A FIRST LOOK AT GRAPH THEORY

A FIRST LOOK AT GRAPH THEORY

John Clark

Derek Allan Holton

Department of Mathematics and Statistics
University of Otago
New Zealand

Published by

World Scientific Publishing Co. Pte. Ltd.

5 Toh Tuck Link, Singapore 596224

USA office: 27 Warren Street, Suite 401-402, Hackensack, NJ 07601

UK office: 57 Shelton Street, Covent Garden, London WC2H 9HE

Library of Congress Cataloging-in-Publication Data

Clark, John.

A first look at graph theory / John Clark and Derek Allan Holton.

xv, 330 p.; 21.5 cm.

Includes bibliographical references and index.

ISBN 9810204892 ISBN 9810204906 (pbk)

1. Graph theory. I. Holton, Derek Allan, 1941– . II. Title.

III. Title: Graph theory.

QA166.C56 1991

511'.5--dc20

91-14633

CIP

British Library Cataloguing-in-Publication Data

A catalogue record for this book is available from the British Library.

First published 1991

Reprinted 1996, 1998, 2001, 2003, 2005

Printed in Singapore.

To our long-suffering wives,
Austina and Marilyn,
with sincere thanks for their patience,
and to
John and Alan Clark,
who should have seen much more of their father
during their summer holidays.

Preface

With the increasing use of computers in society there has been a dramatic growth in all aspects of computer education. At university level, computer science students quickly learn that their subject has many facets, some of which are better appreciated when the student has an appropriate mathematical background. Unfortunately, much of this mathematical background is not the sort found easily in the mathematical education of earlier generations.

Much of the theory of computer science uses an area of mathematics loosely described as "discrete mathematics", this term chosen to emphasise its contrast with the "continuous mathematics" of the more traditional calculus courses. Discrete mathematics covers many topics and this book takes a first look at one of these — Graph Theory. This topic has a surprising number of applications, not just to computer science but to many other sciences (physical, biological and social), engineering and commerce.

From what we have said so far, the reader may have got the impression that this is a book mainly for computer scientists. Not so. Graph Theory is at last being acknowledged as an important subject in the undergraduate *mathematics* curriculum. Perhaps one should expect a more theoretical treatment here than in the computer science setting. However we feel that a blend of the theory with some of its many varied applications is highly desirable for both disciplines — for those mainly concerned with the applications of graphs, the theory helps to strengthen the ideas and point the way to independent applications; conversely, the applications of graph theory to "real world" situations reinforces the theoretical aspects and illustrates one of the many ways in which mathematics is applied. As a result this text is a mixture of both theory and applications and can be used by both the serious mathematics student, her computer science cousin or any other relation keen to learn about one of the most rapidly growing areas of modern mathematics.

The book began in 1985 as a set of notes for a second year course of 40 one-hour lectures in the Department of Mathematics at the University of Otago. The students attending the course, then and in subsequent years, were mainly a mixture of computer science majors and mathematics majors. Not all topics covered in the book were dealt with in these lectures and, indeed, some theoretical aspects may be quite difficult for the average second year university student. However, Graph Theory is particularly suited to selective study and hopefully our treatment here provides material for the individual teacher to tailor to their course's requirements.

We have also provided a plentiful supply of exercises. These are of varying difficulty. Some deal with the algorithmic aspects of the text, some on the theoretical while others introduce some new but related ideas. We encourage the reader to do as many as possible — mathematics is not a spectator sport and is best appreciated with active participation!

We wish to express our sincere thanks to several people involved in the preparation of this book. Jane Hill helped enormously with the typing and typesetting, John Marshall proofread an early version of the text and also made numerous suggestions on style, Gordon Yau prepared some of the diagrams, Maree Watson typed portions of the manuscript, Graeme McKinstry provided very useful LaTeX expertise while Mark Borrie gave frequently-needed computer assistance.

However our main thanks go to our families for their patience during the last few months of the book's preparation.

A Note to the Reader

One of the beauties of Graph Theory is that it depends very little on other branches of mathematics. However in our text we do occasionally rely on the reader having what is often called "mathematical maturity". This means an ability on behalf of the reader to understand and appreciate a mathematical argument or proof. This ability is something that usually is not acquired overnight but is the outcome of an ongoing exposure to mathematics and its accompanying logic. Hopefully, the reader's mathematical maturity will grow as he progresses through the text. If so, then we will have achieved one of the goals of the book.

On a more concrete level, we will assume that the reader knows about the principle of mathematical induction. This is dealt with in many undergraduate textbooks. In particular, the text by Mott, Kandel and Baker, mentioned in the section on Further Reading, has a nice treatment of this.

We also assume that the reader is familiar with the notion of a set. We use $\emptyset$ to denote the empty set and, for two sets A and B, we denote the set difference, consisting of all elements which belong to A but not B, by $A - B$.

Each of the ten chapters of the book is split up into numbered sections, with, for example, Section 2.5 denoting the fifth section of Chapter 2. At the end of most sections there is a collection of numbered exercises. For example, Exercise 2.5.3 refers to the third exercise accompanying Section 2.5. Within each chapter, results such as theorems and corollaries are also numbered consecutively. For example, Theorem 4.3 is the third result of Chapter 4 and it is followed by Corollary 4.4.

The end of a proof of a theorem or corollary is shown by the symbol $\square$.

We have referred to several books and articles throughout the text and details of these are given in the bibliography at the end of the book. Such a reference is given by a number in square brackets with, for example, [8] referring to the eighth item in the bibliography.

Further Reading

We have been much influenced in both our choice and treatment of topics by other Graph Theory texts. Of these we first mention Bondy and Murty's "Graph Theory with Applications" [7]. Sadly, this excellent text is currently out of print, but hopefully your library has a copy. We also learned much from Wilson's enjoyable "Introduction to Graph Theory" [65]. (We have just become aware of, but not yet seen, a new text by Wilson [66], coauthored with Watkins, and on the basis of [65], we feel we can highly recommend it!) On a more advanced level there is also the recent text by Gould [28], "Graph Theory", which provides an up-to-date treatment of the theory with an algorithmic flavour. For an advanced authorative account of the theoretical side of the subject we refer the reader to Behzad, Chartrand and Lesniak-Forster's "Graphs and Digraphs" [4], while Chartrand's "Graphs as Mathematical Models" [13] and Ore's "Graphs and Their Uses" [49] provide more elementary treatments of both the theory and its applications. (Ore's book has just been re-issued in a new edition updated by Wilson.)

For a more algorithmic flavour, we recommend Smith's "Network Optimisation Practise" [57] and Albertson and Hutchinson's "Discrete Mathematics with Algorithms" [1]. We also refer the reader to the more advanced Gibbon's "Algorithmic Graph Theory" [26] and Sysło, Deo and Kowalik's "Discrete Optimization Algorithms with Pascal Programs" [58].

There are several recent texts on the more general area of discrete mathematics. Of those that have substantial treatments of graphs, we mention Mott, Kandel and Baker's "Discrete Mathematics for Computer Scientists" [45] and Polimeni and Straight's "Foundations of Discrete Mathematics" [50]. Finally, as excellent sources of numerous applications of graphs, there are the two books, both titled "Applied Combinatorics", by Roberts [55] and by Tucker [61].

Further Reading

[illegible]

Contents

Preface vii

A Note to the Reader ix

Further Reading xi

1 An Introduction to Graphs
- 1.1 The Definition of a Graph . . . 1
- 1.2 Graphs as Models . . . 3
- 1.3 More Definitions . . . 7
- 1.4 Vertex Degrees . . . 13
- 1.5 Subgraphs . . . 17
- 1.6 Paths and Cycles . . . 25
- 1.7 The Matrix Representation of Graphs . . . 35
- 1.8 Fusion . . . 41

2 Trees and Connectivity
- 2.1 Definitions and Simple Properties . . . 47
- 2.2 Bridges . . . 52
- 2.3 Spanning Trees . . . 57
- 2.4 Connector Problems . . . 62
- 2.5 Shortest Path Problems . . . 69
- 2.6 Cut Vertices and Connectivity . . . 78

3 Euler Tours and Hamiltonian Cycles
- 3.1 Euler Tours . . . 83
- 3.2 The Chinese Postman Problem . . . 96
- 3.3 Hamiltonian Graphs . . . 99
- 3.4 The Travelling Salesman Problem . . . 110

4 Matchings
- 4.1 Matchings and Augmenting Paths . . . 121
- 4.2 The Marriage Problem . . . 129
- 4.3 The Personnel Assignment Problem . . . 135
- 4.4 The Optimal Assignment Problem . . . 143
- 4.5 A Chinese Postman Problem Postscript . . . 155

5 Planar Graphs

5.1 Plane and Planar Graphs 157
5.2 Euler's Formula 162
5.3 The Platonic Bodies 169
5.4 Kuratowski's Theorem 173
5.5 Non-Hamiltonian Plane Graphs 181
5.6 The Dual of a Plane Graph 185

6 Colouring

6.1 Vertex Colouring 191
6.2 Vertex Colouring Algorithms 199
6.3 Critical Graphs 205
6.4 Cliques 208
6.5 Edge Colouring 212
6.6 Map Colouring 219

7 Directed Graphs

7.1 Definitions (and More Definitions) 229
7.2 Indegree and Outdegree 238
7.3 Tournaments 246
7.4 Traffic Flow 254

8 Networks

8.1 Flows and Cuts 261
8.2 The Ford and Fulkerson Algorithm 274
8.3 Separating Sets 282

9 Ramsey Theory

9.1 A Party 291
9.2 A Generalisation of the Party Problem 293
9.3 Another Generalisation of the Party Problem 297
9.4 The Compleat Ramsey 300

10 Reconstruction

10.1 The Reconstruction Conjecture 303
10.2 Reconstruction of Regular and Disconnected Graphs 308
10.3 Edge Reconstruction 313
10.4 The Infinite 316

Bibliography **321**

Index **325**

A FIRST LOOK AT GRAPH THEORY

Chapter 1

An Introduction to Graphs

1.1 The Definition of a Graph

In a hockey league there are eight teams, which we denote by S, T, U, V, W, X, Y and Z. After a few weeks of the season the following games have been played:

S has played X and Z, T has played W, X and Z
U has played Y and Z, V has played W and Y,
W has played T, V and Y, X has played S and T,
Y has played U, V and W, and Z has played S, T and U.

We may illustrate this situation by either of the two diagrams of Figure 1.1, where the teams are represented by (large) dots and two such dots are joined by a line whenever the corresponding teams have played each other.

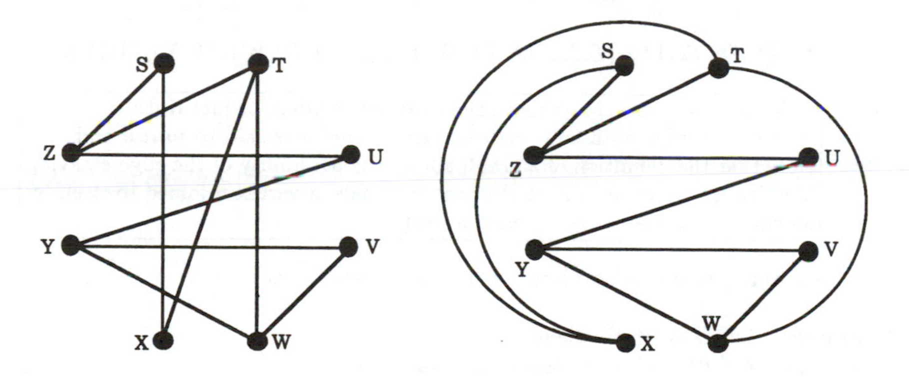

Figure 1.1: Games played in the hockey league.

In the diagram on the left the dots have been joined using straight lines, while in the other diagram three of the lines used are not straight. But since we are simply interested in which games have been played, the manner in which a pair of dots is joined is of no importance and so it does not matter whether the lines are straight or not.

The diagrams may be used to describe other situations. For example, the eight dots may represent eight people, with the lines joining a pair of dots if the two people know each other. Or, the dots could be communication centres with the lines denoting communication links. Indeed, as we will see later, many real-world situations can conveniently be described by means of such drawings, which we call **graphs**.

A graph, then, can be thought of as a drawing or diagram consisting of a collection of vertices (dots or points) together with edges (lines) joining certain pairs of these vertices, as in the diagrams of Figure 1.1. Notice that in the left-hand diagram there are intersecting edges (lines), for example, SX and ZU; the points of intersection of these edges are not however vertices (points) of our graph. To avoid any confusion that could arise in this way, the vertices in any drawing of a graph will normally be drawn as large dots. However, as we will now see from its formal definition, from a mathematical point of view a graph does not need to be drawn.

> A **graph** $G = (V(G), E(G))$ consists of two finite sets:
> $V(G)$, the **vertex set** of the graph, often denoted by just V, which is a nonempty set of elements called **vertices**, and
> $E(G)$, the **edge set** of the graph, often denoted by just E, which is a possibly empty set of elements called **edges**,
> such that each edge e in E is assigned an unordered pair of vertices (u, v), called the **end vertices** of e.

Thus for both graphs of Figure 1.1, the vertex set is

$$V = \{S, T, U, V, W, X, Y, Z\},$$

the edge set E has 10 edges and these edges are assigned the unordered pairs of vertices

$$(S,X),\ (S,Z),\ (T,W),\ (T,X),\ (T,Z),\ (U,Y),\ (U,Z),\ (V,W),\ (V,Y),\ (W,Y).$$

> Vertices are also sometimes called **points, nodes,** or just **dots**.
> If e is an edge with end vertices u and v then e is said to **join** u and v. Note that the definition of a graph allows the possibility of the edge e having identical end vertices, i.e., it is possible to have a vertex u joined to itself by an edge — such an edge is called a **loop**.

We now give two examples to illustrate the above definitions.

Example 1. Let $G = (V, E)$ where
$V = \{a, b, c, d, e\}$, $E = \{e_1, e_2, e_3, e_4, e_5, e_6, e_7, e_8\}$,
and the ends of the edges are given by:

$$e_1 \leftrightarrow (a,b),\quad e_2 \leftrightarrow (b,c),\quad e_3 \leftrightarrow (c,c),\quad e_4 \leftrightarrow (c,d),$$
$$e_5 \leftrightarrow (b,d),\quad e_6 \leftrightarrow (d,e),\quad e_7 \leftrightarrow (b,e),\quad e_8 \leftrightarrow (b,e).$$

We can then represent G diagrammatically as in Figure 1.2.

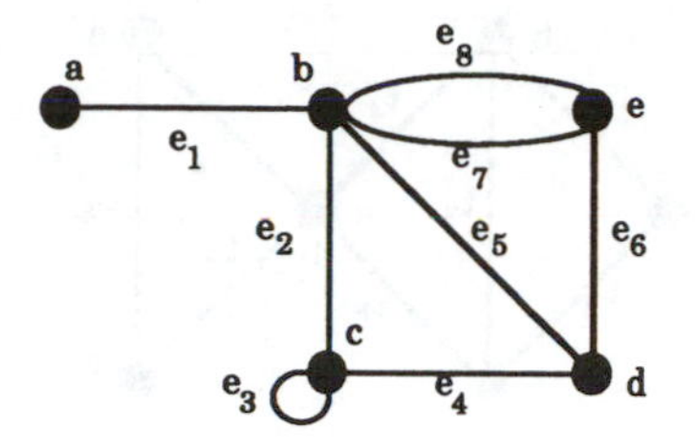

Figure 1.2: A graph G with five vertices and eight edges.

Example 2. Let H be the graph (V, E) where

$$V = \{1, 2, 3, 4, 5, 6\}, \quad E = \{a, b, c, d, e, f, g, h\},$$

and the ends of the edges are given by:

$$\begin{array}{llll} a \leftrightarrow (1,2), & b \leftrightarrow (1,1), & c \leftrightarrow (2,3), & d \leftrightarrow (3,4), \\ e \leftrightarrow (2,4), & f \leftrightarrow (3,4), & g \leftrightarrow (1,4), & h \leftrightarrow (4,5). \end{array}$$

We can represent H diagrammatically as in Figure 1.3.

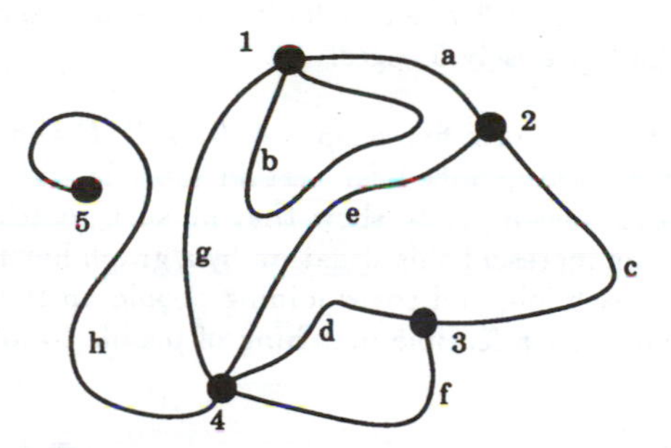

Figure 1.3: A graph H with five vertices and eight edges.

1.2 Graphs as Models

We now give some problems in which graphs provide a natural mathematical model. At this stage we only describe the problems and delay detailed discussion of their solution to later chapters.

Problem 1. Suppose that the graph of Figure 1.4 represents a network of telephone lines and poles. We are interested in the network's vulnerability to accidental disruption. We want to identify those lines and poles that must stay in service to avoid disconnecting the network.

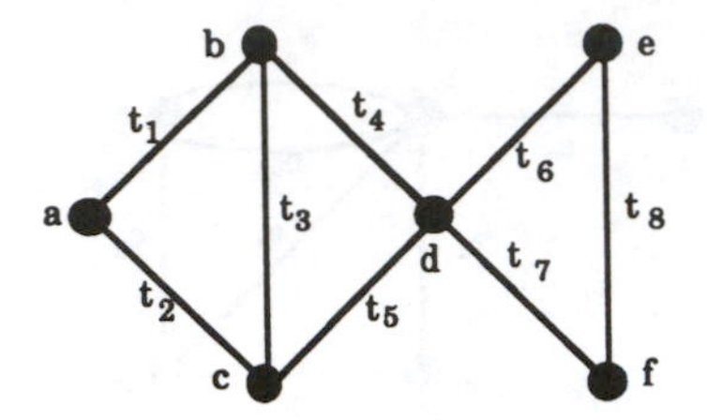

Figure 1.4: A network of telephone lines.

There is no single line whose disruption (removal) will disconnect the graph (network), but the graph will become disconnected if we remove the two lines represented by the edges t_4 and t_5, for example. When it comes to poles, the network is more vulnerable since there is a single vertex, vertex d, whose removal disconnects the graph. This illustrates the notions of edge connectivity and vertex connectivity of a graph, discussed in Sections 8.3 and 2.6.

We may also want to find a smallest possible set of edges needed to connect the six vertices. There are several examples of such minimal sets. One is

$$\{t_1, t_3, t_5, t_6, t_7\}.$$

In fact, as we will see in Chapter 2 when we look at trees and connected graphs, any such minimal set will have precisely 5 edges.

Problem 2. Suppose that we have five people A, B, C, D, E and five jobs a, b, c, d, e and some of these people are qualified for certain jobs. Is there a feasible way of allocating one job to each person, or to show that no such matching up of jobs and people is possible? We can represent this situation by a graph having a vertex for each person and a vertex for each job, and edges joining people up to jobs for which they are qualified. Does there exist a feasible matching of people to jobs for the graph of Figure 1.5?

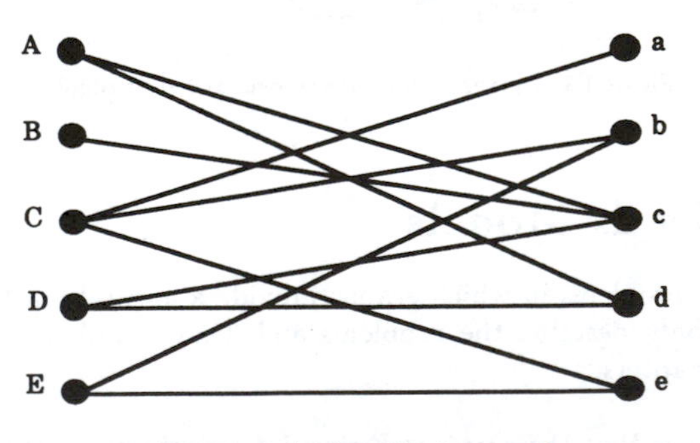

Figure 1.5: A job applications graph.

The answer is no. The reason can be found by considering people A, B, and D. These three people as a set are collectively qualified for only two jobs, c and d, hence there is no feasible matching possible for these three people, much less all five people. We will look at an algorithm for finding a feasible matching, if any exists, later in Section 4.2.

Problem 3. Suppose a salesman's territory includes several cities with highways connecting certain pairs of these cities. His job requires him to visit each city personally. Is it possible for him to schedule a round trip by car enabling him to visit each specified city exactly once? We can represent the transportation system involved by a graph G whose vertices correspond to the cities and such that two vertices are joined by an edge if and only if a highway connects the corresponding cities (and does not pass through any other specified city). The graphs G_1 and G_2 of Figure 1.6 denote two such salesmen's territories. The desired schedule is possible for G_1 but not for G_2. (In G_1, starting at vertex u_1, we can visit each vertex and arrive back at u_1 by taking the edges e_1, e_2, e_3, e_4, and e_6 in turn.)

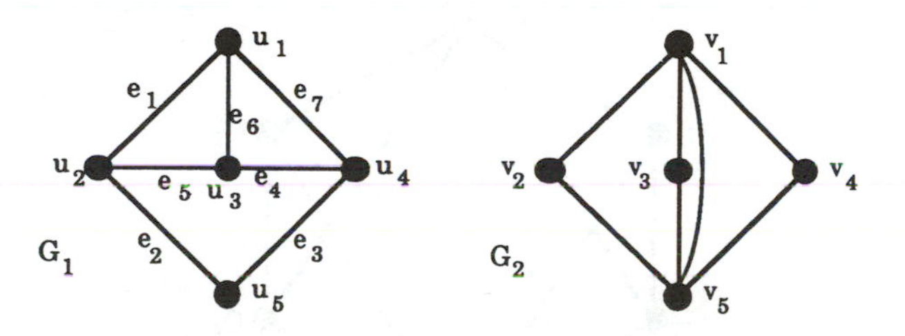

Figure 1.6: Two travelling salesmen's territories.

Problem 4. Suppose we have three houses each of which have to be supplied with electricity, gas and water. Is it possible to connect each utility with each of the three houses without the lines or mains crossing?

We can represent the connection of the three houses to the three utilities by the graph of Figure 1.7. Here we have a graph with six vertices, three of which represent the houses (denoted by H_1, H_2, H_3), the other three represent the utilities (denoted by E, G, H), and an edge joins two vertices if and only if one vertex denotes a house and the other vertex a utility.

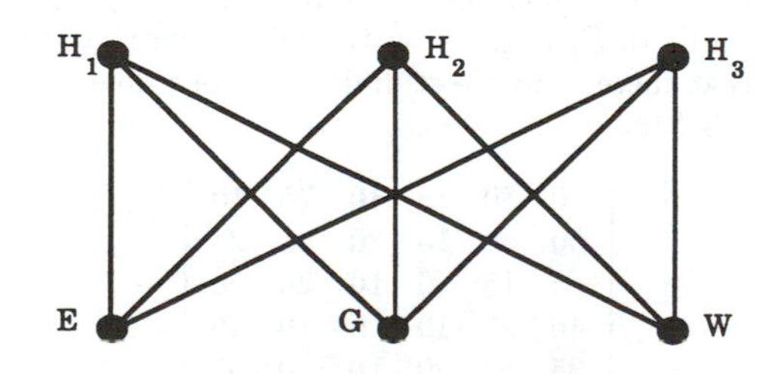

Figure 1.7: The three utilities graph.

The problem then is as to whether or not we can draw this graph in such a way that no two edges intersect. The answer is no — we will see why when we look later at *planar* graphs in Chapter 5.

Problem 5. Six radio broadcasting companies $C_1, \ldots, C_6$ have applied to the Minister of Broadcasting for frequency channels. If two companies' transmitters are within 200 kilometres of each other they can not be assigned the same frequency since there will be too much interference. The Minister wishes to assign as small a number of different frequencies as possible, taking this interference into account.

To illustrate the problem, let G be the graph of Figure 1.8 with vertex set $C_1, \ldots, C_6$ with two vertices C_i and C_j joined by an edge if and only if the two companies C_i and C_j have their transmitters less than 200 kilometres apart. As a particular example, consider the following graph G where, for example, C_2's transmitter is within 200 kilometres of those of C_1, C_3, C_4 and C_5.

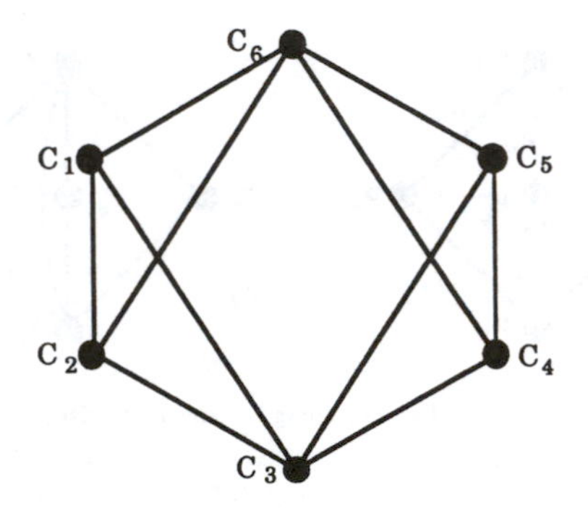

Figure 1.8: Radio transmitters and their interference graph.

Now suppose we assign different colours to the vertices of G in such a way that no two vertices of the same colour are joined by an edge. Then, thinking of the colours as representing the frequency channels, the Minister wants to find the minimum number of colours with which the vertices can be coloured in this way. For our example, the answer is 3 — we can colour C_1 and C_4 red, C_2 and C_5 blue, and C_3 and C_6 yellow.

Problem 6. A company has branches in each of six cities $C_1, C_2, \ldots, C_6$. The airfare for a direct flight from C_i to C_j is given by the (i, j)th entry of the following matrix (where ∞ indicates that there is no direct flight). For example the fare from C_1 to C_4 is \$40, from C_2 to C_3 is \$15.

$$\begin{bmatrix} 0 & 50 & \infty & 40 & 25 & 10 \\ 50 & 0 & 15 & 20 & \infty & 25 \\ \infty & 15 & 0 & 10 & 20 & \infty \\ 40 & 20 & 10 & 0 & 10 & 25 \\ 25 & \infty & 20 & 10 & 0 & 55 \\ 10 & 25 & \infty & 25 & 55 & 0 \end{bmatrix}$$

The company is interested in computing a table of cheapest fares between pairs of cities. (Even if there is a direct flight between two cities this may not be the cheapest route.) We can first represent the situation by a *weighted* graph, i.e., a graph with "weights" attached to the edges according to the airfares, as in Figure 1.9.

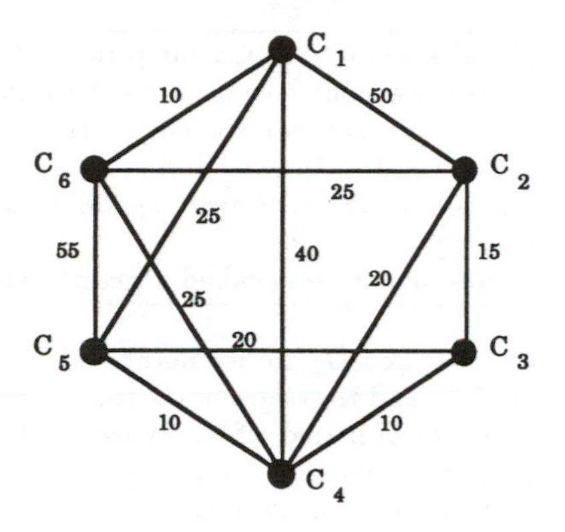

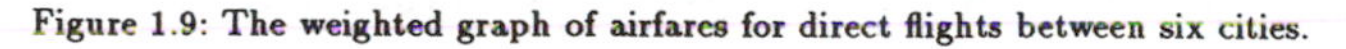
Figure 1.9: The weighted graph of airfares for direct flights between six cities.

The problem can then be solved using *Dijkstra's algorithm* — see Section 2.5. This type of problem is called a *shortest path problem*.

1.3 More Definitions

Let G be a graph. If two (or more) edges of G have the same end vertices then these edges are called **parallel**.

For example, the edges e_7 and e_8 of the graph of Figure 1.2 are parallel.

A vertex of G which is not the end of any edge is called **isolated**.

Two vertices which are joined by an edge are said to be **adjacent** or **neighbours**.

The set of all neighbours of a fixed vertex v of G is called the **neighbourhood set** of v and is denoted by $N(v)$.

Thus, in the graph of Figure 1.9, C_1 and C_4 are adjacent but C_2 and C_5 are not. The neighbourhood set $N(C_3)$ of C_3 is $\{C_2, C_4, C_5\}$.

A graph is called **simple** if it has no loops and no parallel edges.

Much of the graphs that we consider will be simple. (The reader should be warned that other authors may differ from us in their use of the term "graph". For example, in some texts a graph which is not simple is called a **multigraph**.)

It is often the case that two graphs have the same structure, differing only in the way their vertices and edges are labelled or only in the way they are represented geometrically. For many purposes, we can regard the two graphs as *essentially* the same. This essential likeness has a special name and we now define this formally.

> A graph $G_1 = (V_1, E_1)$ is said to be **isomorphic** to the graph $G_2 = (V_2, E_2)$ if there is a one-to-one correspondence between the vertex sets V_1 and V_2 and a one-to-one correspondence between the edge sets E_1 and E_2 in such a way that if e_1 is an edge with end vertices u_1 and v_1 in G_1 then the corresponding edge e_2 in G_2 has its end points the vertices u_2 and v_2 in G_2 which correspond to u_1 and v_1 respectively.
> Such a pair of correspondences is called a **graph isomorphism**.

In other words, the graphs G_1 and G_2 are isomorphic if the vertices of G_1 can be paired off with the vertices of G_2 and the edges of G_1 can be paired off with the edges of G_2 in such a way that the ends of paired off edges are paired off. Thus G_1 is really just the same graph as G_2, apart from a possible change in how the vertices and edges are named (or a possible redrawing of the graphs).

Figure 1.10 shows five fairly obvious pairs of isomorphic graphs.

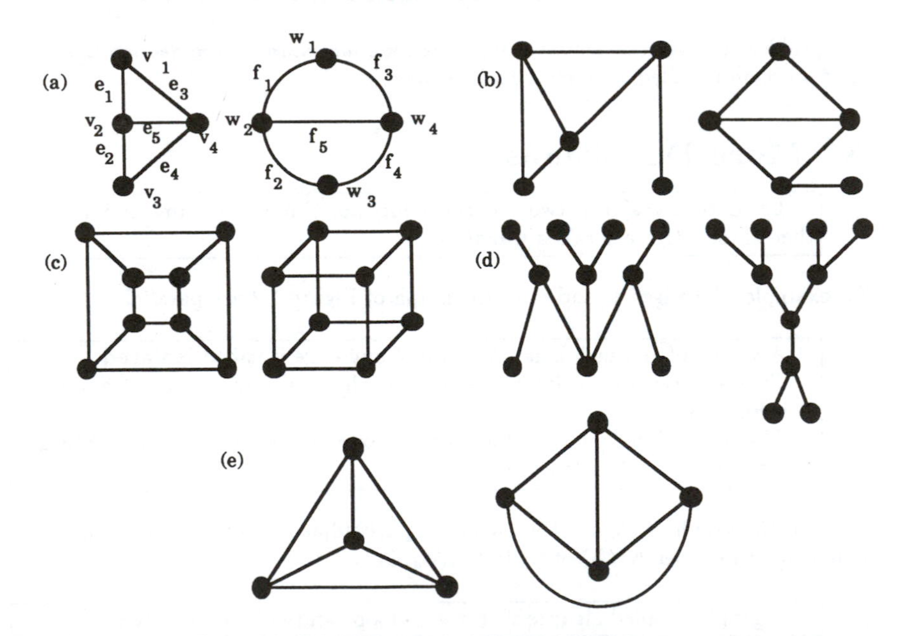

Figure 1.10: Isomorphic pairs of graphs.

Figure 1.11 gives some examples to illustrate that often it can be quite difficult to determine if two graphs are isomorphic.

In Figure 1.11 (a) an isomorphism is given by the following one-to-one correspondence of vertices:

$$u_1 \leftrightarrow v_1,\ u_2 \leftrightarrow v_3,\ u_3 \leftrightarrow v_5,\ u_4 \leftrightarrow v_2,\ u_5 \leftrightarrow v_4, u_6 \leftrightarrow v_6$$

(Note how u_1, u_2, u_3 are only joined to u_4, u_5, u_6 and similarly v_1, v_3, v_5 are only joined to v_2, v_4, v_6.) Isomorphisms for Figure 1.11 (b) and (c) are shown by the labelling of the vertices.

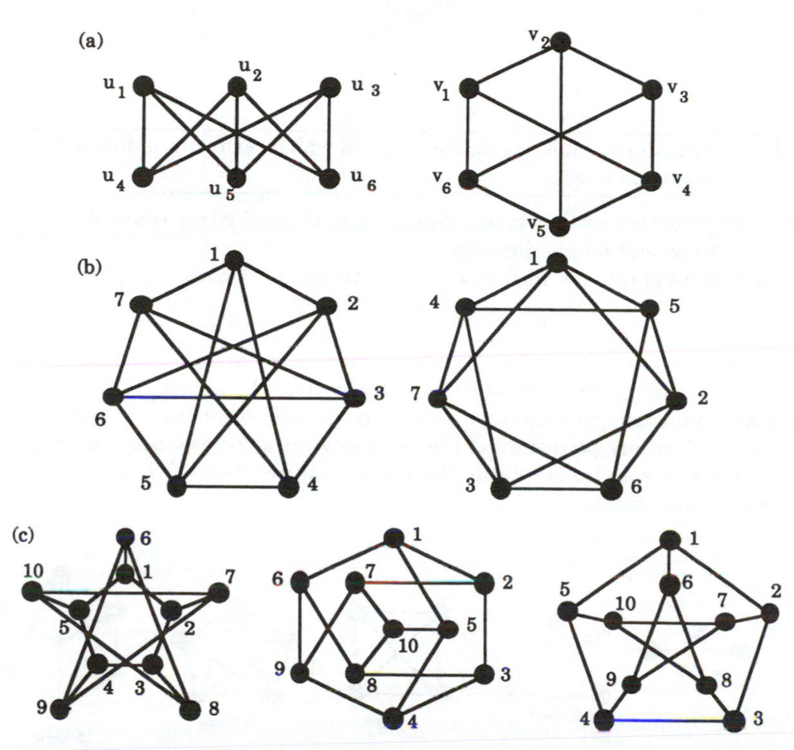

Figure 1.11: Some more groups of isomorphic graphs.

The problem of determining when two graphs are isomorphic gets harder as the number of vertices and edges of the graphs get larger. For example, while there are only 4 non-isomorphic simple graphs on three vertices and 11 on 4 vertices there are 1044 non-isomorphic simple graphs on just seven vertices.

Clearly if two graphs G_1 and G_2 are isomorphic then they must have

(i) the same number of vertices and (ii) the same number of edges.

However, as we will now see, conditions (i) and (ii) are not sufficient. The graph G of Figure 1.12 has the same number of vertices and edges as the graphs in Figure 1.11 (a) but G is not isomorphic to these graphs.

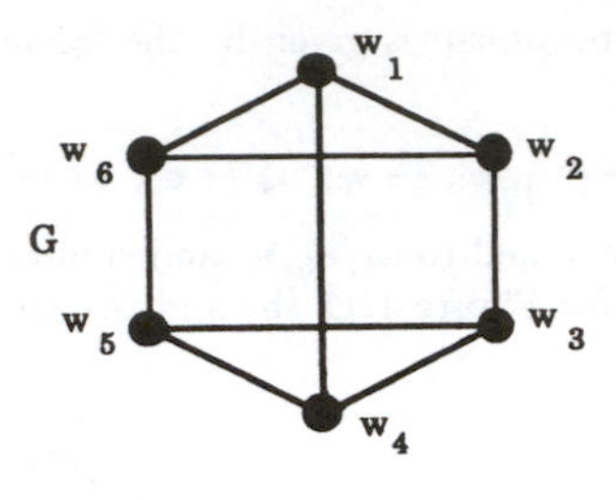

Figure 1.12

A **complete** graph is a simple graph in which each pair of distinct vertices is joined by an edge.

Thus, a graph with n vertices is complete if it has as many edges as possible provided there are no loops and no parallel edges.

If the complete graph has vertices $v_1, \ldots, v_n$ then the edge set can be given by

$$E = \{(v_i, v_j) : v_i \neq v_j;\ \ i, j = 1, \ldots, n.\}$$

It follows that the graph has $\frac{1}{2}n(n-1)$ edges (since there are $n-1$ edges incident with each of the n vertices v_i, so a total of $n \times (n-1)$, but divide by 2 since $(v_j, v_j) = (v_j, v_i)$).

Given any two complete graphs with the same number of vertices, n, then they are isomorphic. In fact any pairing off of the vertices gives a corresponding pairing off of the edges and hence an isomorphism. For this reason we speak of *the* complete graph on n vertices. It is denoted by K_n.

Figure 1.13 shows $K_1, \ldots, K_6$.

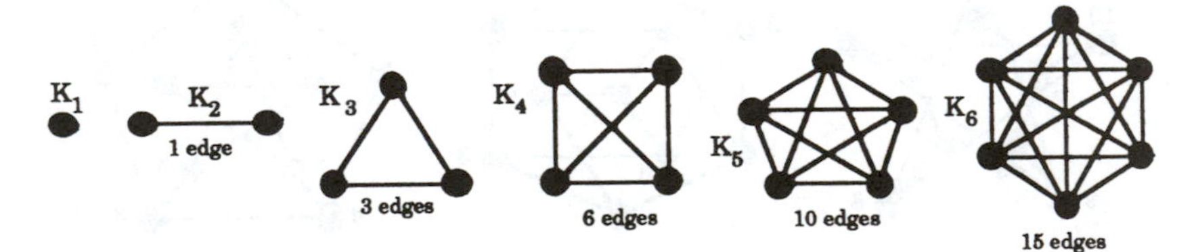

Figure 1.13: The complete graphs on at most six vertices.

An **empty** (or **trivial**) graph is a graph with no edges.

Let G be a graph. If the vertex set V of G can be partitioned into two nonempty subsets X and Y (i.e., $X \cup Y = V$ and $X \cap Y = \emptyset$) in such a way that each edge of G has one end in X and one end in Y then G is called **bipartite**. The partition $V = X \cup Y$ is called **a bipartition** of G.

A **complete bipartite graph** is a simple bipartite graph G, with bipartition $V = X \cup Y$, in which every *vertex* in X is joined to *every* vertex of Y. If X has m vertices and Y has n vertices, such a graph is denoted by $K_{m,n}$.

Any complete bipartite graph with a bipartition into two sets of m and n vertices is isomorphic to $K_{m,n}$ — in fact any pairing off of the two sets of m vertices together with any pairing off of the two sets of n vertices will give an isomorphism. In particular, $K_{m,n}$ is (of course) isomorphic to $K_{n,m}$.

Since each of the m vertices in the partition set X of $K_{m,n}$ is adjacent to each of the n vertices in the partition set Y, $K_{m,n}$ has $m \times n$ edges.

Note that there is now an unfortunate ambiguity in the use of the word *complete*, since a complete bipartite graph will not in general be complete. Indeed, as the reader should easily verify, the only complete bipartite graph which is complete is $K_{1,1}$.

Figure 1.14 shows two bipartite graphs. They are not complete bipartite. However, the graphs of Figure 1.15 *are* complete bipartite.

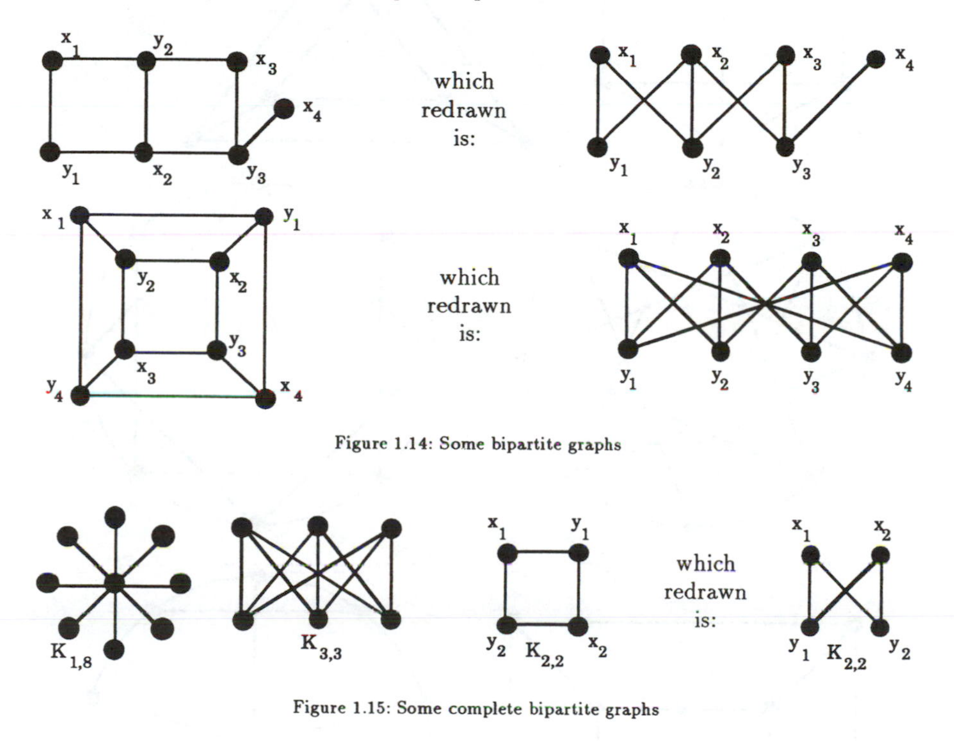

Figure 1.14: Some bipartite graphs

Figure 1.15: Some complete bipartite graphs

Exercises for Section 1.3

1.3.1 Make a list of drawings of all the graphs with n vertices and e edges for all n, e with $n + e \leq 6$. The list should not include any isomorphic pairs of graphs. Determine which of the graphs on your list are simple. (You should get a total of 65 graphs of which 14 are simple.)

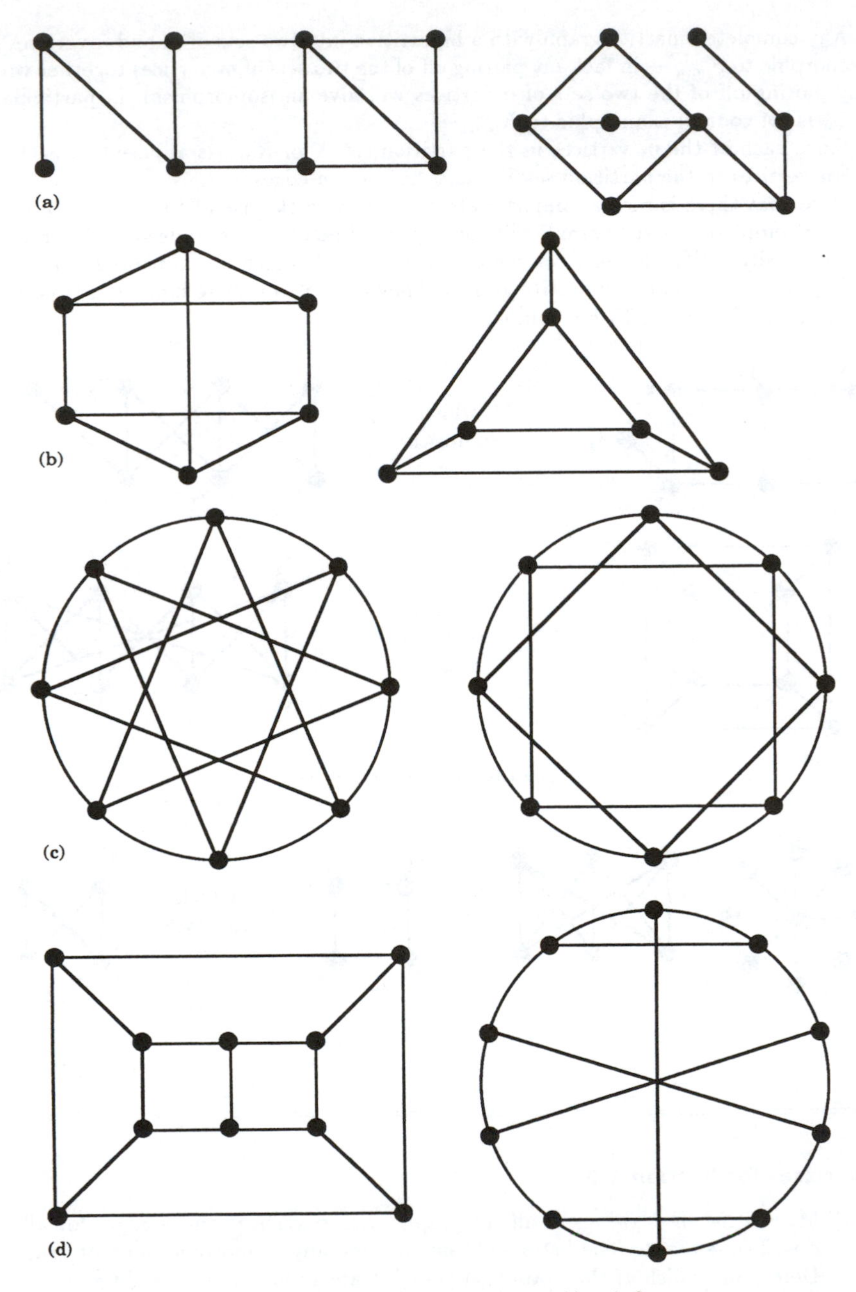

Figure 1.16: Which of these are isomorphic pairs?

1.3.2 Determine which of the pairs of graphs in Figure 1.16 are isomorphic pairs. (Give an argument justifying your answer.)

1.3.3 In each collection of three graphs shown in Figure 1.17 there is exactly one isomorphic pair. Find each pair and justify that your answer is correct.

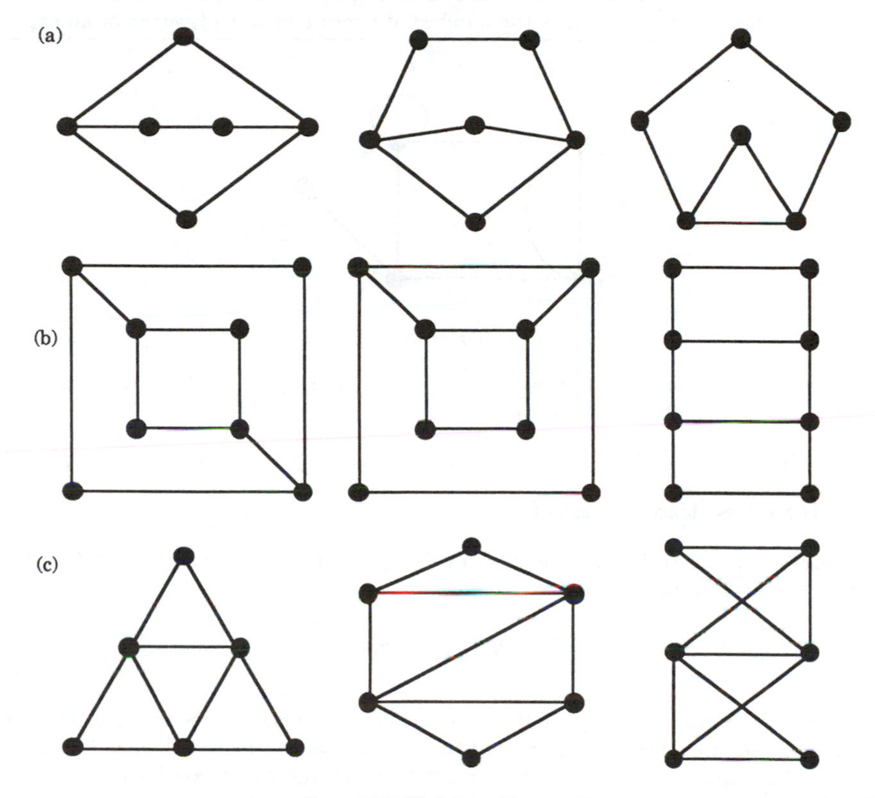

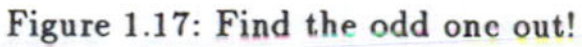

Figure 1.17: Find the odd one out!

1.3.4 Find all nonisomorphic complete bipartite graphs with at most 7 vertices.

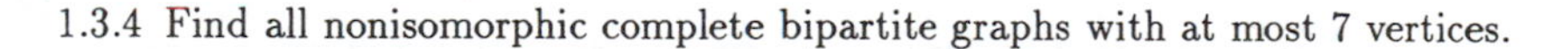

1.4 Vertex Degrees

An edge e of a graph G is said to be **incident** with the vertex v if v is an end vertex of e. In this case we also say that v is **incident** with e. Two edges e and f which are incident with a common vertex v are said to be **adjacent**.

Note that an edge is incident with either 1 or 2 vertices (1 if it is a loop), whereas a vertex may be incident with any finite number, including 0, of edges.

Let v be a vertex of the graph G. The **degree** $d(v)$ (or $d_G(v)$ if we want to emphasize G) of v is the number of edges of G incident with v, counting each loop twice, i.e., it is the number of times v is an end vertex of an edge.

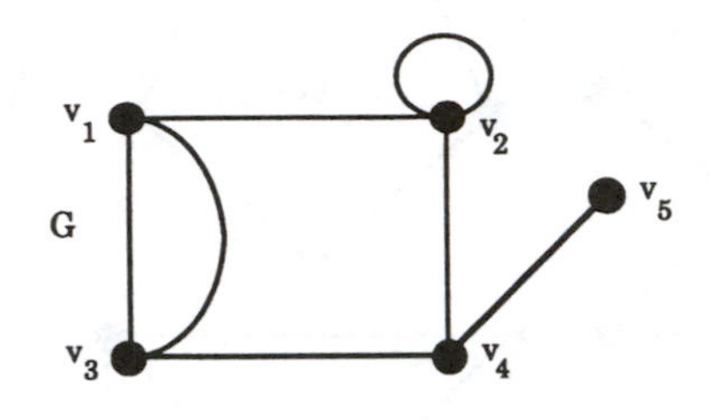

Figure 1.18

In the graph of Figure 1.18 we have $d(v_1) = 3$, $d(v_2) = 4$, $d(v_3) = 3$, $d(v_4) = 3$, $d(v_5) = 1$. Note that in this example

$$d(v_1) + d(v_2) + d(v_3) + d(v_4) + d(v_5) = 14 = 2 \times (\text{number of edges in } G).$$

This is no coincidence because of:

Theorem 1.1 (The First Theorem of Graph Theory) *For any graph G with e edges and n vertices, $v_1, \ldots, v_n$,*

$$\sum_{i=1}^{n} d(v_i) = 2e.$$

Proof Each edge, since it has two end vertices, contributes precisely 2 to the sum of the degrees, i.e., when the degrees of the vertices are summed each edge is counted twice. (Note that even a loop contributes 2 although the 2 ends are identical.) □

A vertex of a graph is called **odd** or **even** depending on whether its degree is odd or even.

In the graph G of Figure 1.18 there is an even number of odd vertices (these vertices being v_1, v_3, v_4, v_5). Again this is a consequence of a general result:

Corollary 1.2 *In any graph G there is an even number of odd vertices.*

Proof Let W be the set of odd vertices of G and let U be the set of even vertices of G. Then, for each $u \in U$, $d(u)$ is even and so $\sum_{u \in U} d(u)$, being a sum of even numbers, is even. However

$$\sum_{u \in U} d(u) + \sum_{w \in W} d(w) = \sum_{v \in V} d(v) = 2e,$$

by Theorem 1.1, where V is the vertex set of G and e is the number of its edges. Thus

$$\sum_{w \in W} d(w) = 2e - \sum_{u \in U} d(u)$$

is even (being the difference of two even numbers). As all the terms in $\sum_{w \in W} d(w)$ are odd and their sum is even there must be an even number of them (because the sum of an odd number of odd numbers is odd). □

Note that it is not true in general that a graph must have an *odd* number of *even* vertices, e.g., the graph of Figure 1.19 has four even vertices (and we leave the reader to construct a graph which has exactly n even vertices for *any* given n).

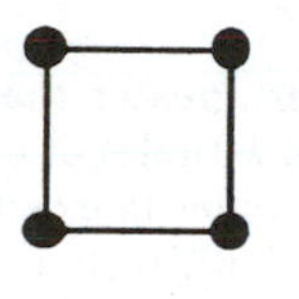

Figure 1.19

> If for some positive integer k, $d(v) = k$ for *every* vertex v of the graph G, then G is called **k-regular**.
> A **regular** graph is one that is k-regular for some k.

The graph drawn above is 2-regular. The complete graph K_n is $(n-1)$-regular. The complete bipartite graph $K_{n,n}$ on $2n$ vertices is n-regular. The graph G_2 of Figure 1.12 is 3-regular, as is the graph of Figure 1.11 (c).

Exercises for Section 1.4

1.4.1 List the degrees of each of the vertices of the graph G of Figure 1.20. How many even vertices does G have? How may odd vertices does G have?

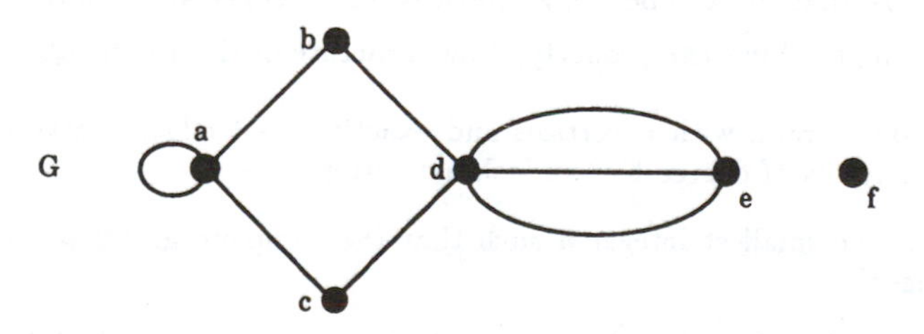

Figure 1.20: Find the degree of each vertex of this graph.

1.4.2 Let G be a graph in which there is no pair of adjacent edges. What can you say about the degrees of the vertices in G?

1.4.3 Let G be a graph with n vertices and e edges and let m be the smallest positive integer such that $m \geq 2e/n$. Prove that G has a vertex of degree at least m.

1.4.4 Prove that it is impossible to have a group of nine people at a party such that each one knows exactly five of the others in the group.

1.4.5 Let G be a graph with n vertices, t of which have degree k and the others have degree $k+1$. Prove that $t = (k+1)n - 2e$, where e is the number of edges in G.

1.4.6 Let G be a k-regular graph, where k is an odd number. Prove that the number of edges in G is a multiple of k.

1.4.7 Let k be some positive integer, greater than 1. The **k-cube**, Q_k, is the graph whose vertices are the ordered k-tuples of 0's and 1's, two vertices being joined by an edge if and only if they differ in exactly one position. Thus, for example, for $k = 3$ the vertices are $(0,0,0)$, $(0,0,1)$, $(0,1,0)$, $(1,0,0)$, $(1,0,1)$, $(1,1,0)$, $(0,1,1)$, $(1,1,1)$ and, for example, $(0,0,0)$ is joined to $(0,0,1)$, $(0,1,0)$ and $(1,0,0)$ but not to any other vertex. The 3-cube Q_3 is shown in Figure 1.21.

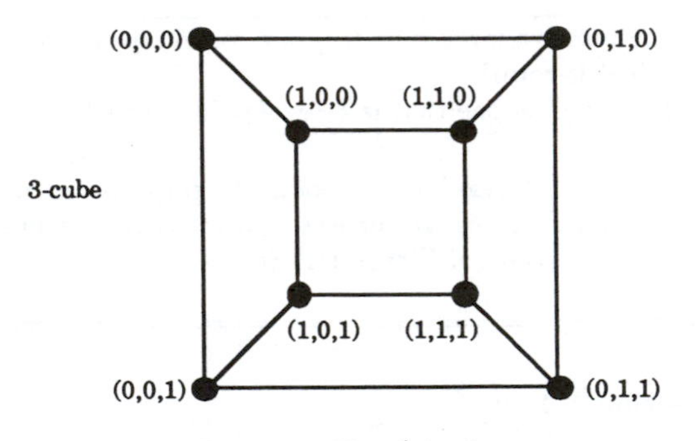

Figure 1.21: The 3-cube Q_3.

(a) Show that the k-cube has 2^k vertices, $k2^{k-1}$ edges and is bipartite.

(b) Using the bipartite property, draw a picture of the 4-cube Q_4.

1.4.8 Let G be a graph with n vertices and exactly $n-1$ edges. Prove that G has either a vertex of degree 1 or an isolated vertex.

1.4.9 What is the smallest integer n such that the complete graph K_n has at least 500 edges?

1.4.10 Prove that there is no simple graph with six vertices, one of which has degree 2, two have degree 3, three have degree 4 and the remaining vertex has degree 5.

1.4.11 Prove that there is no simple graph on four vertices, three of which have degree 3 and the remaining vertex has degree 1.

1.4.12 Let G be a simple regular graph with n vertices and 24 edges. Find all possible values of n and give examples of G in each case.

1.5 Subgraphs

It is often the case that a graph under study is contained within some larger graph also being investigated.

> Let H be a graph with vertex set $V(H)$ and edge set $E(H)$ and, similarly, let G be a graph with vertex set $V(G)$ and edge set $E(G)$. Then we say that H is a **subgraph** of G if $V(H) \subseteq V(G)$ and $E(H) \subseteq E(G)$. In such a case, we also say that G is a **supergraph** of H.

For example, in Figure 1.22, G_1 is a subgraph of both G_2 and G_3 but G_3 is not a subgraph of G_2.

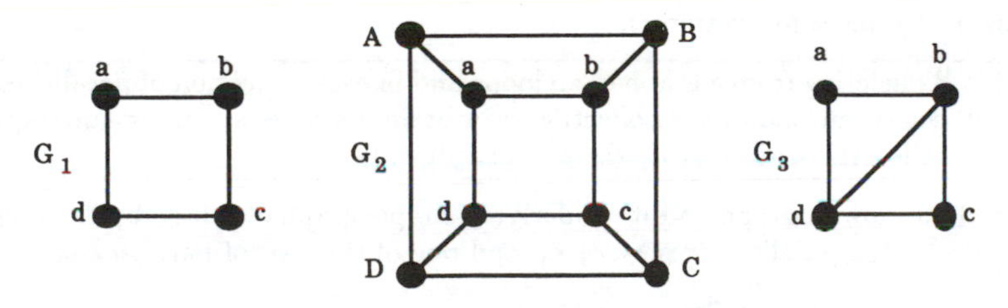

Figure 1.22: $G_1 \subseteq G_2$, $G_1 \subseteq G_3$ but $G_3 \not\subseteq G_2$.

> Any graph *isomorphic* to a subgraph of G is also referred to as a **subgraph** of G.
>
> If H is a subgraph of G then we write $H \subseteq G$. When $H \subseteq G$ but $H \neq G$, i.e., $V(H) \neq V(G)$ or $E(H) \neq E(G)$, then H is called a **proper subgraph** of G.
>
> A **spanning subgraph** (or **spanning supergraph**) of G is a subgraph (or supergraph) H with $V(H) = V(G)$, i.e., H and G have exactly the same vertex set.

It follows easily from the definitions that any *simple* graph on n vertices is a subgraph of the complete graph K_n.

In Figure 1.22, G_1 is a proper spanning subgraph of G_3.

The simplest types of subgraph of a graph G are those obtained by the **deletion** of a vertex or an edge and we now define these.

If $G = (V, E)$ and V has at least two elements (i.e., G has at least two vertices), then for any vertex v of G, $G - v$ denotes the subgraph of G with vertex set $V - \{v\}$ and whose edges are all those of G which are not incident with v, i.e., $G - v$ is obtained from G by removing v and all the edges of G which have v as an end. $G - v$ is referred to as a **vertex deleted subgraph**.

If $G = (V, E)$ and e is an edge of G then $G - e$ denotes the subgraph of G having V as its vertex set and $E - \{e\}$ as its edge set, i.e., $G - e$ is obtained from G by removing the edge e, (but not its endpoint(s)). $G - e$ is referred to as an **edge deleted subgraph**.

We extend the above definition to cater for the deletion of several vertices or edges.

If $G = (V, E)$ and U is a proper subset of V then $G - U$ denotes the subgraph of G with vertex set $V - U$ and whose edges are all those of G which are not incident with any vertex in U.

If F is a subset of the edge set E then $G - F$ denotes the subgraph of G with vertex set V and edge set $E - F$, i.e., obtained by deleting all the edges in F, but not their endpoints.

$G - U$ and $G - F$ are also referred to as **vertex deleted** and **edge deleted subgraphs** (respectively).

Figure 1.24 gives some examples.

By deleting from a graph G all loops and in each collection of parallel edges all edges but one in the collection we obtain a simple spanning subgraph of G, called the **underlying simple graph** of G.

Figure 1.23 shows a graph and its underlying simple graph obtained by deleting the loop e_6, two of the parallel edges e_2, e_3, e_4, and one of the pair of parallel edges e_7, e_8.

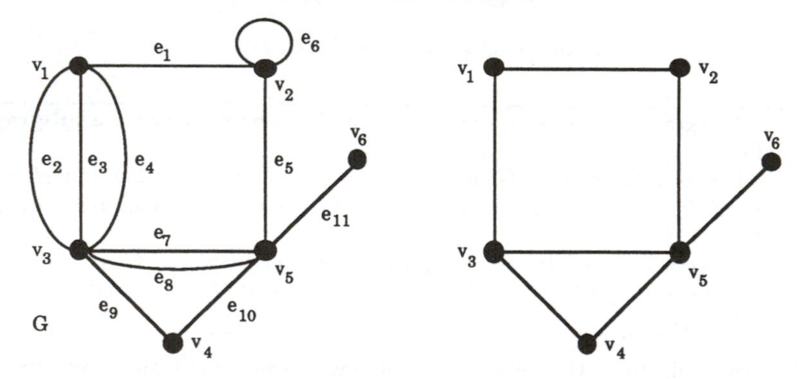

Figure 1.23: A graph and its underlying simple graph.

Some of the more important subgraphs we shall encounter are the *induced subgraphs* and we now define these.

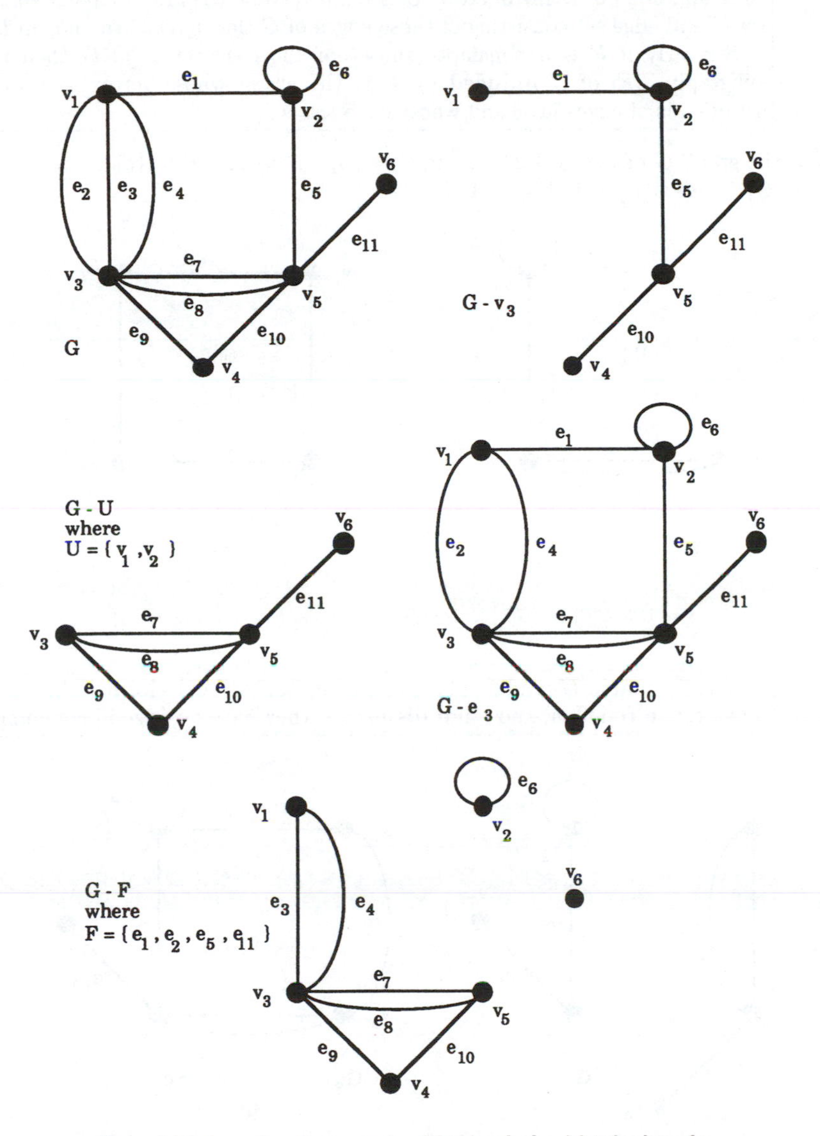

Figure 1.24: A graph and some vertex deleted and edge deleted subgraphs.

If U is a nonempty subset of the vertex set V of the graph G then the subgraph $G[U]$ of G **induced** by U is defined to be the graph having vertex set U and edge set consisting of those edges of G that have both ends in U.

Similarly if F is a nonempty subset of the edge set E of G then the subgraph $G[F]$ of G **induced** by F is the graph whose vertex set is the set of ends of edges in F and whose edge set is F.

For the graph G of Figure 1.24, taking $U = \{v_2, v_3, v_5\}$ and $F = \{e_1, e_3, e_5, e_7, e_9\}$, we get $G[U]$ and $G[F]$ as in Figure 1.25.

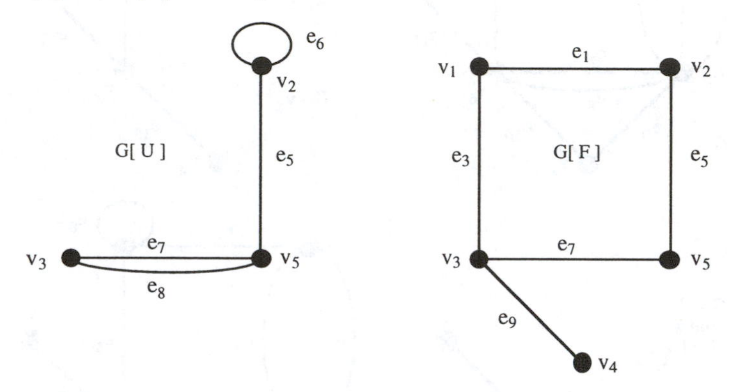

Figure 1.25: $G[U]$ and $G[F]$ for $U = \{v_2, v_3, v_5\}$ and $F = \{e_1, e_3, e_5, e_7, e_9\}$.

Two subgraphs G_1 and G_2 of a graph G are said to be **disjoint** if they have no vertex in common, and **edge disjoint** if they have no edge in common.

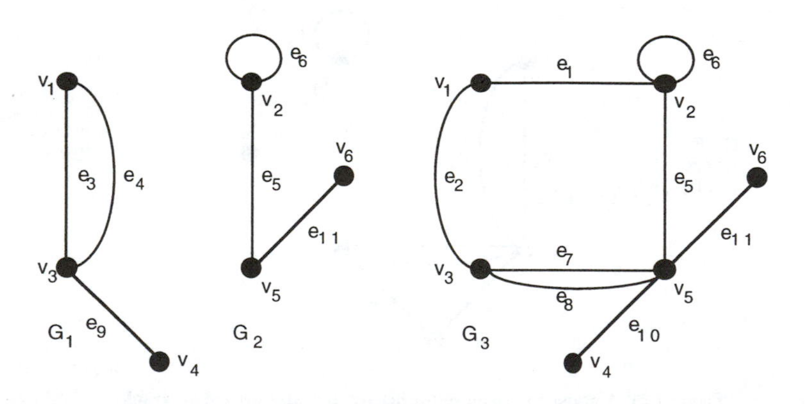

Figure 1.26: G_1 and G_2 are disjoint and G_1 and G_3 are edge disjoint.

For example, Figure 1.26 shows three subgraphs G_1, G_2 and G_3 of the graph G of Figure 1.24. Of these, G_1 and G_2 are disjoint and G_1 and G_3 are edge disjoint.

If two subgraphs G_1 and G_2 are disjoint then they must also be edge disjoint — if not then there would be an edge e of G in both G_1 and G_2 and then the end(s) of e would also be in both G_1 and G_2.

Given two subgraphs G_1 and G_2 of G, the **union** $G_1 \cup G_2$ is the subgraph of G with vertex set consisting of all those vertices which are in either G_1 or G_2 (or both) and with edge set consisting of all those edges which are in either G_1 or G_2 (or both); symbolically

$$V(G_1 \cup G_2) = V(G_1) \cup V(G_2),$$

$$E(G_1 \cup G_2) = E(G_1) \cup E(G_2).$$

For example, Figure 1.27 shows $G[U] \cup G[F]$ for the subgraphs $G[U]$ and $G[F]$ of Figure 1.25, while, for G_1 and G_3 of Figure 1.26, we have $G_1 \cup G_3 = G$.

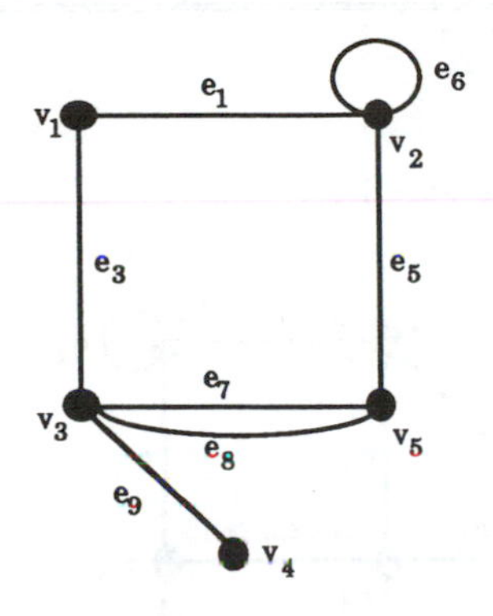

Figure 1.27: $G[U] \cup G[F]$.

If G_1 and G_2 are two subgraphs of G with at least one vertex in common then the **intersection** $G_1 \cap G_2$ is given by

$$V(G_1 \cap G_2) = V(G_1) \cap V(G_2),$$

$$E(G_1 \cap G_2) = E(G_1) \cap E(G_2).$$

Notice the requirement that the two subgraphs G_1 and G_2 must have at least one vertex in common before we can form their intersection — if there is no vertex belonging to both G_1 and G_2 we get $V(G_1) \cap V(G_2) = \emptyset$, denying us the possibility of any reasonable definition of an intersection, since any graph must have at least one vertex.

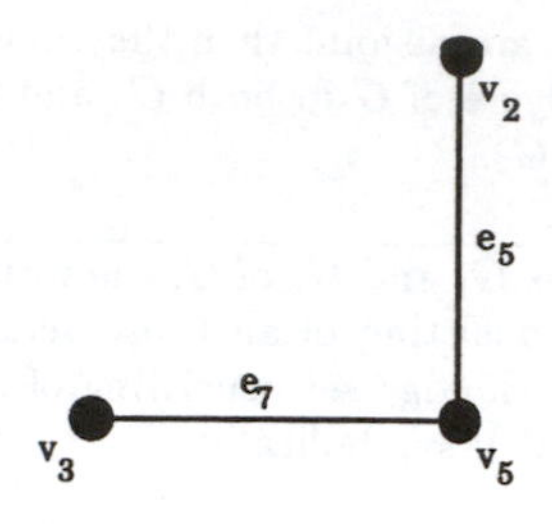

Figure 1.28: $G[U] \cap G[F]$.

Figure 1.28 shows $G[U] \cap G[F]$ for the subgraphs $G[U]$ and $G[F]$ of Figure 1.25, while, for G_1 and G_3 of Figure 1.26, we have $G_1 \cap G_3$ consists of three isolated vertices, namely v_1, v_3 and v_4.

Exercises for Section 1.5

1.5.1 Let G be the graph of Figure 1.29.

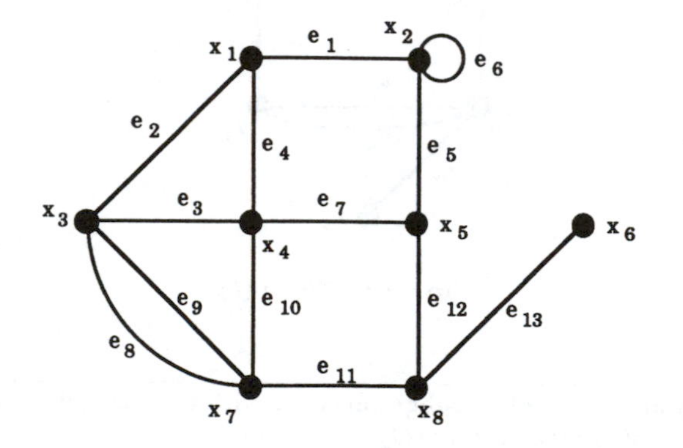

Figure 1.29

(a) Find $G - U$ where $U = \{x_1, x_3, x_5, x_7\}$.

(b) Find $G - F$ where $F = \{e_2, e_4, e_6, e_8, e_{10}, e_{12}\}$.

(c) Find $G[U]$ where $U = \{x_2, x_3, x_4, x_7\}$.

(d) Find $G[F]$ where $F = \{e_1, e_2, e_8, e_{11}\}$.

(e) Find a subgraph H of G isomorphic to K_3.

(f) Is there a subgraph of G isomorphic to K_4?

(g) What is the underlying simple graph of G? In how many ways can this be obtained?

(h) What is the intersection of the two subgraphs you found in (a) and (b)?

(i) What is the union of the two subgraphs you found in (c) and (d)?

1.5.2 Let G be a simple graph with n vertices. The **complement** $\overline{G}$ of G is defined to be the simple graph with the same vertex set as G and where two vertices u and v are adjacent precisely when they are *not* adjacent in G. Roughly speaking then, the complement of G can be obtained from the complete graph K_n by "rubbing out" all the edges of G. Figure 1.30 shows a graph G on 6 vertices and its complement $\overline{G}$.

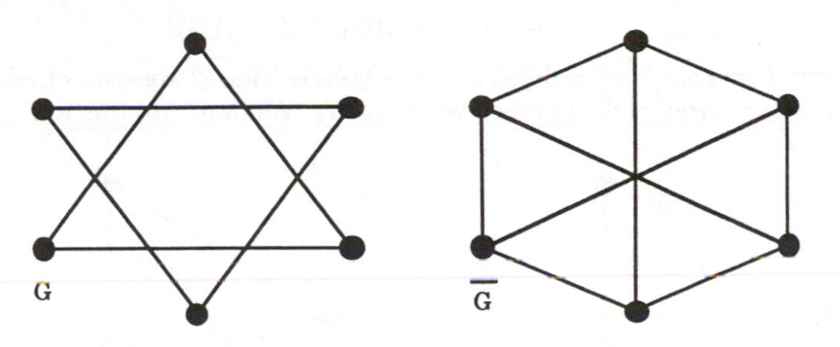

Figure 1.30: A graph and its complement.

Find the complements of the graphs in Figure 1.31.

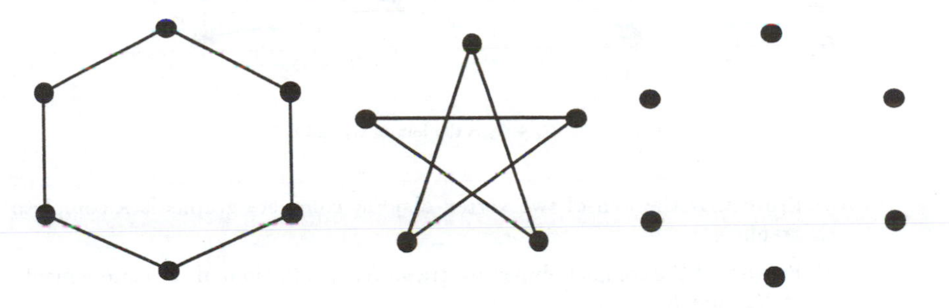

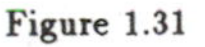

Figure 1.31

1.5.3 A simple graph is called **self-complementary** if it is isomorphic to its own complement.

(a) Find which of the graphs of Figure 1.31 are self-complementary.

(b) Prove that if G is a self-complementary graph with n vertices then n is either $4t$ or $4t + 1$ for some integer t. (Hint: consider the number of edges in K_n.)

(c) List all the self-complementary graphs with 4 or 5 vertices.

1.5.4 Let G be a simple graph with n vertices and let $\overline{G}$ be its complement.

(a) Prove that, for each vertex v in G, $d_G(v) + d_{\overline{G}}(v) = n - 1$.

(b) Suppose that G has exactly one even vertex. How many odd vertices does $\overline{G}$ have?

1.5.5 Let G_1 and G_2 be two graphs with no vertex in common. We define the **join** of G_1 and G_2, denoted by $G_1 + G_2$, to be the graph with vertex set and edge set given as follows:

$$V(G_1 + G_2) = V(G_1) \cup V(G_2),$$

$$E(G_1 + G_2) = E(G_1) \cup E(G_2) \cup J$$

where $J = \{x_1x_2 : x_1 \in V(G_1), x_2 \in V(G_2)\}$. Thus J consists of edges which join every vertex of G_1 to every vertex of G_2. We illustrate this in Figure 1.32.

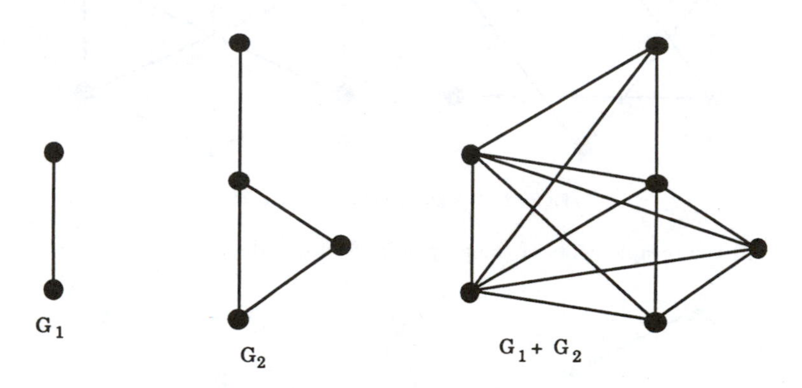

Figure 1.32: $G_1 + G_2$ is the join of G_1 and G_2.

(a) Prove that the join of two vertex disjoint complete graphs is a complete graph.

(b) Prove that the complete bipartite graph $K_{m,n}$ is the join of the complements of K_m and K_n.

(c) Let G_1, G_2 and G_3 be three graphs with no vertex common to any pair. Prove that $(G_1+G_2)+G_3 = G_1+(G_2+G_3)$. (This means that the expression $G_1 + G_2 + G_3$ is unambiguous.)

(d) Prove that if G_1 and G_2 are disjoint simple graphs then the complement of their join is the union of their complements.

1.6 Paths and Cycles

A **walk** in a graph G is a finite sequence

$$W = v_0 e_1 v_1 e_2 v_2 \dots v_{k-1} e_k v_k$$

whose terms are alternately vertices and edges such that, for $1 \leq i \leq k$, the edge e_i has ends v_{i-1} and v_i.

Thus each edge e_i is immediately preceded and succeeded by the two vertices with which it is incident.

We say that the above walk W is a $v_0 - v_k$ **walk** or a **walk from** v_0 **to** v_k. The vertex v_0 is called the **origin** of the walk W, while v_k is called the **terminus** of W. Note that v_0 and v_k need not be distinct.

The vertices $v_1, \dots, v_{k-1}$ in the above walk W are called its **internal vertices**. The integer k, the number of edges in the walk, is called the **length** of W.

Note that in a walk there may be repetition of vertices and edges.

In a *simple* graph, a walk $v_0 e_1 v_1 e_2 v_2 \dots v_{k-1} e_k v_k$ is determined by the sequence $v_0 v_1 \dots v_k$ of its vertices, since for each pair $v_{i-1} v_i$ there is only one possible edge with ends determined by the pair. In fact, even in graphs that are not simple, a walk is often simply denoted by a sequence of vertices

$$v_0 v_1 v_2 \dots v_{k-1} v_k$$

where consecutive vertices are adjacent. When this is done, it is to be understood that the discussion is valid for every walk with that vertex sequence.

A **trivial walk** is one containing no edges.

Thus, for any vertex v of G, $W = v$ gives a trivial walk. It has length 0.

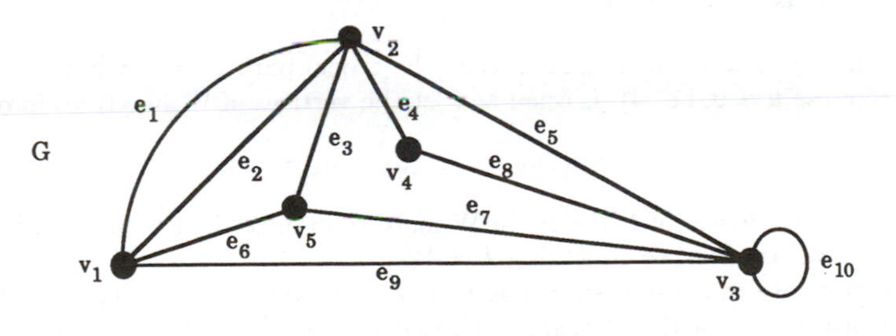

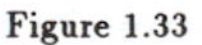
Figure 1.33

In Figure 1.33, $W_1 = v_1\ e_1\ v_2\ e_5\ v_3\ e_{10}\ v_3\ e_5\ v_2\ e_3\ v_5$ and $W_2 = v_1\ e_1\ v_2\ e_1\ v_1\ e_1\ v_2$ are both walks, of length 5 and 3 respectively, from v_1 to v_5 and from v_1 to v_2 respectively.

> Given two vertices u and v of a graph G, a $u-v$ walk is called **closed** or **open** depending on whether $u = v$ or $u \neq v$.

The two examples W_1 and W_2 above are both open while $W_3 = v_1\ v_5\ v_2\ v_4\ v_3\ v_1$ is closed.

> If the edges $e_1, e_2, \ldots, e_k$ of the walk $W = v_0e_1v_1e_2v_2 \ldots e_kv_k$ are distinct then W is called a **trail**.

In other words, a trail is a walk in which no edge is repeated. The examples W_1 and W_2 above are not trails, since, for example, e_5 is repeated in W_1, while e_1 is repeated in W_2. However, W_3 is a trail.

> If the vertices $v_0, v_1, \ldots, v_k$ of the walk $W = v_0e_1v_1e_2v_2 \ldots e_kv_k$ are distinct then W is called a **path**.
>
> Clearly any two paths with the same number of vertices are isomorphic.
>
> A path with n vertices will sometimes be denoted by P_n. Note that P_n has length $n-1$.

In other words, a path is a walk in which no vertex is repeated. Thus in a path no edge can be repeated either, so every path is a trail. Not every trail is a path, though. For example, W_3 above is not since v_1 is repeated. However $W_4 = v_2\ v_4\ v_3\ v_5v_1$ *is* a path in the graph G (of Figure 1.33).

By definition, every path is a walk. Although the converse of this statement is not true in general, we do have the following.

Theorem 1.3 *Given any two vertices u and v of a graph G, every $u-v$ walk contains a $u-v$ path, i.e., given any walk*

$$W = ue_1v_1 \ldots v_{k-1}e_kv$$

then, after some deletion of vertices and edges if necessary, we can find a subsequence P of W which is a $u-v$ path.

Proof If $u = v$, i.e., if W is closed, then the trivial path $P = u$ will do.

Now suppose $u \neq v$, i.e., W is open and let the vertices of W be given, in order, by

$$u = u_0, u_1, u_2, \ldots, u_{k-1}, u_k = v.$$

If none of the vertices of G occurs in W more than once then W is already a $u-v$ path and so we are finished by taking $P = W$.

So now suppose that there are vertices of G that occur in W twice or more. Then there are distinct i, j, with $i < j$, say, such that $u_i = u_j$. If the terms $u_i, u_{i+1}, \ldots, u_{j-1}$ (and the preceding edges) are deleted from W then we obtain a $u-v$ walk W_1 having fewer vertices than W. If there is no repetition of vertices in W_1, then W_1 is a $u-v$ path and setting $P = W_1$ finishes the proof.

If this is not the case, then we repeat the above deletion procedure until finally arriving at a $u-v$ walk that is a path, as required. □

To illustrate the above proof let us look at the walk

$$W = v_1\ e_1\ v_2\ e_5\ v_3\ e_{10}\ v_3\ e_5\ v_2\ e_3\ v_5$$

in our example G of Figure 1.33. Here, (see Figure 1.34),

$$u = u_1 = v_1,\ u_2 = v_2,\ u_3 = v_3,\ u_4 = v_3,\ u_5 = v_2,\ u_6 = v_5 = v.$$

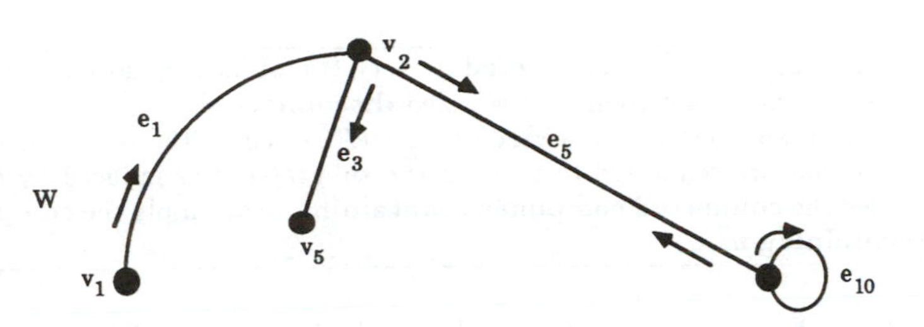

Figure 1.34: The walk W.

Since $u_3 = u_4$ the deletion procedure deletes u_3 (and the edge e_{10}) to give the walk

$$W_1 = v_1\ e_1\ v_2\ e_5\ v_3\ e_5\ v_2\ e_3\ v_5.$$

shown in Figure 1.35. The next stage is to delete $v_2e_5v_3e_5$ to give

$$W_2 = v_1\ e_1\ v_2\ e_3\ v_5,$$

also shown in Figure 1.35, and this is a path, as required.

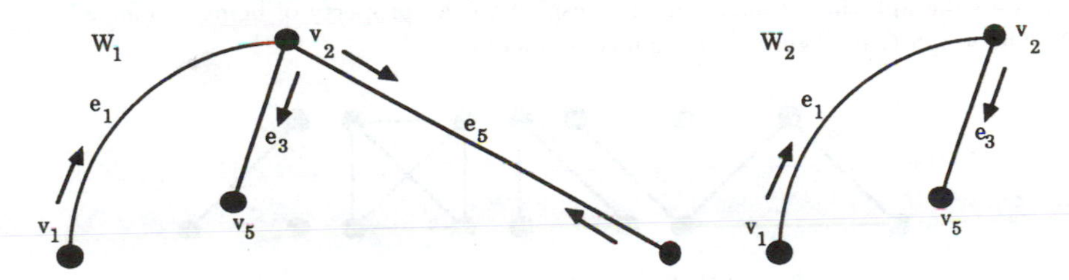

Figure 1.35: The deletion procedure applied to the walk W of Figure 1.34.

A vertex u is said to be **connected** to a vertex v in a graph G if there is a path in G from u to v.

Clearly if u is connected to v then v is connected to u — just reverse the path. Also any vertex u is connected to itself by the trivial path $P = u$.

Moreover, if u is connected to v and v is connected to w, then u is connected to w. To see this, suppose that $W_1 = ue_1 \dots e_k v$ is a $u - v$ path and $W_2 = vf_1 \dots f_t w$ is a

$v - w$ path. Then, joining the two paths together in the obvious way, we get a $u - w$ walk

$$W = ue_1 \dots e_k v f_1 \dots f_t w$$

and, by Theorem 1.3, this contains the required $u - w$ path.

The above process of joining two walks which have a common "end point" to form a longer walk is called **concatenation** — meaning "stringing together".

> A graph G is called **connected** if every two of its vertices are connected. A graph that is not connected is called **disconnected**.
>
> Given any vertex u of a graph G, let $C(u)$ denote the set of all vertices in G that are connected to u. Then the subgraph of G induced by $C(u)$ is called the **connected component containing** u, or simply the **component containing** u.

If u and v are two connected vertices in the graph G, i.e., if there is a path from u to v, then, by the remarks above, $C(u) = C(v)$ and so u and v have the same connected component. Conversely, if u and v have the same component then v is in $C(u)$ so v and u must be connected.

Another way of describing a component is as follows: it is a connected subgraph C of the graph G which is not properly contained in any other connected subgraph of G. For if there was a connected subgraph D of G such that $C \subseteq D$ then, since D is connected, every vertex w of D is connected to every vertex of C, showing, from above, that C and D have the same vertex set, and by the definition of component, C has got all the edges of G that have both ends in C and so D can not have any more edges than C, i.e., $C = D$. We also express this by saying that a component of G is a subgraph that is maximal with respect to the property of being connected.

The graph G of Figure 1.36 has 6 components.

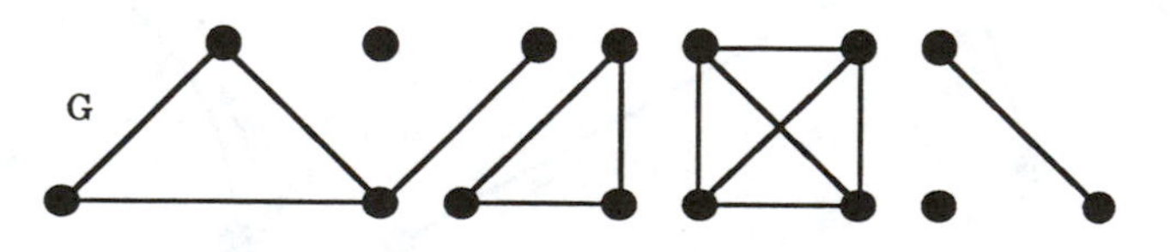

Figure 1.36: A graph with six connected components.

> The number of components of a graph G is denoted by $\omega(G)$.

Thus $\omega(G) = 6$ for the graph G of Figure 1.36. Of course a graph G is connected if and only if $\omega(G) = 1$.

It is sometimes not obvious what $\omega(G)$ is or what the components of G look like. For example, in Figure 1.37, $\omega(G) = 2$ but this is not seen immediately.

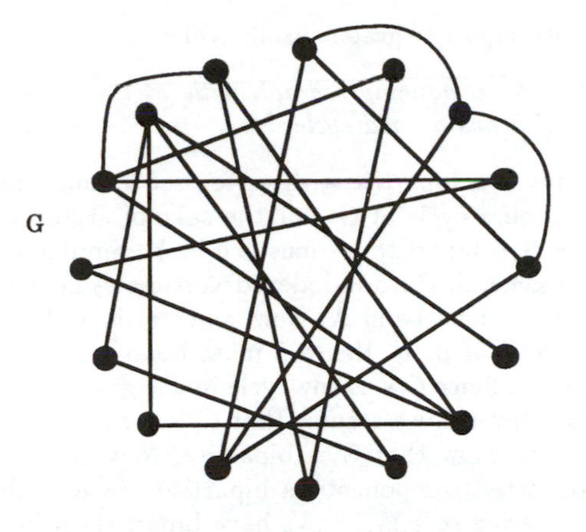

Figure 1.37: $\omega(G) = 2$ — can you see why?

A nontrivial closed trail in a graph G is called a **cycle** if its origin and internal vertices are distinct. In detail, the closed trail $C = v_1v_2 \ldots v_nv_1$ is a cycle if

(i) C has at least 1 edge and

$v_1, v_2, \ldots, v_n$ are n distinct vertices.

A cycle of length k, i.e., with k edges, is called a **k-cycle**. A k-cycle is called **odd** or **even** depending on whether k is odd or even.

A 3-cycle is often called a **triangle**.

Clearly any two cycles of the same length are isomorphic. An n-cycle, i.e., a cycle with n vertices, will sometimes be denoted by C_n.

For example in Figure 1.38, $C = v_1\ v_2\ v_3\ v_4\ v_1$ is a 4-cycle, $T = v_1\ v_2\ v_5\ v_3\ v_4\ v_5\ v_1$ is a nontrivial closed trail which is not a cycle (because v_5 occurs twice as an internal vertex), and $C' = v_1\ v_2\ v_5\ v_1$ is a triangle.

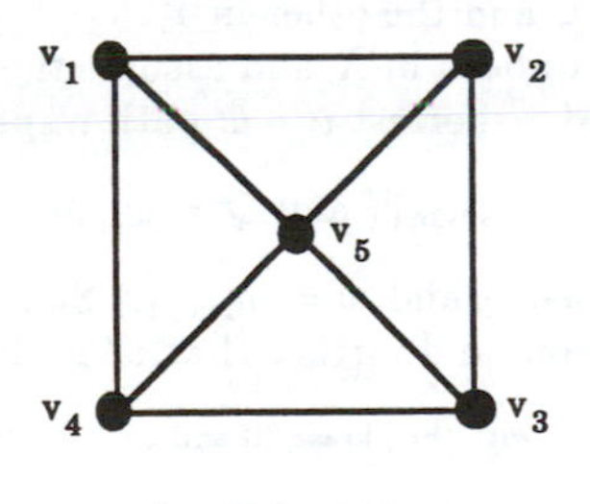

Figure 1.38

Note that a loop is just a 1-cycle. Also, given a pair of parallel edges e_1 and e_2 with distinct end vertices v_1 and v_2, we can form the cycle $v_1\ e_1\ v_2\ e_2\ v_1$ of length 2. Conversely, the two edges of any cycle of length 2 are a pair of parallel edges.

We now characterize bipartite graphs using cycles.

Theorem 1.4 *Let G be a nonempty graph with at least two vertices. Then G is bipartite if and only if it has no odd cycles.*[1]

Proof Suppose that G is bipartite with vertex set V and bipartition $V = X \cup Y$. Let $C = v_0 v_1 \ldots v_k v_0$ be a cycle of G. For the sake of argument, assume that v_0 is in X. Then, because G is bipartite, v_1 must be in Y. Similarly v_2 must be in X, v_3 in Y, etc. In fact, in general, the odd-indexed vertices v_{2i+1} must be in Y while the even-indexed vertices v_{2i} must be in X. Now, since v_0 is in X, we must have (at the "other end" of the cycle) v_k in Y. Hence k must be an odd number. Thus the cycle $C = v_0 v_1 \ldots v_k v_0$ is even. Since C was any cycle in the graph G, G has no odd cycles.[2]

Now, to prove the converse, we assume that G is a nonempty graph which has no odd cycles. We wish to show that G is bipartite. Now G will be bipartite if each of its nonempty connected components is bipartite, since, if these components are $C_1, \ldots, C_n$ and their vertex sets $V_1, \ldots, V_n$ have bipartitions $V_1 = X_1 \cup Y_1, \ldots, V_n = X_n \cup Y_n$, then the vertex set V of G has bipartition $V = X \cup Y$ where

$$X = X_0 \cup X_1 \cup X_2 \cup \ldots \cup X_n \text{ and } Y = Y_1 \cup Y_2 \cup \ldots \cup Y_n,$$

where X_0 is the set of isolated vertices in G. (The details of this argument are asked for in Exercise 1.6.15.) As a result of this it is enough to show that if G is a nonempty *connected* graph with no odd cycles then G is bipartite.

With this assumption, let u be a fixed vertex of G. We define two subsets of the vertex set V of G as follows:

X is the set of all vertices v of G with the property that any shortest $u - v$ path of G has even length,

Y is the set of all vertices w of G with the property that any shortest $u - w$ path of G has odd length,

i.e., X consists of those vertices of G an "even distance" from u, while Y consists of those vertices of G an "odd distance" from u.

Note that u itself is in X. Then, clearly, $V = X \cup Y$ and X and Y have no element in common. We show that $V = X \cup Y$ is a bipartition of G by showing that any edge of G must have one end in X and the other in Y.

Let v and w be two vertices both in X and assume they are adjacent. Let P and Q be a shortest $u - v$ path and a shortest $u - w$ path respectively, say

$$P = u_1, u_2, \ldots, u_{2n+1} \text{ and } Q = w_1, w_2, \ldots, w_{2m+1},$$

(so that $u = u_1 = w_1$, $v = u_{2n+1}$ and $w = w_{2m+1}$.) Suppose that w' is a vertex that the two paths have in common, and further that w' is the last such vertex. (If $v = w$

---

[1] Some readers may not be familiar with the phrase "if and only if". It involves two implications. Here for example it says:

(1) *if* a nonempty graph is bipartite *then* it has no odd cycles, *and also the converse statement, namely:*

(2) *if* a nonempty graph has no odd cycles *then* it is bipartite.

Another way of describing this is to say that the following two statements are equivalent:

(i) the nonempty graph G is bipartite ;

(ii) the nonempty graph G has no odd cycles.

[2] We are not finished the proof — in fact only half-finished; we still have to prove "the other way".

then of course $w' = v = w$. Moreover, there is always such a vertex w' since the vertex u is common to both paths.) Then that part of P from u to w' is a shortest path from u to w' and that part of Q from u to w' is a shortest path from u to w' also. In other words, we have two shortest paths from u to w'. It follows that, since these two paths have the same length, there exists an i such that $w' = u_i = w_i$. However this produces an odd cycle in G:

$$C = \underbrace{u_i u_{i+1} \dots u_{2n+1}}_{*} \underbrace{w_{2m+1} w_{2m} \dots w_i}_{**}$$

since if i is odd then the above parts $*$ and $**$ are both of even length while if i is even then they are both of odd length, giving the total length of C as odd + 1 + odd or even + 1 + even, in either case odd. Since G has no odd cycles, something is wrong (!), namely the assumption that v and w are adjacent.

Hence there are no edges in G joining vertices of X. A similar argument shows that there are no edges of G joining vertices of Y. Hence G is bipartite, as required. □

Exercises for Section 1.6

1.6.1 Let G be the graph of Figure 1.39.

(a) Find a closed walk of length 6. Is your walk a trail?

(b) Find an open walk of length 12. Is your walk a path?

(c) Find a closed trail of length 6. Is your trail a cycle?

(d) What is the length of the longest cycle in G?

(e) What is the length of a longest path in G? How many paths in G are there of this length?

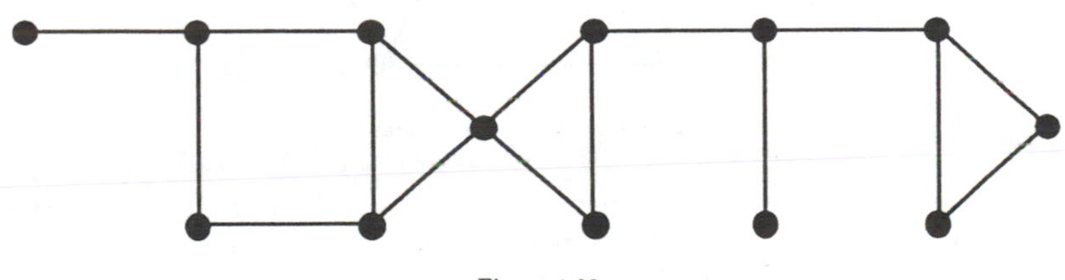

Figure 1.39

1.6.2 Let G be a graph with 15 vertices and 4 connected components. Prove that G has at least one component with at least 4 vertices. What is the largest number of vertices that a component of G can have?

1.6.3 Let v be a vertex in the graph G and let T be a closed trail with origin (and terminus) v. Modify the proof of Theorem 1.3 to show that T contains a cycle with origin v.

1.6.4 For each $n \geq 4$, the **wheel** graph, W_n, with n vertices, is defined to be the join $K_1 + C_{n-1}$, of an isolated vertex with a cycle of length $n-1$. (See Exercise 1.5.5 for the definition of *join*.)

(a) Draw the graphs W_4, W_5, W_6 and W_7 in such a way that you can see why the name *wheel* is given to this family of graphs.

(b) Show that W_n contains a cycle of length k for each k, $4 \leq k \leq n$.

1.6.5 Prove that if u is an odd vertex in a graph G then there must be a path in G from u to another odd vertex v of G.

1.6.6 Give an example of a graph in which the length of the longest cycle is 9 and the length of the shortest cycle is 4.

1.6.7 The graph G of Figure 1.40 is known as the **Petersen graph**. Find in G

(a) a trail of length 5,

(b) a path of length 9,

(c) cycles of lengths 5 , 6 , 8 and 9.

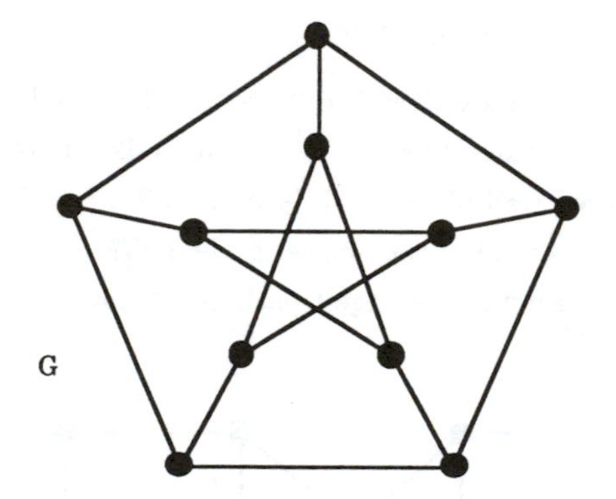

Figure 1.40: The Petersen graph.

1.6.8 For any two vertices u and v connected by a path in a graph G, we define the **distance** between u and v, denoted by $d(u,v)$, to be the length of a shortest u-v path. If there is no path connecting u and v we define $d(u,v)$ to be infinite. (This idea of distance was used in the proof of Theorem 1.4.)

(a) Prove that for any vertices u, v, w in G we have

$$d(u,w) \leq d(u,v) + d(v,w).$$

(b) Prove that if $d(u,v) \geq 2$, then there is a vertex z in G such that

$$d(u,v) = d(u,z) + d(z,v).$$

(c) Prove that in the Petersen graph of Exercise 1.6.7 above, $d(u,v) \leq 2$ for any pair of vertices u, v. (By using the symmetry of the graph you need not look at *every* pair of vertices.)

1.6.9 Let G be a connected graph with vertex set V. For each $v \in V$, the **eccentricity** of v, denoted by $e(v)$, is defined by

$$e(v) = \max\{d(u,v) : u \in V, u \neq v\}.$$

The **radius** of G, denoted by rad G , is defined by

$$\text{rad } G = \min\{e(v) : v \in V\},$$

while the **diameter** of G, denoted by diam G, is defined by

$$\text{diam } G = \max\{e(v) : v \in V\}.$$

Thus the diameter of G is given by $\max\{d(u,v) : u, v \in V\}$.

(a) Find the radius and the diameter of the graph of Figure 1.39 and the Petersen graph (Figure 1.40).

(b) Prove that for any connected graph G,

$$\text{rad } G \leq \text{diam } G \leq 2 \text{ rad } G.$$

(c) Find the radius and the diameter of the wheel graphs W_n of Exercise 1.6.4.

(d) Which simple graphs have diameter 1?

1.6.10 Let G be a simple connected graph. The **square** of G, denoted by G^2, is defined to be the graph with the same vertex set as G and in which two vertices u and v are joined by an edge if and only if in G we have $1 \leq d(u,v) \leq 2$. An example of a graph G and its square is shown in Figure 1.41.

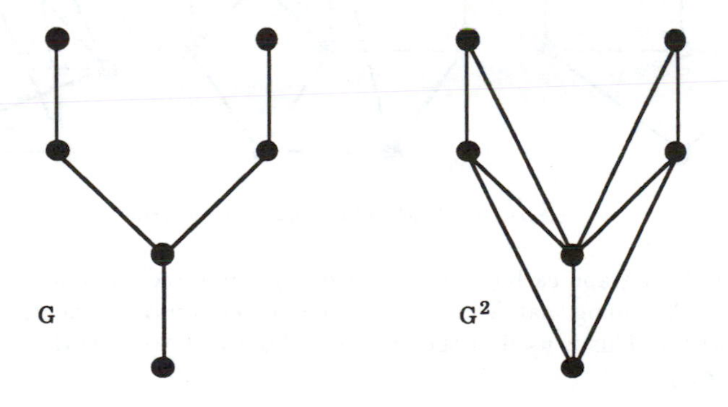

Figure 1.41: A graph and its square.

(a) Show that the square of $K_{1,3}$ is K_4. Can you find two more graphs whose square is K_4?

(b) Draw the squares of the paths P_4, P_5, P_6, the cycles C_5, C_6 and the wheels W_5 and W_6.

1.6.11 Let G be a simple graph with n vertices, where $n \geq 2$. Prove that G has two vertices u and v with $d(u) = d(v)$. (Hint: if G is nonempty, consider $G - e$ where e is an edge of G, and use induction on the number of edges of G.)

1.6.12 Let G be a simple graph. Show that if G is not connected then its complement $\overline{G}$ is connected.

1.6.13 A **complete tripartite graph** G is a simple graph G in which the vertex set V is the union of three nonempty subsets V_1, V_2 and V_3 where $V_i \cap V_j = \emptyset$ for $i \neq j$ and an edge joins two vertices u, v of G if and only if u and v do not belong to the same V_i. If V_1, V_2 and V_3 have r,s and t elements respectively where $r \leq s \leq t$ we denote G by $K_{r,s,t}$.

(a) Draw $K_{1,2,2}$, $K_{2,2,2}$, $K_{2,3,3}$.

(b) How many edges are there in $K_{r,s,t}$?

(c) Formulate a definition of a **complete n-partite graph** for any $n \geq 3$.

1.6.14 Which of the graphs in Figure 1.42 are bipartite? Justify your answer using Theorem 1.4 and redraw those that are bipartite showing the bipartite property more clearly.

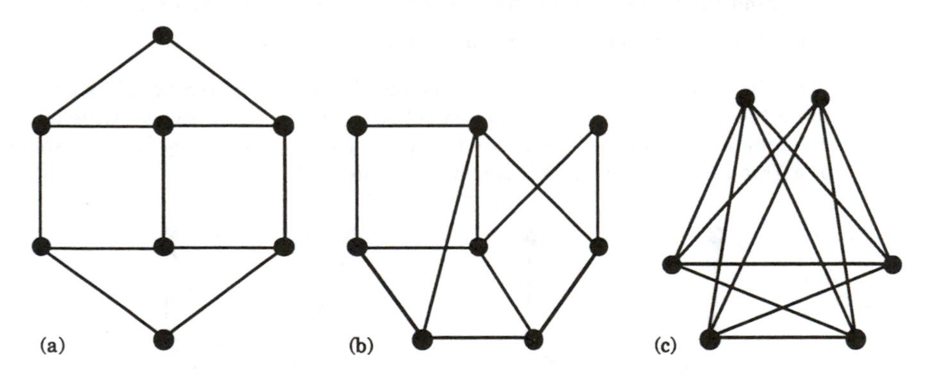

Figure 1.42: Which of these graphs are bipartite?

1.6.15 Let G be a graph each of whose nonempty connected components is a bipartite graph. Assuming that G has at least one nonempty component, prove that G is bipartite. (This is used in the proof of Theorem 1.4 — see there on how to get started.)

1.7 The Matrix Representation of Graphs

There are essentially two different ways of representing a graph inside a computer, namely by using the adjacency matrix or the incidence matrix of a graph.

> Let G be a graph with n vertices, listed as $v_1, \ldots, v_n$. The **adjacency matrix** of G, with respect to this particular listing of the n vertices of G, is the $n \times n$ matrix $A(G) = (a_{ij})$ where the (i,j)th entry a_{ij} is the number of edges joining the vertex v_i to the vertex v_j.

Figure 1.43 shows a graph G with vertices listed as $v_1, \ldots, v_4$ and its adjacency matrix $A(G)$ with respect to this listing.

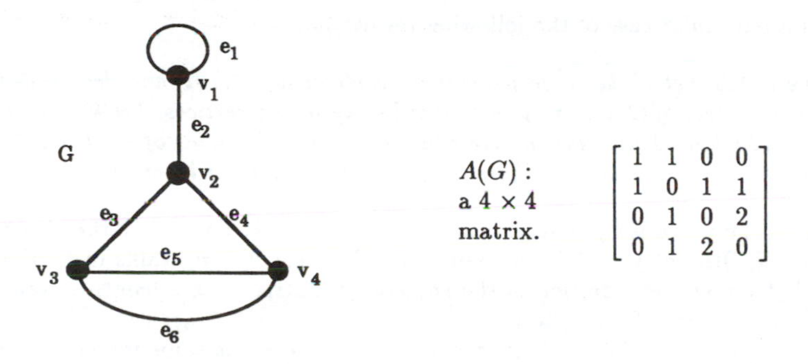

Figure 1.43: A graph and its adjacency matrix.

Note that in $A(G)$ we have $a_{ij} = a_{ji}$ for each i and j. A matrix with this property is called **symmetric**. Note also that if G has no loops then all the entries of the main diagonal of $A(G)$ are 0, while if G has no parallel edges then the entries of $A(G)$ are either 0 or 1.

Given an $n \times n$ symmetric matrix $A = (a_{ij})$ in which all the entries are non-negative integers, we can associate with it a graph G whose adjacency matrix is A, simply by letting G have n vertices, labelled 1 to n, say, and joining vertex i to vertex j by a_{ij} edges. Figure 1.44 shows such a symmetric matrix A and a graph produced from it.

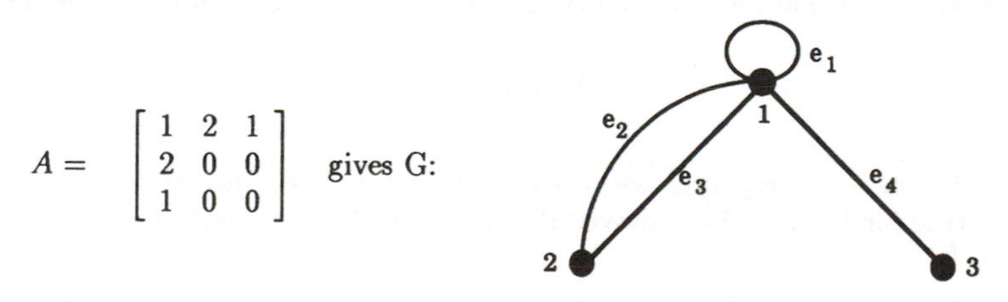

Figure 1.44: A symmetric matrix A of non-negative integers and a graph G with $A(G) = A$.

In this example the matrix multiplication $A \times A = A^2$ gives

$$A^2 = \begin{bmatrix} 6 & 2 & 1 \\ 2 & 4 & 2 \\ 1 & 2 & 1 \end{bmatrix} = (b_{ij}), \text{ say,}$$

and in fact b_{ij} gives the number of walks of length 2 from vertex i to vertex j. For example, $b_{11} = 6$ and the 6 walks of length 2 from vertex 1 to itself are

$$1e_1 1e_1 1,\ 1e_2 2e_2 1,\ 1e_3 2e_3 1,\ 1e_2 2e_3 1,\ 1e_3 2e_2 1,\ 1e_4 3e_4 1.$$

As a further example, $b_{23} = b_{32} = 2$ and the 2 walks of length 2 from vertex 2 to vertex 3 are

$$2e_2 1e_4 3,\ 2e_3 1e_4 3.$$

This is a particular case of the following result:

Theorem 1.5 *Let G be a graph with n vertices $v_1, \ldots, v_n$ and let A denote the adjacency matrix of G with respect to this listing of the vertices. Let k be any positive integer and let A^k denote the matrix multiplication of k copies of A. Then the (i,j)th entry of A^k is the number of different $v_i - v_j$ walks in G of length k.*

Proof The proof is by mathematical induction on k. For $k = 1$ the theorem says that the (i,j)th entry of A is the number of different $v_i - v_j$ walks in G of length 1, which is true by the definition of the adjacency matrix since a length 1 walk from v_i to v_j is just an edge from v_i to v_j.

Now suppose that the result is true for A^{k-1}, where k is some integer greater than 1. We wish to prove it true for A^k. Thus, setting $A^{k-1} = (b_{ij})$, we are assuming that b_{ij} is the number of different walks of length $k-1$ from v_i to v_j and we want to prove that if $A^k = (c_{ij})$ then c_{ij} is the number of different walks of length k from v_i to v_j. Set $A = (a_{ij})$. Since $A^k = A^{k-1} \times A$, from the definition of matrix multiplication we get

$$c_{ij} = \sum_{t=1}^{n} ((i,t)\text{th element of } A^{k-1}) \times ((t,j)\text{th element of } A) = \sum_{t=1}^{n} b_{it}\, a_{tj}.$$

Now every $v_i - v_j$ walk of length k consists of a $v_i - v_t$ walk of length $k-1$, where v_t is adjacent to v_j, followed by an edge $v_t v_j$. Since there are b_{it} such walks of length $k-1$ and a_{tj} such edges for each vertex v_t, the total number of all $v_i - v_j$ walks is

$$\sum_{t=1}^{n} b_{it}\, a_{tj}.$$

Since this, by the above, is just c_{ij} we have established the result for A^k. Thus assuming the result true for $k-1$ we have proved it true for k. Hence by induction the proof is complete. □

We can use the above result to determine whether or not a graph is connected:

Theorem 1.6 *Let G be a graph with n vertices $v_1, \ldots, v_n$ and let A denote the adjacency matrix of G with respect to this listing of the vertices. Let $B = (b_{ij})$ be the matrix*

$$B = A + A^2 + \cdots + A^{n-1}.$$

Then G is a connected graph if and only if for every pair of distinct indices i, j we have $b_{ij} \neq 0$, i.e., if and only if B has no zero entries off the main diagonal.

Proof Let $a_{ij}^{(k)}$ denote the (i,j)th entry of the matrix A^k for each $k = 1, \ldots, n-1$. Then

$$b_{ij} = a_{ij}^{(1)} + a_{ij}^{(2)} + \cdots + a_{ij}^{(n-1)}.$$

However, by Theorem 1.5, $a_{ij}^{(k)}$ denotes the number of distinct walks of length k from v_i to v_j and so

$$\begin{aligned} b_{ij} &= (\text{number of different } v_i - v_j \text{ walks of length } 1) \\ &+ (\text{number of different } v_i - v_j \text{ walks of length } 2) \\ &\ \ \vdots \\ &+ (\text{number of different } v_i - v_j \text{ walks of length } (n-1)), \end{aligned}$$

i.e., b_{ij} is the number of different $v_i - v_j$ walks of length less than n.

Now suppose that G is connected. Then for every pair of distinct indices i, j there is a path from v_i to v_j. Since G has only n vertices this path goes through at most n vertices and so it has length less than n, i.e., there is at least 1 path from v_i to v_j of length less than n. Hence $b_{ij} \neq 0$ for each i, j with $i \neq j$, as required.

Conversely, suppose that for each distinct pair i, j we have $b_{ij} \neq 0$. Then, from above, there is at least 1 walk (of length less than n) from v_i to v_j. In particular, v_i is connected to v_j. Thus G is a connected graph, as required, since i and j were an arbitrary pair of distinct vertices. □

As an illustration, we use Theorem 1.6 to find whether a particular graph G is connected or not using its adjacency matrix $A(G)$. Suppose that $A(G)$ is the following matrix A:

$$A = \begin{bmatrix} 0 & 0 & 1 & 0 & 0 \\ 0 & 0 & 0 & 1 & 0 \\ 1 & 0 & 0 & 0 & 1 \\ 0 & 1 & 0 & 0 & 1 \\ 0 & 0 & 1 & 1 & 0 \end{bmatrix}.$$

Here $n = 5$ so the theorem tells us to look at $B = A + A^2 + A^3 + A^4$. Now

$$A^2 = \begin{bmatrix} 1 & 0 & 0 & 0 & 1 \\ 0 & 1 & 0 & 0 & 1 \\ 0 & 0 & 2 & 1 & 0 \\ 0 & 0 & 1 & 2 & 1 \\ 1 & 1 & 0 & 0 & 2 \end{bmatrix} \text{ so } A + A^2 = \begin{bmatrix} 1 & 0 & 1 & 0 & 1 \\ 0 & 1 & 0 & 1 & 1 \\ 2 & 0 & 2 & 1 & 1 \\ 0 & 1 & 1 & 2 & 1 \\ 1 & 1 & 1 & 1 & 2 \end{bmatrix},$$

$$A^3 = \begin{bmatrix} 0 & 0 & 2 & 1 & 0 \\ 0 & 0 & 1 & 2 & 0 \\ 2 & 1 & 0 & 0 & 3 \\ 1 & 2 & 0 & 0 & 3 \\ 0 & 0 & 3 & 3 & 0 \end{bmatrix} \text{ so } A + A^2 + A^3 = \begin{bmatrix} 1 & 0 & 3 & 1 & 1 \\ 0 & 1 & 1 & 3 & 1 \\ 3 & 1 & 2 & 1 & 4 \\ 1 & 3 & 1 & 2 & 4 \\ 1 & 1 & 4 & 4 & 2 \end{bmatrix}, \text{ and}$$

$$A^4 = \begin{bmatrix} 2 & 1 & 0 & 0 & 3 \\ 1 & 2 & 0 & 0 & 3 \\ 0 & 0 & 5 & 4 & 0 \\ 0 & 0 & 4 & 5 & 0 \\ 3 & 3 & 0 & 0 & 6 \end{bmatrix} \text{ so } B = A + A^2 + A^3 + A^4 = \begin{bmatrix} 3 & 1 & 3 & 1 & 4 \\ 1 & 3 & 1 & 3 & 4 \\ 3 & 1 & 7 & 5 & 4 \\ 1 & 3 & 5 & 7 & 4 \\ 4 & 4 & 4 & 4 & 8 \end{bmatrix}.$$

Since this last matrix, B, has no nonzero entries off the main diagonal we conclude that the graph is connected. Figure 1.45 gives a drawing of the graph G.

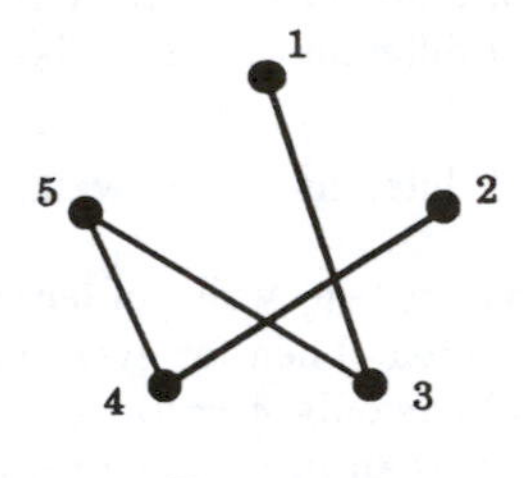

Figure 1.45

Often we do not have to go up to A^{n-1}. For example, for the graph with adjacency matrix

$$A = \begin{bmatrix} 0 & 1 & 1 & 1 & 1 & 1 \\ 1 & 0 & 1 & 0 & 0 & 0 \\ 1 & 1 & 0 & 1 & 0 & 0 \\ 1 & 0 & 1 & 0 & 1 & 0 \\ 1 & 0 & 0 & 1 & 0 & 1 \\ 1 & 0 & 0 & 0 & 1 & 0 \end{bmatrix} \text{ we have } A^2 = \begin{bmatrix} 5 & 1 & 2 & 2 & 2 & 1 \\ 1 & 2 & 1 & 2 & 1 & 1 \\ 2 & 1 & 3 & 1 & 2 & 1 \\ 2 & 2 & 1 & 3 & 1 & 2 \\ 2 & 1 & 2 & 1 & 3 & 1 \\ 1 & 1 & 1 & 2 & 1 & 1 \end{bmatrix}$$

and here, by Theorem 1.5, since every entry is nonzero there is at least one walk of length 2 from every vertex in the graph to every other vertex and so G is connected. Notice however that in our worked example above, we *did* have to go up to A^{n-1}.

For the graph G with adjacency matrix

$$A = \begin{bmatrix} 0 & 1 & 0 & 0 & 0 \\ 1 & 0 & 0 & 0 & 1 \\ 0 & 0 & 0 & 1 & 0 \\ 0 & 0 & 1 & 0 & 0 \\ 0 & 1 & 0 & 0 & 0 \end{bmatrix}$$

we have $n = 5$ and (check) $B = A + A^2 + A^3 + A^4$ is

$$\begin{bmatrix} 3 & 3 & 0 & 0 & 3 \\ 3 & 6 & 0 & 0 & 3 \\ 0 & 0 & 2 & 2 & 0 \\ 0 & 0 & 2 & 2 & 0 \\ 3 & 3 & 0 & 0 & 3 \end{bmatrix}.$$

So, since B has zeros off the diagonal, G is not connected. In fact, looking at B we can identify two components, one with vertices v_1, v_2, v_5, the other with v_3, v_4.

Another matrix associated with a graph G is given as follows.

> Suppose that G has n vertices, listed as $v_1, \ldots, v_n$, and t edges, listed as $e_1, \ldots, e_t$. The **incidence matrix** of G, with respect to these particular listings of the vertices and edges of G, is the $n \times t$ matrix $M(G) = (m_{ij})$ where m_{ij} is the number of times that the vertex v_i is incident with the edge e_j, i.e.,
>
> $$m_{ij} = \begin{cases} 0 & \text{if } v_i \text{ is not an end of } e_j \\ 1 & \text{if } v_i \text{ is an end of the non-loop } e_j \\ 2 & \text{if } v_i \text{ is an end of the loop } e_j. \end{cases}$$

Figure 1.46 shows a graph G, with four vertices $v_1, \ldots, v_4$ and six edges $e_1, \ldots, e_6$, and its incidence matrix $M(G)$ with respect to these listings of the vertices and edges.

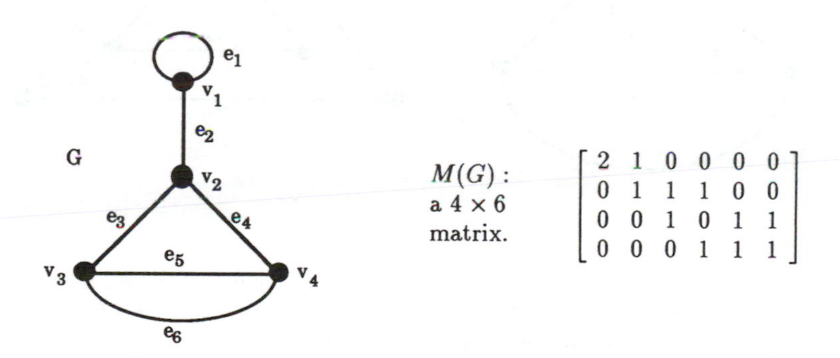

Figure 1.46: A graph and its incidence matrix.

Notice that the sum of the elements in the ith row of $M(G)$ gives us the degree of the vertex v_i, while the sum of the elements in each column is 2 (corresponding to the 2 ends of the edge).

Exercises for Section 1.7

1.7.1 Draw the graphs having the following matrices as their adjacency matrices.

(a) $\begin{bmatrix} 1 & 1 & 0 & 1 & 0 \\ 1 & 0 & 1 & 0 & 0 \\ 0 & 1 & 0 & 1 & 0 \\ 1 & 0 & 1 & 0 & 0 \\ 0 & 0 & 0 & 0 & 1 \end{bmatrix}$ (b) $\begin{bmatrix} 1 & 1 & 1 & 0 & 0 \\ 1 & 1 & 1 & 0 & 0 \\ 1 & 1 & 1 & 0 & 0 \\ 0 & 0 & 0 & 0 & 1 \\ 0 & 0 & 0 & 1 & 0 \end{bmatrix}$

(c) $\begin{bmatrix} 0 & 1 & 0 & 0 \\ 1 & 0 & 2 & 2 \\ 0 & 2 & 1 & 2 \\ 0 & 2 & 2 & 1 \end{bmatrix}$ (d) $\begin{bmatrix} 0 & 1 & 2 & 3 \\ 1 & 0 & 3 & 2 \\ 2 & 3 & 0 & 1 \\ 3 & 2 & 1 & 0 \end{bmatrix}$.

1.7.2 Write down the adjacency matrix and the incidence matrix for each of the graphs in Figure 1.47 using the ordering of the vertices and edges given.

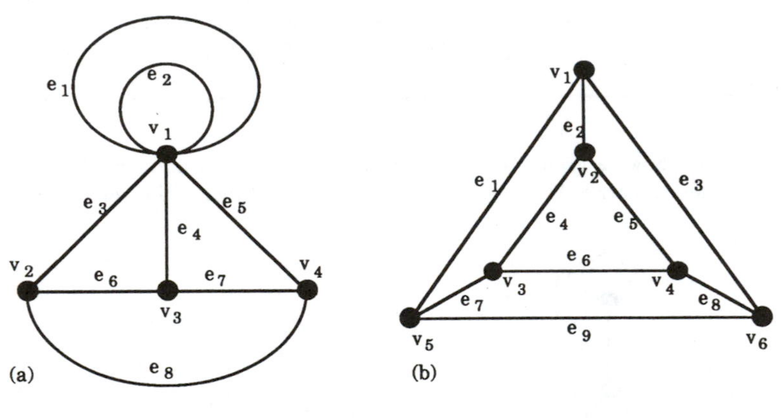

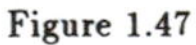
Figure 1.47

1.7.3 Let G be a simple graph and A be its adjacency matrix. Prove that the entries on the main diagonal of A^2 give the degrees of the vertices of G. Does this remain true if we drop the "simple" condition?

1.7.4 Let G be a bipartite graph. Show that we can list the vertices of G so that the corresponding adjacency matrix of G has the form

$$A[G] = \left[\begin{array}{c|c} O & C \\ \hline D & O \end{array}\right]$$

where the O's, C and D are submatrices, the submatrices O have entries all zero, and C is the "mirror-image" of D. (Formally, C is the matrix transpose of D.)

1.7.5 Use the powers of the adjacency matrix to determine if the graphs of Exercise 1.7.1 are connected or not.

1.8 Fusion

Let u and v be distinct vertices of a graph G. We can construct a new graph G_1 by **fusing** (or **identifying**) the two vertices, namely by replacing them by a single new vertex x such that every edge that was incident with either u or v in G is now incident with x, i.e., the end u and the end v become end x.

Thus the new graph G_1 has one less vertex than G but the same number of edges as G and the degree of the new vertex x is the sum of the degrees of u and v.

We illustrate the process in Figure 1.48.

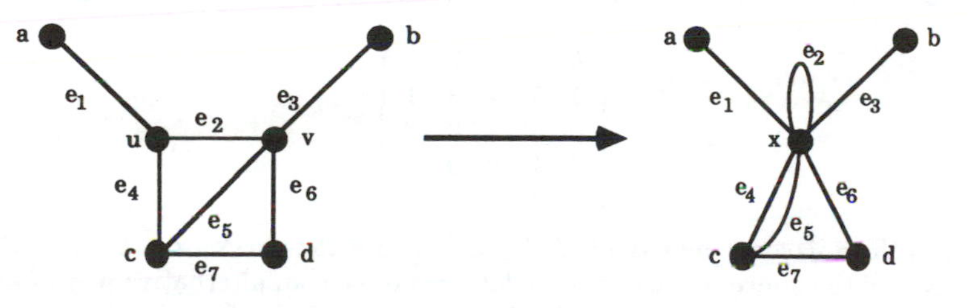

Figure 1.48: The fusion of vertices u and v.

The fusion process when applied to adjacent vertices u and v does not alter the number of connected components of the graph — the component containing u (and v) in G gets changed to the component containing the fused vertex x in G_1, while all other components of G remain unaltered.

If v_i is fused to its neighbour v_j to form the new vertex w then, since each edge of the form v_iv_k or of the form v_jv_k gets changed to one of the form wv_k, it follows that in the adjacency matrix of the new graph G_1 the entries in the row (and column) corresponding to w are just the sum of the corresponding entries given by v_i and v_j in the adjacency matrix for G.

One can more precisely describe this as the following two step process:

The adjacency matrix after fusion of two adjacent vertices u and v.

Step 1. Change u's row to the sum of u's row with v's row and (symmetrically) change u's column to the sum of u's column with v's column.

Step 2. Delete the row and column corresponding to v. The resulting matrix is the adjacency matrix of the new graph G_1.

As an example, we consider the fusion of u and v given in Figure 1.48. Listing the vertices of G as a, u, b, v, c, d, the corresponding adjacency matrix is

$$\begin{bmatrix} 0 & 1 & 0 & 0 & 0 & 0 \\ 1 & 0 & 0 & 1 & 1 & 0 \\ 0 & 0 & 0 & 1 & 0 & 0 \\ 0 & 1 & 1 & 0 & 1 & 1 \\ 0 & 1 & 0 & 1 & 0 & 1 \\ 0 & 0 & 0 & 1 & 1 & 0 \end{bmatrix}.$$

Since u and v are the second and fourth vertex in the list, Step 1 of the process gives the matrix

$$\begin{bmatrix} 0 & \mathbf{1} & 0 & 0 & 0 & 0 \\ \mathbf{1} & \mathbf{1} & \mathbf{1} & \mathbf{1} & \mathbf{2} & \mathbf{1} \\ 0 & \mathbf{1} & 0 & 1 & 0 & 0 \\ 0 & \mathbf{1} & 1 & 0 & 1 & 1 \\ 0 & \mathbf{2} & 0 & 1 & 0 & 1 \\ 0 & \mathbf{1} & 0 & 1 & 1 & 0 \end{bmatrix}.$$

Step 2 now deletes the fourth row and column (corresponding to v) of this matrix to give

$$\begin{bmatrix} 0 & 1 & 0 & 0 & 0 \\ 1 & 1 & 1 & 2 & 1 \\ 0 & 1 & 0 & 0 & 0 \\ 0 & 2 & 0 & 0 & 1 \\ 0 & 1 & 0 & 1 & 0 \end{bmatrix}$$

and this is the adjacency matrix of G_1 (as the reader can check from Figure 1.48).

We now use the above remarks and techniques to give an alternative way of finding whether or not a graph G is connected. The problem with the $B = A + A^2 + \cdots + A^{n-1}$ method described earlier is that it involves much matrix multiplication and this can be very time-consuming when dealing with a large number of vertices (i.e., when n is large). The following method is more efficient and also tells us how many connected components the graph has.

The fusion algorithm for connectedness.

Step 1. Replace G by its underlying simple graph — it is easy to see that the underlying simple graph has exactly the same number of connected components as G and it has the advantage of having all entries in its adjacency matrix as either 0 or 1. To get the adjacency matrix of the new graph just replace all nonzero entries off the diagonal by 1 and make all entries on the diagonal 0. We denote the underlying simple graph of G also by G.

Step 2. Fuse vertex v_1 to the first of the vertices $v_2, \ldots, v_n$ with which it is adjacent to give a new graph, also denoted by G, in which the new vertex is also denoted by v_1. (The above two step process gives the adjacency matrix $A(G)$.)

Step 3. Carry out step 1 on the new graph G.

Step 4. Carry out steps 2 and 3 repeatedly with v_1 and the vertices of the new graphs until v_1 is not adjacent to any of the other vertices.

Step 5. Carry out steps 2–4 on the vertex v_2 (instead of v_1) of the latest graph and then on all the remaining vertices of the resulting graphs in turn. The final graph is empty and the number of its (isolated) vertices is the number of connected components of the initial graph G.

We illustrate the algorithm in Figures 1.49 and 1.50 starting with a graph G with seven vertices listed as $v_1, \ldots, v_7$. The resulting graphs are shown in pairs G and G_0, where G_0 is the underlying simple graph of G, with their adjacency matrices shown immediately below. Following through the algorithm, after the initial simplification, vertex v_1 is first fused with v_4 (and the new vertex labelled v_1), as shown in Figure 1.49. Then (see Figure 1.50) v_1 is fused with v_5 and then with v_2. Next v_3 is fused with v_6 and v_7 in turn — the new vertices are labelled v_3 in both cases.

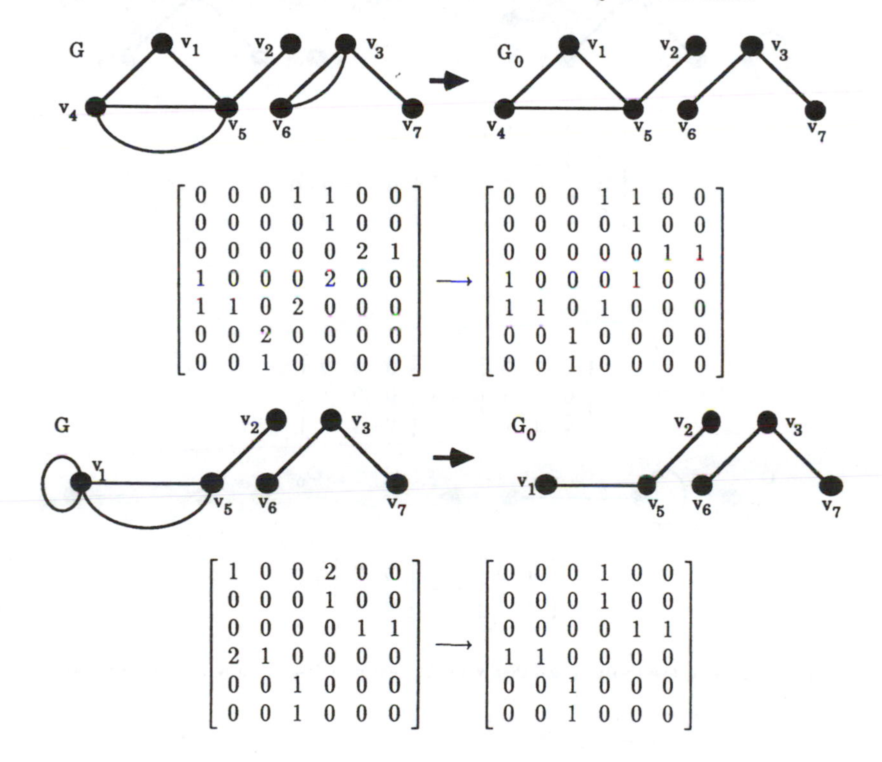

Figure 1.49: v_1 is fused with v_4.

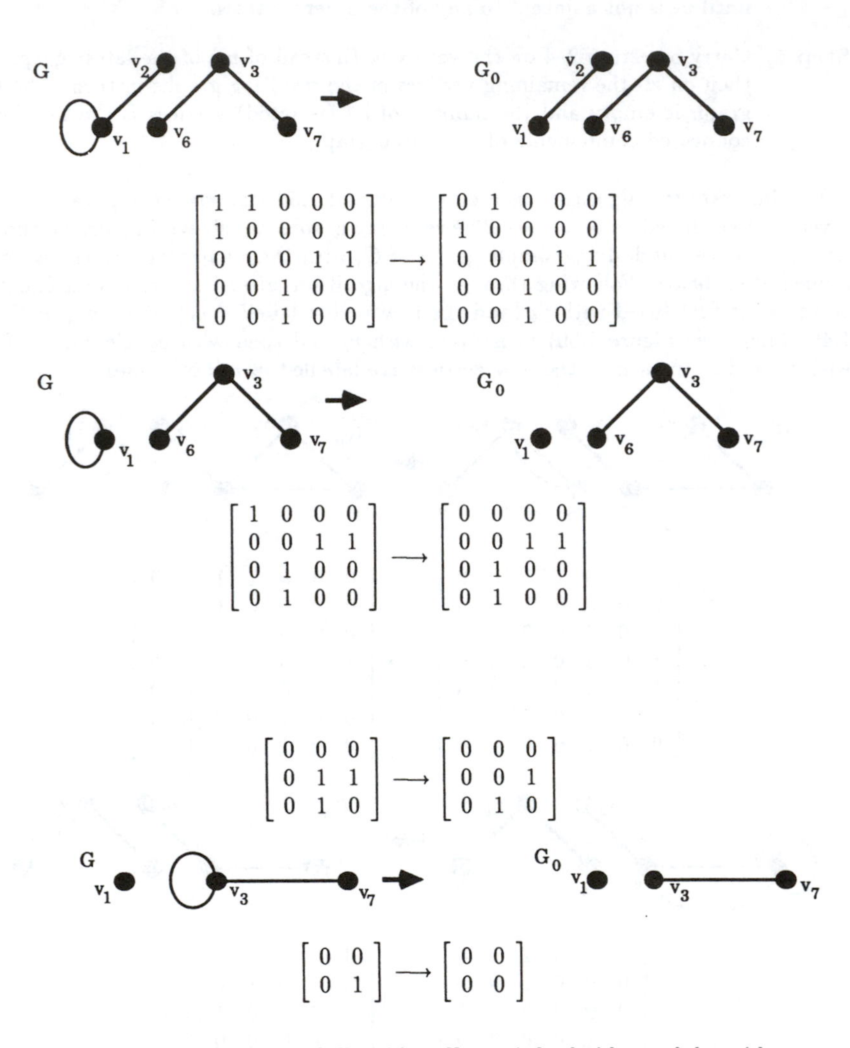

Figure 1.50: v_1 is fused with v_5 and then with v_2. Next v_3 is fused with v_6 and then with v_7.

Since the final adjacency matrix $A(G)$ in Figure 1.50 is a 2×2 matrix we can conclude that the original graph G has 2 connected components. Of course this is obvious from our drawing of the original G in Figure 1.49. However usually a computer has no such drawing to help it but only a knowledge of the graph in matrix form.

Exercises for Section 1.8

1.8.1 Use the fusion process to determine whether the graphs of Exercise 1.7.1, specified by their adjacency matrix, are connected or not. At each stage of the process, give both the corresponding graph and its adjacency matrix.

1.8.2 Give a formal proof of the remark made in the main text that if two adjacent vertices u and v of the graph G are fused to produce the graph G_1 then $\omega(G) = \omega(G_1)$.

Chapter 2

Trees and Connectivity

2.1 Definitions and Simple Properties

A graph G is called **acyclic** if it contains no cycles.

Since loops are cycles of length one while a pair of parallel edges produces a cycle of length two, any acyclic graph must be simple.

A graph G is called a **tree** if it is a connected acyclic graph.

Figure 2.1 shows all trees with at most five vertices.

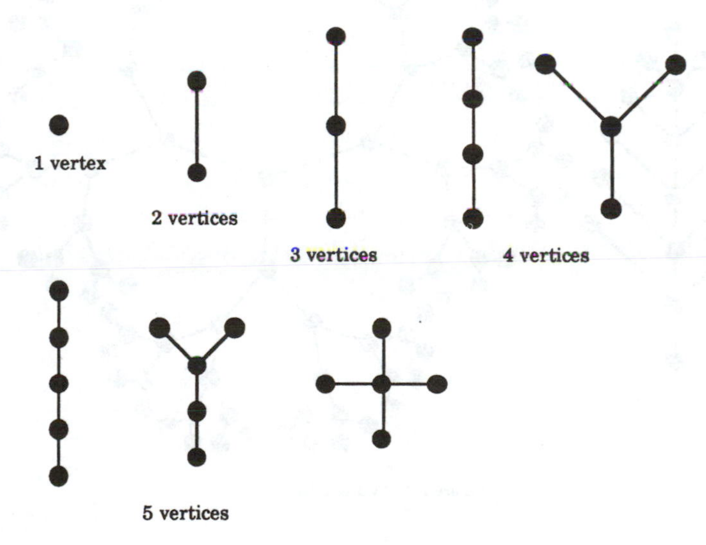

Figure 2.1: The trees with at most five vertices.

Figure 2.2 shows all trees with six vertices while Figure 2.3 shows two more luxuriant trees.

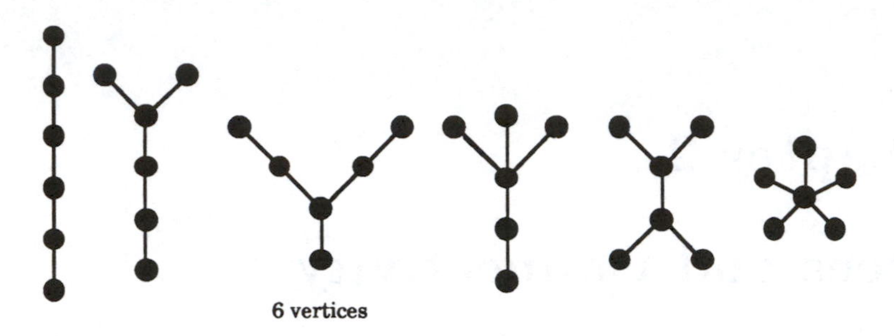

Figure 2.2: The trees with six vertices.

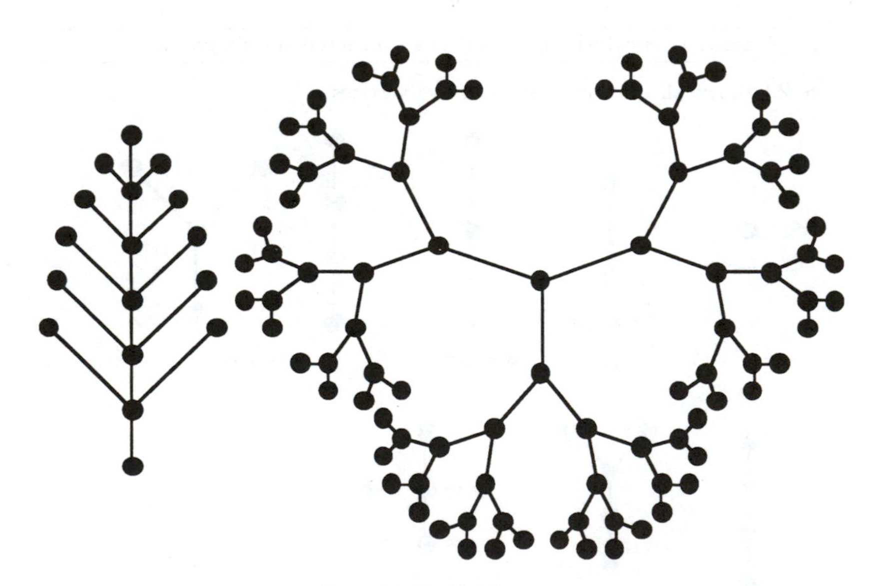

Figure 2.3: Two fancier trees.

Part (a) of the first result in this chapter gives an important property of trees, while part (b) shows that this property almost characterises trees.

Theorem 2.1 *(a) Let u and v be distinct vertices of a tree T. Then there is precisely one path from u to v.*

(b) Let G be a graph without any loops. If for every pair of distinct vertices u and v of G there is precisely one path from u to v, then G is a tree.

Proof (a) Suppose that the result is false. Then there are two different paths from u to v, say $P = u_0u_1u_2 \ldots u_mv$ and $P' = v_0v_1v_2 \ldots v_nv$ where $u_0 = u = v_0$. There is then a first index k greater than 0 for which $u_k \neq v_k$. Let $x = u_{k-1} = v_{k-1}$ and let w be the first vertex after x which belongs to both P and P'. (The vertex x might well be u and the vertex w might well be v but at least there are such vertices x and w.) Then $w = u_i = v_j$ for some indices i and j both greater than $k - 1$. (See Figure 2.4.)

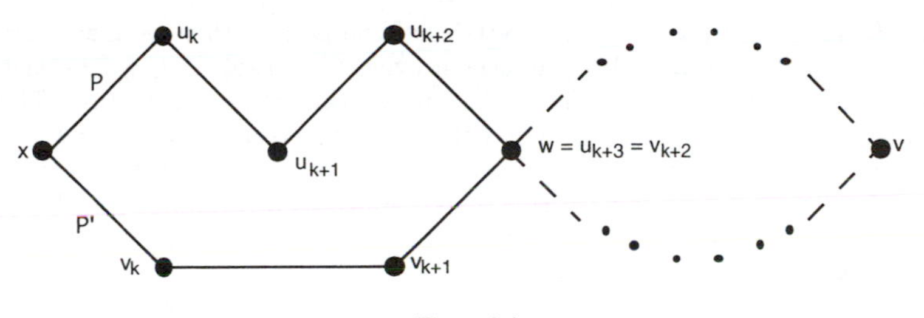

Figure 2.4

This produces a cycle $C = xu_{k+1} \ldots u_iv_{j-1} \ldots v_{k+1}x$ since no two of the "u" terms are repeated (as P is a path), no two of the "v" terms are repeated (as P' is a path), and, by the definition of w, no "u" term in C can be a "v" term. Since T is a tree, T has no cycles. This contradiction means that our initial assumption (the first sentence of the proof) must be false. Thus there is precisely one path from u to v.

(b) Since there is a path between each pair of its vertices, G must be connected. Thus it remains to show that G has no cycles. Firstly, since G has no loops it has no cycles of length one. Now suppose that G has a cycle of length at least two, say $C = v_1v_2 \ldots v_nv_1$ where $n \geq 2$. Since a cycle is a trail, the edge v_nv_1 does not appear in the path $v_1v_2 \ldots v_n$. Thus $P = v_1v_n$ and $P' = v_1v_2 \ldots v_n$ are two different paths from v_1 to v_n. (See Figure 2.5.) This contradicts our assumptions. Hence G has no cycles and so is a tree. □

The next result shows that, with one exception, all trees have at least two "leaves".

Theorem 2.2 *Let T be a tree with at least two vertices and let $P = u_0u_1 \ldots u_n$ be a longest path in T (so that there is no path in T of length greater than n). Then both u_0 and u_n have degree 1, i.e., $d(u_0) = 1 = d(u_n)$.*

Proof Suppose that $d(u_0) > 1$. The edge $f = u_0u_1$ contributes 1 to the degree of u_0 and so there must be another edge e from u_0 to a vertex v of T (which is different

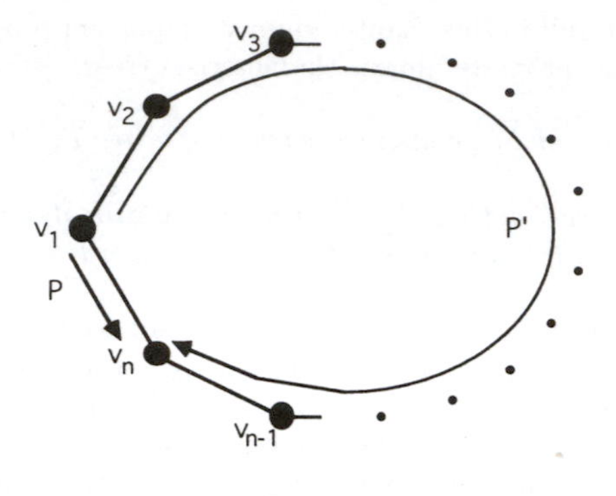

Figure 2.5

from f). If this vertex v is one of the vertices of the path P then we may set $v = u_i$ for some $i = 0, 1, \ldots, n$ and this produces a cycle $C = u_0 u_1 \ldots u_i u_0$ (the last edge being e). Since T is a tree it has no cycles and so this is a contradiction. Thus the remaining possibility for v is that it is not one of the vertices of the path P. But then $P_1 = v u_0 u_1 \ldots u_n$, where the first edge is e, gives a path of length $n+1$ in T, in contradiction to our assumption that P is a longest path (and of length n). This final contradiction shows that there is no such edge e and so $d(u_0) = 1$, as required. Similarly $d(u_n) = 1$, as required. □

Corollary 2.3 *A tree with at least two vertices has more than one vertex of degree 1.*

Proof In such a tree T there is a longest path P (of length greater than 0) and so the Theorem produces at least two vertices of degree 1. □

Note that in Figure 2.2 all our trees with 6 vertices had 5 edges. We now use Corollary 2.3 to prove a more general result:

Theorem 2.4 *If T is a tree with n vertices then it has precisely $n - 1$ edges.*

Proof We use induction on n. When $n = 1$, i.e., T has only 1 vertex, then, since it has no loops, T can not have any edges, i.e., it has $n - 1 = 0$ edges. This establishes that the result is true for $n = 1$.

Now suppose that the result is true for $n = k$ where k is some positive integer. We use this to show that the result is true for $n = k+1$. Let T be a tree with $k+1$ vertices and let u be a vertex of degree 1 in T. (Note that such a vertex exists by Corollary 2.3.) Let $e = uv$ denote the unique edge of T which has u as an end. Then if x and y are vertices in T both different from u, any path P joining x to y does not go through the vertex u since if it did it would involve the edge e twice. Thus the subgraph $T - u$, obtained from T by deleting the vertex u (and the edge e), is connected. Moreover if C is a cycle in $T - u$ then C would be a cycle in T — impossible, since T is a tree. Thus the subgraph $T - u$ is also acyclic. Hence $T - u$ is a tree. However $T - u$ has k vertices (since T has $k + 1$) and so, by our induction assumption, $T - u$ has $k - 1$

edges. Since $T - u$ has exactly 1 edge less than T (the edge e), it follows that T has k edges, as required. In other words, assuming the result is true for k, we have shown that it is true for $k+1$. Thus, by the principle of mathematical induction, it is true for all positive integers k. □

Let G be an acyclic graph. Then any subgraph of G must also contain no cycles. In particular, the connected components of G are also acyclic and so they are trees. For this reason an acyclic graph is also called a **forest**. Figure 2.6 gives an example of a forest.

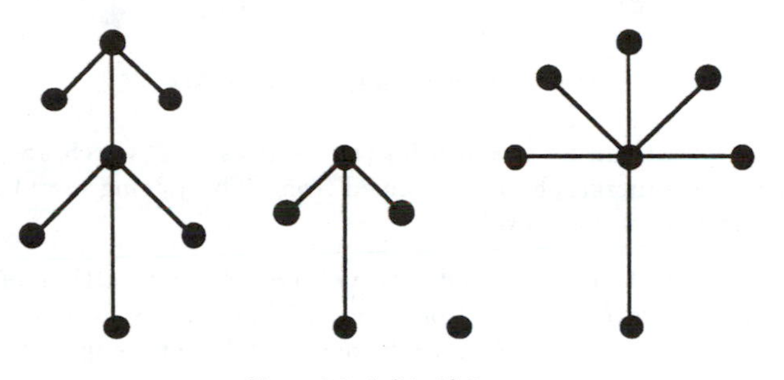

Figure 2.6: A (black) forest.

The next result extends Theorem 2.4, a result on trees, to cover forests.

Theorem 2.5 *Let G be an acyclic graph with n vertices and k connected components, i.e., $\omega(G) = k$. Then G has $n - k$ edges.*

Proof Denote the k components of G by $C_1, \ldots, C_k$ and suppose that for each $i, 1 \leq i \leq k$, the ith component C_i has n_i vertices. Then $n = n_1 + n_2 + \cdots + n_k$. Also, since each C_i is a tree, by Theorem 2.4 it has $n_i - 1$ edges and so since each edge of G belongs to precisely one component of G the total number of edges in G is $(n_1 - 1) + (n_2 - 1) + \cdots + (n_k - 1)$. Thus G has $(n_1 + n_2 + \cdots + n_k) - k$ edges, i.e., $n - k$ edges, as required. □

Exercises for Section 2.1

2.1.1 Give a list of all trees with 7 vertices and all trees with 8 vertices. Your lists should not contain any pair of trees which are isomorphic. (There are 11 non-isomorphic trees on 7 vertices and 23 non-isomorphic trees on 8 vertices.)

2.1.2 The complete bipartite graphs $K_{1,n}$, known as the **star graphs**, are trees. Figure 2.7 shows the star graphs $K_{1,4}$ and $K_{1,6}$. Prove that the star graphs are the only complete bipartite graphs which are trees.

2.1.3 Prove that any tree with at least two vertices is a bipartite graph.

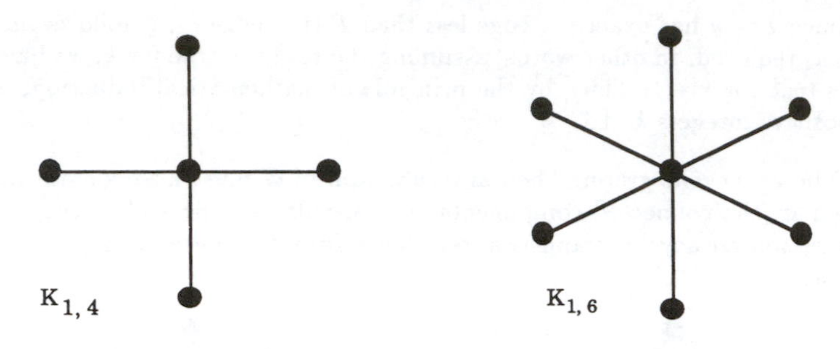

Figure 2.7: The star graphs $K_{1,4}$ and $K_{1,6}$.

2.1.4 Let T be a tree and let u and v be two vertices of T which are not adjacent. Let G be the supergraph of T obtained from T by joining u and v by an edge. Prove that G contains a cycle.

2.1.5 Let F be a forest and let u and v be two nonadjacent vertices of F. Let G be the supergraph of F obtained from F by joining u and v by an edge. Prove that G is a forest if and only if u and v belong to different components of F.

2.1.6 Let T be a tree with n vertices, where $n \geq 4$, and let v be a vertex of maximum degree in T.

(a) Show that T is a path if and only if $d(v) = 2$.

(b) Prove that T is isomorphic to a star graph (see Exercise 2.1.2 above) if and only if $d(v) = n - 1$.

(c) Prove that if $d(v) = n-2$ then any other tree with n vertices and maximum vertex degree $n - 2$ is isomorphic to T.

(d) Prove that if $n \geq 6$ and $d(v) = n-3$ then there are exactly 3 non-isomorphic trees which T can be.

2.1.7 Let T be a tree with n vertices, where $n \geq 3$. Show that there is a vertex v in T with $d(v) \geq 2$ such that every vertex adjacent to v, except possibly for one, has degree 1.

2.1.8 Let T be a tree and let v be a vertex of maximum degree in T, say $d(v) = k$. Prove that T has at least k vertices of degree 1.

2.2 Bridges

The first theorem of this section shows that if we remove an edge from a graph then the number of connected components of the graph either remains unchanged or it increases by exactly 1. Recall that $\omega(G)$ denotes the number of connected components of G.

Theorem 2.6 *Let e be an edge of the graph G and, as usual, let $G-e$ be the subgraph obtained by deleting e. Then $\omega(G) \leq \omega(G-e) \leq \omega(G)+1$.*

Proof Let e have end vertices u and v and let C be the connected component of G to which e (and u and v) belong. First suppose that u and v are distinct. Then the edge e forms, by itself, a path from u to v.

Suppose now that there is another path P from u to v. Then P can not involve the edge e and so must also be a path in the subgraph $G-e$. Thus, in this case, u and v are connected in $G-e$ (as well as in G). Moreover if x and y are any two vertices in the component C then there is still a walk from x to y in $G-e$ (since given a walk from x to y in G, we can simply replace any occurrence of the edge e in this walk by the path P to produce a walk in $G-e$). This shows that in this case the vertices of C still give one connected component in the subgraph $G-e$. Since any other connected component of G is unaffected by the removal of e, it follows that $G-e$ has the same number of components as G, i.e., $\omega(G-e)=\omega(G)$.

We now consider the remaining possibility,namely that the edge e gives the only path from u to v. Let x be any vertex in the component C. Then either

(i) there is a path from x to u which does not involve e or

(ii) every path from x to u must involve e.

If (i) holds then x is in the same component of $G-e$ as u, because the deletion of e does not affect such a path.

If (ii) holds then in any path P from x to u, since it involves e, e must in fact be the last edge in the path and so the second last vertex must be v. This then produces a path from x to v which does not involve e and so x is in the same component of $G-e$ as v. Moreover since e is the only path from u to v in G, u and v are in different components in $G-e$. The above arguments then show that the deletion of e in this case has split up C into two components in $G-e$, namely one where every vertex is connected to u and another in which every vertex is connected to v. Thus we have increased the number of components by 1, i.e., in this case $\omega(G-e)=\omega(G)+1$.

The above argument assumes that u is different from v. If $u=v$ then e is a loop and its removal does not change the number of components, i.e., if e is a loop then $\omega(G-e)=\omega(G)$. □

An edge e of a graph G is called a **bridge** (or a **cut edge** or an **isthmus**) if the subgraph $G-e$ has more connected components than G has.

Thus, if e is a bridge in G, by Theorem 2.6 we have $\omega(G-e)=\omega(G)+1$, i.e., $G-e$ has one more component than G has. The bridges of the graph G of Figure 2.8 are the edges e and f.

Roughly speaking a bridge is an edge which is the only link between two parts of a graph. Its deletion will cut up the graph into more disjoint parts. We illustrate this in Figure 2.8.

The proof of Theorem 2.6 shows that an edge e with end vertices u and v is a bridge in the graph G if and only if e is not a loop and e gives the only path in G from u to v. Another way of saying this is given in the following theorem.

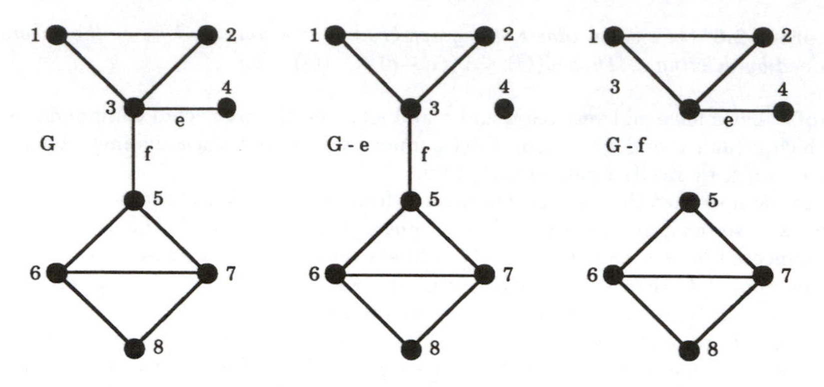

Figure 2.8: A graph and two bridge deletions.

Theorem 2.7 *An edge e of a graph G is a bridge if and only if e is not part of any cycle in G.*

Proof Let e have end vertices u and v. If e is not a bridge then, by the above remarks, it is either a loop or there is a path $P = uu_1 \ldots u_n v$ from u to v, different from the edge e. If it is a loop then it forms a cycle (by itself). If there is such a path P then $C = uu_1 \ldots u_n vu$, the concatenation of P with e, is a cycle in G. This shows that if e is not a bridge then it is part of a cycle. This is equivalent to saying that if e is not part of any cycle then e must be a bridge.

Conversely, suppose that e is part of some cycle $C = u_0 u_1 \ldots u_m$ in G. Let $e = u_i u_{i+1}$. In the case where $m = 1$, $C = u_0 u_1$ and so C is just the edge e and e is a loop. On the other hand, if $m > 1$ then $P = u_i u_{i-1} \ldots u_0 u_{m-1} \ldots u_{i+1}$ is a path from u to v different from e. (See Figure 2.9.) Thus, by the remarks preceding the proof, e is not a bridge. This shows that if e is a bridge then it is not part of any cycle in G, completing the proof. □

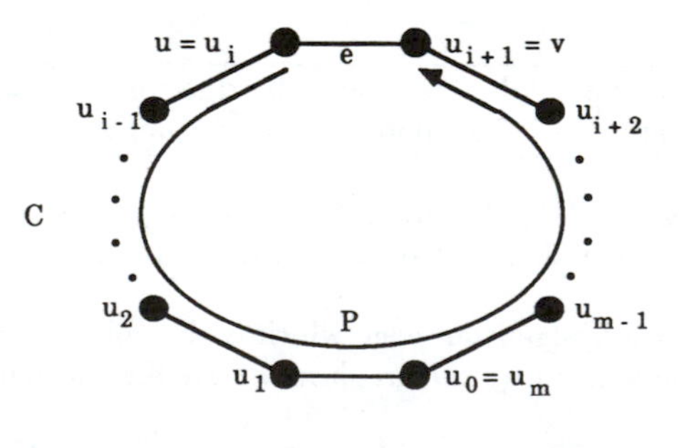

Figure 2.9

We can now describe trees using bridges by the following result.

Theorem 2.8 *Let G be a connected graph. Then G is a tree if and only if every edge of G is a bridge, i.e., if and only if for every edge e of G the subgraph $G-e$ has two components.*

Proof Suppose that G is a tree. Then G is acyclic, i.e., it has no cycles, and so no edge of G belongs to a cycle. In other words, if e is any edge of G then, by Theorem 2.7, it is a bridge, as required.

Conversely suppose that G is connected and that every edge e of G is a bridge. Then G can have no cycles since any edge belonging to a cycle is not a bridge, by Theorem 2.7. Hence G is acyclic and so is a tree as required. □

The corollary of the next result will enable us to give yet another characterisation of trees.

Theorem 2.9 *Let G be a graph with n vertices and q edges and, as usual, let $\omega(G)$ denote the number of connected components of G. Then G has at least $n-\omega(G)$ edges, i.e., $q \geq n-\omega(G)$.*

Proof We use induction on q, starting with $q=0$. If $q=0$ then G has no edges and so $\omega(G)=n$, since in this case each vertex constitutes a component. Thus for $q=0$ we have $q=0 \geq 0=n-\omega(G)$, establishing the result in this case.

Although we do not need to prove the result for $q=1$ at this stage let us do so to give peace of mind to those worried about induction starting at 0 instead of 1. If $q=1$ then there is only one edge in G, say with end(s) u and v. Then u and v are in the same component and all other vertices each give another component. There being at least $n-2$ other such vertices ($n-1$ if $u=v$), we get $\omega(G) \geq n-1$ and so $n-\omega(G) \leq 1=q$, as required.

Now suppose the result is true for some value k of q where $k \geq 0$. Using this assumption we will prove that it is true for a graph of $k+1$ edges. Thus let G be a graph with $k+1$ edges and select one of these edges, e, say. Then the subgraph $G-e$ has k edges and so by our assumption

$$k \geq n-\omega(G-e). \quad (*)$$

(Note that $G-e$ has still n vertices.) Now, by Theorem 2.6, we have $\omega(G-e) \leq \omega(G)+1$ so, multiplying by -1, $-\omega(G-e) \geq -\omega(G)-1$. Substituting back into $(*)$ gives $k \geq n-\omega(G)-1$ and so $k+1 \geq n-\omega(G)$, i.e., our graph G with $k+1$ edges satisfies the required inequality. Hence, assuming the result true for k, we have shown it to be true for $k+1$. Thus, by mathematical induction, the result is true in general. □

Corollary 2.10 *A connected graph G with n vertices has at least $n-1$ edges. (In other words, a graph with n vertices and less than $n-1$ edges can not be connected.)*

Proof If G is connected then $\omega(G)=1$ so, by the Theorem, if q denotes the number of edges in G we have $q \geq n-1$. □

We now provide the promised characterisation of trees:

Theorem 2.11 *Let G be a graph with n vertices. Then the following three statements are equivalent:*
(i) G is a tree,
(ii) G is an acyclic graph with $n-1$ edges,
(iii) G is a connected graph with $n-1$ edges.

Proof We prove (i) $\Rightarrow$ (ii) $\Rightarrow$ (iii) $\Rightarrow$ (i).

(i) $\Rightarrow$ (ii): Suppose that G is a tree. Then by definition G is an acyclic graph and by Theorem 2.4 it has $n-1$ edges. Thus (ii) holds.

(ii) $\Rightarrow$ (iii): Assume that G is an acyclic graph with $n-1$ edges and, as usual, let $\omega(G)$ denote the number of connected components of G. Then, by Theorem 2.5, G has $n-\omega(G)$ edges. Thus $\omega(G)=1$, in other words, G is connected. This establishes (iii).

(iii) $\Rightarrow$ (i): Assume that G is a connected graph with $n-1$ edges. To prove (i), i.e., that G is a tree, we must show that G is acyclic. We do this by contradiction. Thus, assume that G is not acyclic. Then G contains a cycle and every edge of this cycle can not be a bridge, by Theorem 2.7. Choose such an edge e. Then, since e is not a bridge, $G-e$ is still connected. However $G-e$ has $n-2$ edges and n vertices, which is impossible by the above corollary. This contradiction has arisen from our assumption that G is not acyclic. Hence G is acyclic and so a tree as required. □

Exercises for Section 2.2

2.2.1 Find all bridges in the graph of Figure 2.10.

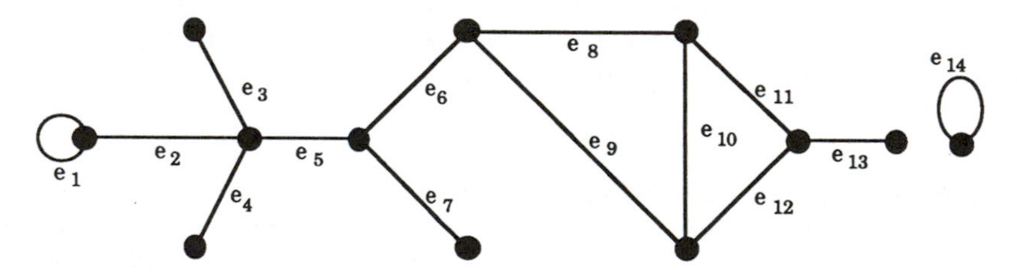

Figure 2.10: Which edges are bridges?

2.2.2 Let G be a connected graph.

(a) If G has 17 edges what is the maximum possible number of vertices in G?

(b) If G has 21 vertices what is the minimum possible number of edges in G?

2.2.3 Let G be a graph with 4 connected components and 24 edges. What is the maximum possible number of vertices in G?

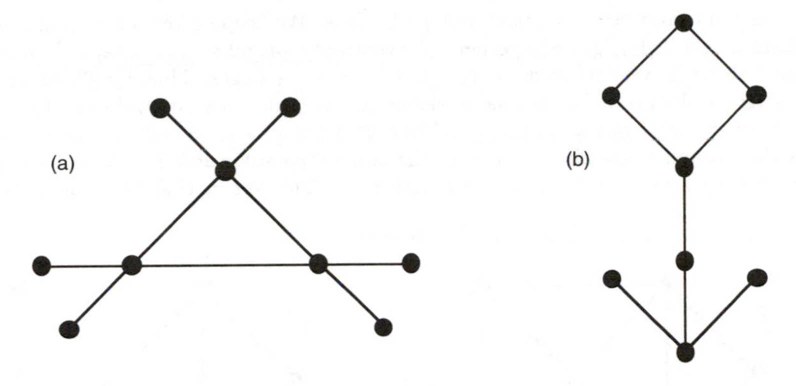

Figure 2.11: Two unicyclic graphs.

2.2.4 A graph G is called **unicyclic** if it is connected and contains precisely one cycle. The graphs of Figure 2.11 are unicyclic.

Prove that a connected graph G with n vertices and e edges is unicyclic if and only if $n = e$.

2.3 Spanning Trees

Let G be a graph. Recall from Section 1.4 that a subgraph H of G is called a spanning subgraph of G if the vertex set of H is the same as the vertex set of G.

> A **spanning tree** of a graph G is a spanning subgraph of G that is a tree.

Our first result of this section shows that the graphs which have spanning trees are easily described.

Theorem 2.12 *A graph G is connected if and only if it has a spanning tree.*

Proof Suppose that G is connected with n vertices and q edges. Then, by Corollary 2.10, we have $q \geq n - 1$. If $q = n - 1$ then, by (iii) $\Rightarrow$(i) of Theorem 2.11, G is a tree and so we can take $T = G$ as a spanning tree of G.

If $q > n - 1$ then, by Theorem 2.4 (or by (i) $\Rightarrow$ (iii) of Theorem 2.11), G is not a tree and so G must contain a cycle. Let e_1 be an edge of such a cycle. Then the subgraph $G - e_1$ is connected (since e_1 is not a bridge), has n vertices, and has $q - 1$ edges. If $q - 1 = n - 1$ then, repeating the above argument gives $T = G - e_1$ as a spanning tree of G.

If $q - 1 > n - 1$ then $G - e_1$ is not a tree so, as before, there is a cycle in $G - e_1$. Removing an edge e_2 from such a cycle gives a subgraph $G - \{e_1, e_2\} = (G - e_1) - e_2$

which is connected, has n vertices and $q-2$ edges. We keep on repeating this process, deleting $q-n+1$ edges altogether, to eventually produce a subgraph T which is connected, has n vertices and $q-(q-n+1)=n-1$ edges. Thus by Theorem 2.11, T is a tree and since it has the same vertex set as G it is a spanning tree of G.

Conversely, if G has a spanning subtree T, then given any two vertices u and v of G then u and v are also vertices of the connected subgraph T. Thus u and v are connected by a path in T and so by a path in G. This shows that G is connected. □

Figures 2.12 and 2.13 illustrate the Theorem.

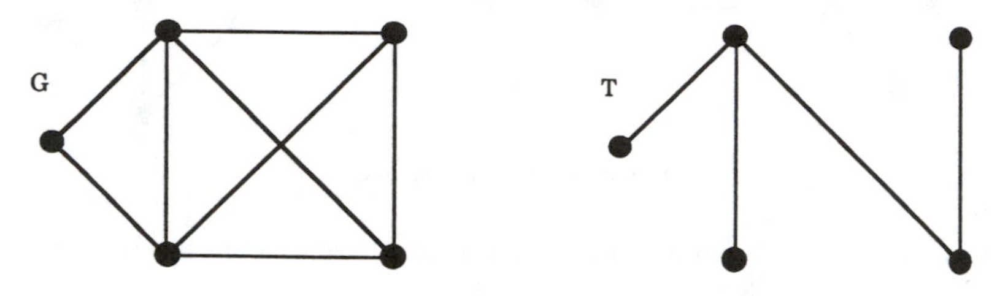

Figure 2.12: A connected graph and a spanning tree.

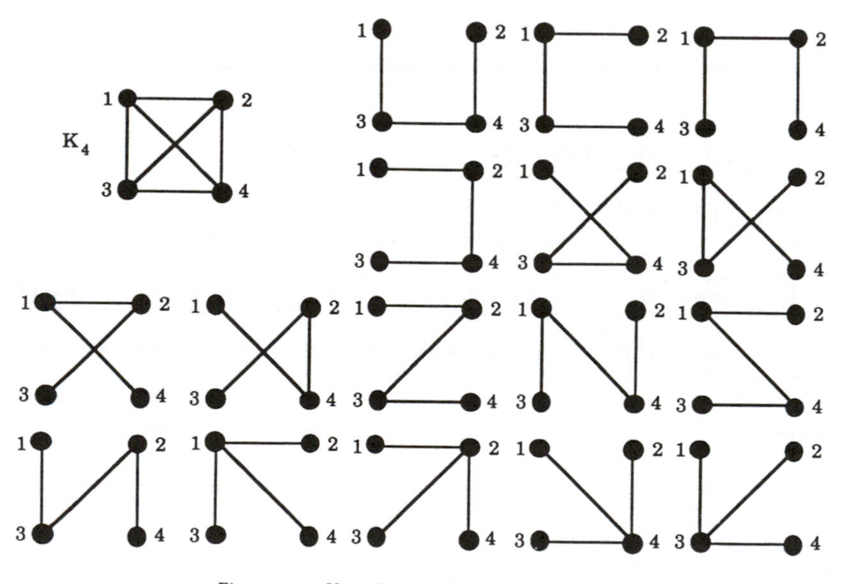

Figure 2.13: K_4 and its 16 different spanning trees.

Note that in the 16 different spanning trees of K_4 shown in Figure 2.13 there are only two non-isomorphic ones — the first 12 shown are isomorphic to each other, while the last four are also isomorphic to each other. K_6, the complete graph on 6 vertices, has 1296 *different* spanning trees, but just 6 *non-isomorphic* ones. Another way of saying this is, given 6 vertices, then there are 1296 different ways of joining these vertices to form a tree if we label the vertices $1, 2, 3, 4, 5, 6$, but if we drop these labels then there are only 6 different ways.

The subject of counting how many spanning trees and non-isomorphic spanning trees there are for a given graph was probably initiated by the English mathematician Arthur Cayley, who used trees to try to count the number of saturated hydrocarbons C_nH_{2n+2} containing a given number of carbon atoms. Cayley was the first person to use the term "tree" (in 1857) and in 1889 [12] he proved the following result which tells us that given n vertices, labelled $1, \ldots, n$, then there are n^{n-2} different ways of joining them to form a tree.

Theorem 2.13 (Cayley, 1889) *The complete graph K_n has n^{n-2} different spanning trees.*

Proof We omit the proof but, for those interested, see pages 32–35 of Bondy and Murty [7] or pages 50–52 of Wilson [65]. □

Exercises for Section 2.3

2.3.1 Give a list of all spanning trees, including isomorphic ones, of the connected graphs of Figure 2.14. How many non-isomorphic spanning trees are there in each case?

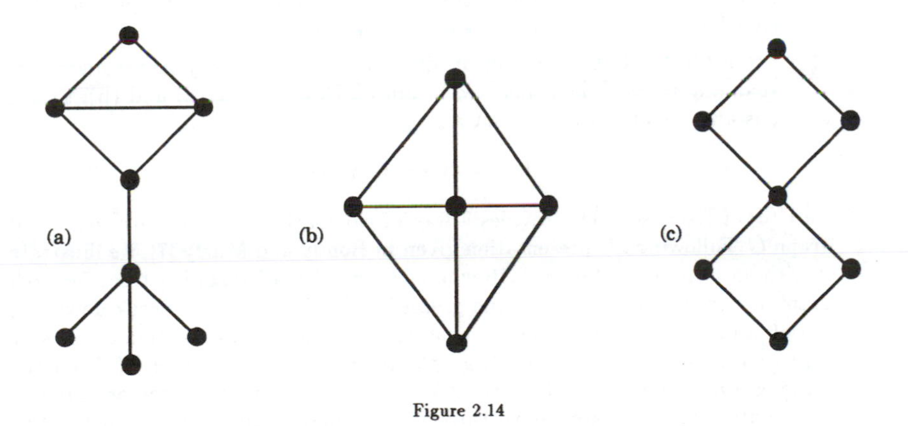

Figure 2.14

2.3.2 Let T be a tree with at least k edges, $k \geq 2$. How many connected components are there in the subgraph of T obtained by deleting k edges of T?

2.3.3 Let G be a connected graph which is not a tree and let C be a cycle in G. Prove that the complement of any spanning tree of G contains at least one edge of C.

2.3.4 Let e be an edge of the connected graph G.

(a) Prove that e is a bridge if and only if it is in every spanning tree of G.

(b) Prove that e is a loop if and only if it is in no spanning tree of G.

2.3.5 Let G be a graph with exactly one spanning tree. Prove that G is a tree.

2.3.6 An edge e (not a loop) of a graph G is said to be **contracted** if it is deleted and then its end vertices are fused (identified). The resulting graph is denoted by $G * e$. Figure 2.15 illustrates this.

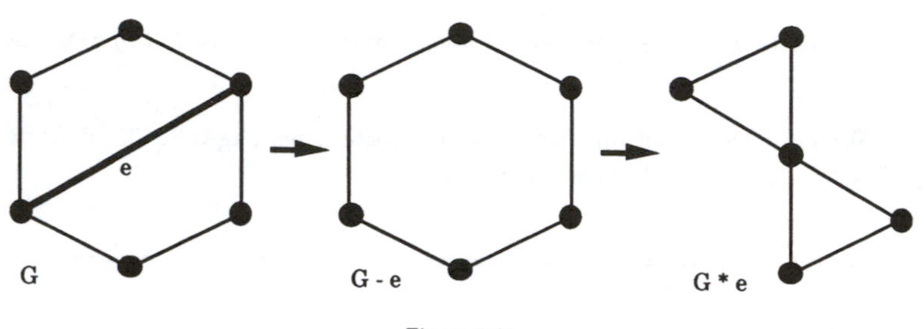

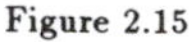

Figure 2.15

(a) Prove that if T is a spanning tree of G which contains e then $T * e$ is a spanning tree of $G * e$.

(b) Prove that if U is a spanning tree of $G * e$ then there is a unique spanning tree T of G which contains e and is such that $U = T * e$.

(c) Let $\tau(G)$ denote the number of different (not necessarily non-isomorphic) spanning trees of the connected graph G. Prove, using (a) and (b), that if e is an edge of G which is not a loop then

$$\tau(G) = \tau(G - e) + \tau(G * e).$$

2.3.7 Part (c) of Exercise 2.3.6 provides a way of calculating $\tau(G)$ for any connected graph G. Following the presentation given in Bondy and Murty [7], we illustrate it with an example in Figure 2.16 where the number of spanning trees for each graph is denoted pictorially by the graph itself. The idea is to break down the initial graph by a series of edge contractions to produce a collection of trees or "trees with loops". The number of graphs in this collection is then $\tau(G)$ for the initial graph G. Part (c) of Exercise 2.3.6 is used in each step of the breakdown. The contracted edges are shown thicker. The final "expression" consists of 11 graphs which are either trees or "trees with loops" and so we can conclude that $\tau(G) = 11$.

Use this method to find $\tau(G)$ for the graph G of Figure 2.15.

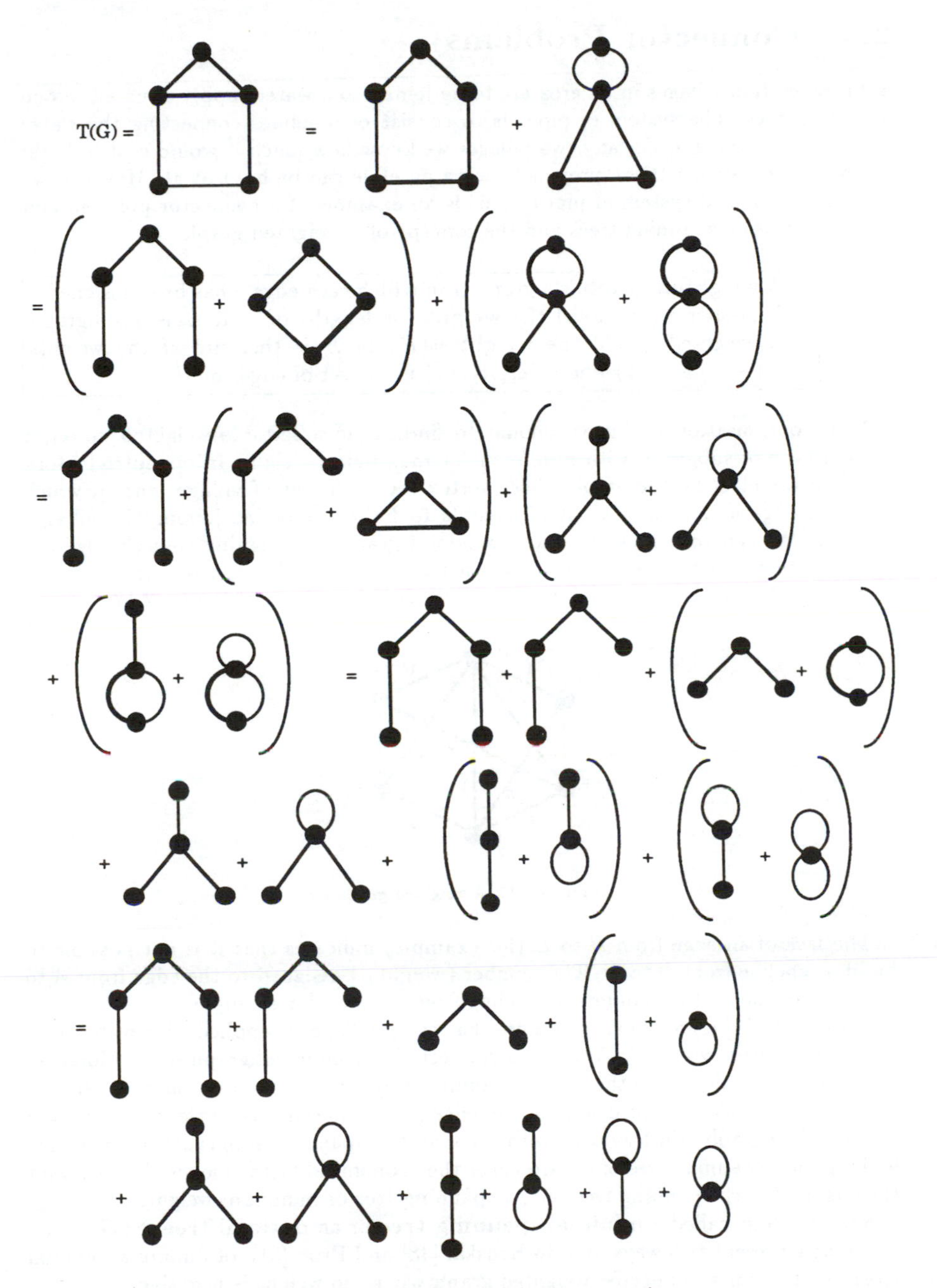

Figure 2.16: Calculation of $\tau(G)$ using contractions.

2.4 Connector Problems

Suppose certain villages in an area are to be joined to a water supply situated in one of the villages. The system of pipes is to consist of pipelines connecting the water towers of two villages. For any two villages we know how much it would cost to build a pipeline connecting them, provided such a pipeline can be built at all. How can we find an economical system of pipes? This is an example of a connector problem and we solve it using spanning trees and the concept of a weighted graph.

> A **weighted graph** is a graph G in which each edge e has been assigned a real number $w(e)$, called the **weight** (or length) of e. If H is a subgraph of a weighted graph, the **weight** $w(H)$ of H is the sum of the weights $w(e_1) + \cdots + w(e_k)$ where $\{e_1, \ldots, e_k\}$ is the set of edges of H.

Many optimisation problems amount to finding, in a suitable weighted graph, a certain type of subgraph with minimum (or maximum) weight. In our introductory problem we let G be the graph whose vertex set is the set of villages and in which xy is an edge if (and only if) it is possible to build a pipeline joining the villages x and y. We can then make G into a weighted graph by assigning to each edge the cost of constructing the corresponding pipeline. For example suppose that there are 6 villages, A, B, C, D, E, F, and we get the weighted graph G of Figure 2.17.

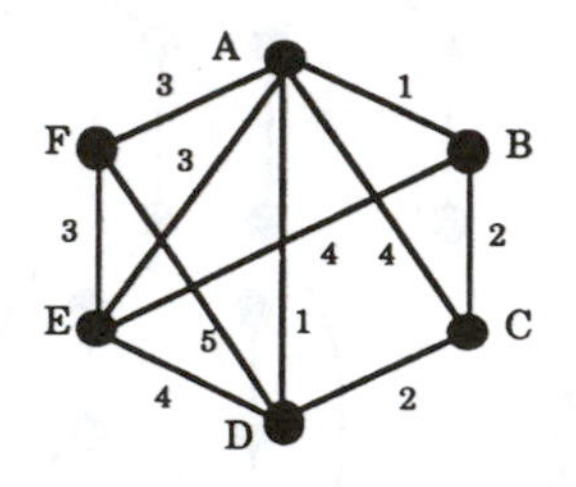

Figure 2.17: A weighted graph.

The lack of an edge from B to D (for example) indicates that it is not possible to build a pipeline from B to D. The number (weight) 4 assigned to the edge from A to C indicates the cost of building a pipeline from A to C, for example.

Since part of the problem is to ensure that every village is supplied with water from the source village we are looking for a connected spanning subgraph of G. Moreover since we want to do this in the most economical way, such a spanning subgraph should have no cycles, because the deletion of an edge (a pipeline) from a cycle in a connected spanning subgraph still leaves us with a connected spanning subgraph. Thus we are looking for a spanning tree of G. Moreover the economical factor means that we want the cheapest such spanning tree, i.e., a spanning tree of minimum weight.

Such a tree is called a **minimal spanning tree** or an **optimal tree** for G.

We now present two ways, due to Kruskal [38] and Prim [51], of finding a minimal spanning tree for a connected weighted graph where no weight is negative.

(1) *Kruskal's algorithm.*

In this algorithm we first choose an edge of G which has smallest weight among the edges of G which are not loops. Next, again avoiding loops, we choose from the remaining edges one of smallest weight possible which does not form a cycle with the edge already chosen. We repeat this process taking edges of smallest weight among those not already chosen, provided no cycle is formed with those that have been chosen. If the graph has n vertices we stop after having chosen $n-1$ edges. These edges form an acyclic subgraph T of G and we will prove below that T is a minimal spanning tree of G. This gives the following three step procedure:

Kruskal's Algorithm

Step 1. Choose e_1, an edge of G, such that $w(e_1)$ is as small as possible and e_1 is not a loop.

Step 2. If edges $e_1, e_2, \ldots, e_i$ have been chosen, then choose an edge e_{i+1}, not already chosen, such that

(i) the induced subgraph $G[\{e_1, \ldots, e_{i+1}\}]$ is acyclic and

(ii) $w(e_{i+1})$ is as small as possible (subject to condition (i)).

Step 3. If G has n vertices, stop after $n-1$ edges have been chosen. Otherwise repeat Step 2.

In Figure 2.18 we perform the algorithm on our example G of Figure 2.17, indicating at each stage the edges chosen by shaded lines. The weight, $w(T)$, of T is $1+1+2+3+3$, i.e., $w(T) = 10$.

We still have to prove that Kruskal's algorithm does produce a minimal spanning tree. We do this now:

Theorem 2.14 *Let G be a weighted connected graph in which the weights of the edges are all non-negative numbers. Let T be a subgraph of G obtained by Kruskal's algorithm. Then T is a minimal spanning tree of G.*

Proof As noted in the description of the algorithm, T is an acyclic subgraph of G with $n-1$ edges. If T has m vertices and k connected components then, by Theorem 2.5, it has $m-k$ edges, i.e., $n-1 = m-k$. Since $m \leq n$ and $k \geq 1$ this can only happen when $n = m$ and $k = 1$ (because $n - m = 1 - k$). Thus T is connected and a spanning subgraph, i.e., T is indeed a spanning tree of G.

It remains then to show that the weight of T is at a minimum. In order to do this, we suppose that S is a spanning tree of G with less weight than T, i.e., $w(S) < w(T)$, and obtain a contradiction. Let $e_1, e_2, \ldots, e_{n-1}$ be the edges of T in the order that they were produced by Kruskal's algorithm. Since S is different from T there will be a first edge, e_k say, in this sequence which does not lie in S. Then the subgraph H of G obtained by adding the edge e_k to S has n edges and so is no longer a tree (by Theorem 2.4). Thus this subgraph H must contain a cycle C, say. C must contain the edge e_k, since otherwise C would be in the acyclic S. Moreover C must contain

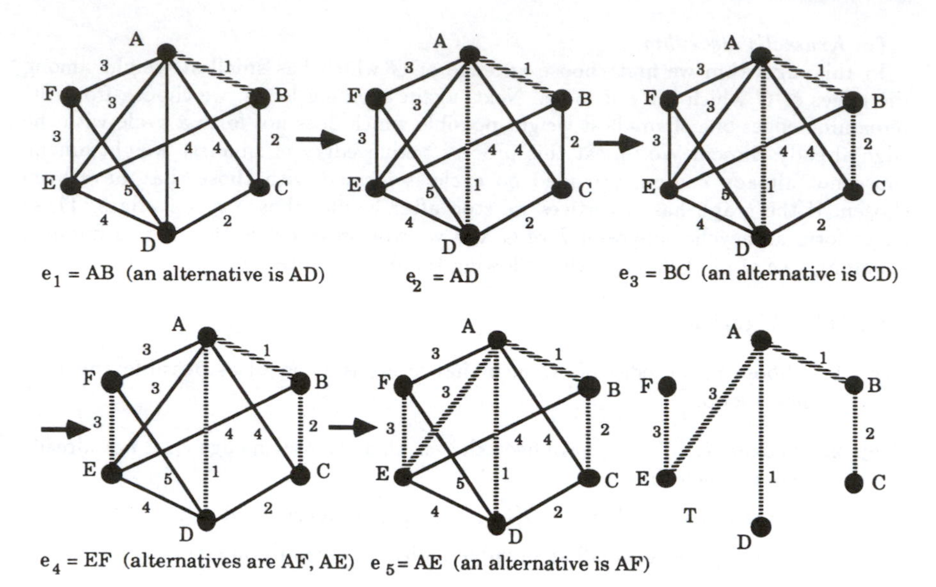

Figure 2.18: A minimal spanning tree construction using Kruskal's algorithm.

an edge e of S that is not in T, since otherwise C would be in the acyclic T. The subgraph $H - e$ is then still connected (since e belongs to a cycle in H) and so, since it has $n - 1$ edges, is also a spanning tree of G. However, since e is not an edge in T and e_k is the first edge in T which is not in S, by the algorithm it follows that $w(e_k) \leq w(e)$. (If $w(e) < w(e_k)$ we would have chosen e instead of e_k as the kth edge.) Since $H - e$ has been formed by replacing e with e_k and $w(e_k) \leq w(e)$ we have that $w(H - e) \leq w(S)$. Moreover $H - e$ has one more edge in common with T than S has because e_k is in T while e is not..

The procedure that we just performed on S to produce $H - e$ is now performed repeatedly, on $H - e$ and its successors, one step at a time, to gradually change S into T, each stage giving a spanning tree with weight at most $w(S)$. Thus at the final stage we get $w(T) \leq w(S)$, a contradiction to our earlier assumption. Hence $w(T)$ is after all at a minimum and so T is a minimum spanning tree, as required. □

(2) *Prim's algorithm.*

In this algorithm for finding a minimal spanning tree we first choose a vertex v_1 of the connected graph G — any vertex v_1 will do. We next choose one of the edges of smallest weight in G which is not a loop and which is incident with v_1, say $e_1 = v_1v_2$. Next we choose an edge of smallest weight in G which is incident with either v_1 or v_2 but with the other end point neither of these, i.e., we choose $e_2 = v_iv_3$ where $i \in \{1, 2\}$ but $v_3 \neq v_1, v_2$. We repeat this process of taking edges of smallest weight one of whose ends is a vertex previously chosen and the other end becoming involved for the first time, until we have chosen $n - 1$ edges (assuming the graph has n vertices). Then at this stage we have involved each of the n vertices of G and by the construction the resulting subgraph is connected. Hence it is a tree by Theorem 2.11 and so a spanning tree of G. We shall prove below that it is a minimal spanning tree. (Note: as in Kruskal we do assume all weights are non-negative.) The algorithm then involves four steps:

Prim's Algorithm

Step 1. Choose any vertex v_1 of G.

Step 2. Choose an edge $e_1 = v_1v_2$ of G such that $v_2 \neq v_1$ and e_1 has smallest weight among the edges of G incident with v_1.

Step 3. If edges $e_1, e_2, \ldots, e_i$ have been chosen involving end points $v_1, v_2, \ldots, v_{i+1}$ choose an edge $e_{i+1} = v_jv_k$ with $v_j \in \{v_1, \ldots, v_{i+1}\}$ and $v_k \notin \{v_1, \ldots, v_{i+1}\}$ such that e_{i+1} has smallest weight among the edges of G with precisely one end in $\{v_1, \ldots, v_{i+1}\}$.

Step 4. Stop after $n-1$ edges have been chosen. Otherwise repeat Step 3.

In Figure 2.19 we perform the algorithm on our example G of Figure 2.17 (the same example used for Kruskal's algorithm), indicating at each stage the edges chosen by shaded lines. Figure 2.20 shows the resulting spanning tree.

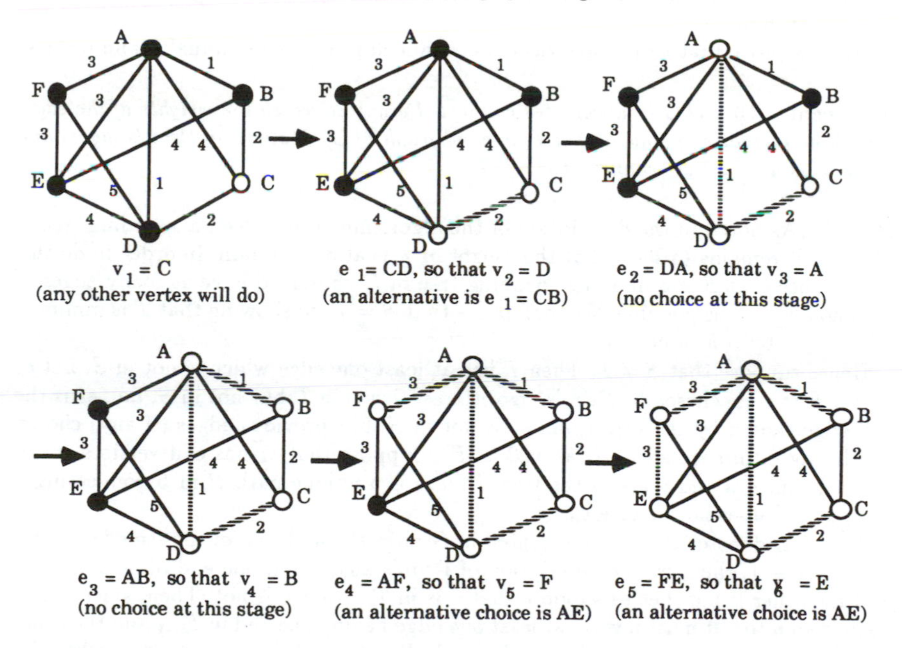
A
B
C
D
E
F
$v_1 = C$
(any other vertex will do)
e_1 = CD, so that v_2 = D
(an alternative is e_1 = CB)
e_2 = DA, so that v_3 = A
(no choice at this stage)
e_3 = AB, so that v_4 = B
(no choice at this stage)
e_4 = AF, so that v_5 = F
(an alternative choice is AE)
e_5 = FE, so that v_6 = E
(an alternative choice is AE)

Figure 2.19: A minimal spanning tree construction using Prim's algorithm.

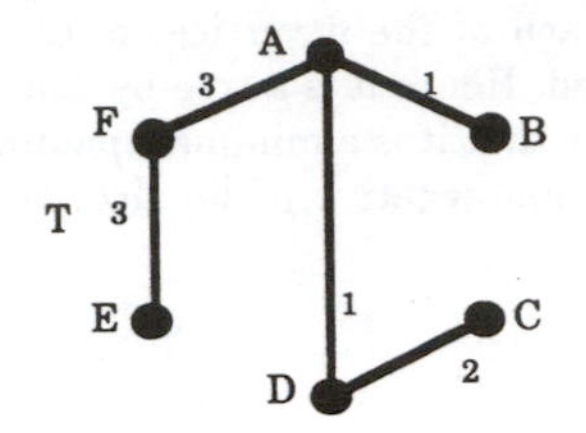

Figure 2.20: The minimal spanning tree.

The weight, $w(T)$, of T is $w(e_1)+w(e_2)+w(e_3)+w(e_4)+w(e_5) = 2+1+1+3+3 = 10$ (which is, as expected, the weight of the minimal spanning tree given by Kruskal's algorithm).

The main differences between Kruskal's algorithm and Prim's algorithm are that

(i) Kruskal's may lead to several subtrees being grown "simultaneously" and then being joined together whereas Prim's leads to one subtree growing steadily, from one initial vertex and

(ii) Kruskal's depends on being able to detect cycles whereas Prim's depends on not choosing a previously chosen vertex.

Difference (ii) means that in computer implementation Prim's algorithm is usually faster than Kruskal's.

We now prove that Prim's algorithm does indeed produce a minimal spanning tree.

Theorem 2.15 *Let G be a weighted connected graph in which the weights of the edges are all non-negative numbers. Let T be a subgraph of G obtained by Prim's algorithm. Then T is a minimal spanning tree of G.*

Proof As noted in the description of the algorithm, T is indeed a spanning tree of G. Thus it remains to show that the weight of T is at a minimum. In order to do this we suppose that S is a minimal spanning tree of G chosen to have as many edges in common with T as possible. We shall prove that $S = T$, so showing that T is minimal. We do this by contradiction.

Hence suppose that $S \neq T$. Then T has at least one edge which is not in S. Let e_k be the first edge chosen by Prim's algorithm which is in T but not in S, i.e., e_k is the kth edge chosen by the algorithm, e_k is not in S, but previous edges (if any) chosen by the algorithm are all in S (as well as T). Suppose that e_k has end vertices u and v. Then, since u and v are in the tree S there is a unique path P in S connecting u to v, and P does not involve e_k.

Now if T_i denotes the subtree created in G after the addition of the ith edge e_i, for $1 \leq i \leq n-1$, then, by the description of Prim's algorithm, one end of e_k is in T_{k-1} and the other is not. Let us suppose that u is in T_{k-1} but v is not. Then, since P is a path from u to v it must involve at least one edge having one end in T_{k-1} and the other not. Let e^* denote such an edge in the path P. Then $w(e^*) \geq w(e)$ since otherwise e^* has less weight than e_k and Prim's algorithm would then have incorporated e^* and not e_k as the kth edge.

Now the path P in S together with the edge e_k gives a cycle in G so if we replace the edge e^* in S with the edge e_k we still have a connected subgraph with n vertices and $n-1$ edges. In other words, replacing e^* in S with e_k gives a new spanning tree R. Since $w(e^*) \geq w(e_k)$, the weight of R is not greater than that of S and so R must also be a minimal spanning tree. However R has one more edge in common with T than S has, namely the edge e_k. This contradicts the assumption that S was chosen to be a minimal spanning tree with as many edges in common with T as possible. This contradiction has arisen from the supposition that $S \neq T$. Hence $S = T$ and T is a minimal spanning tree, as required. □

For computer implementation a weighted graph is usually presented in matrix form: if G has n vertices and w_{ij} denotes the weight of an edge from vertex v_i to v_j, then, assuming that G has no parallel edges, we may present G as the $n \times n$ matrix with (i,j)th entry w_{ij}, taking w_{ij} as ∞ if there is no edge from v_i to v_j.

Thus, for example, taking $v_1 = A, v_2 = B$, etc. in our example above, our graph G is represented by the 6×6 matrix

$$\begin{bmatrix} \infty & 1 & 4 & 1 & 3 & 3 \\ 1 & \infty & 2 & \infty & 4 & \infty \\ 4 & 2 & \infty & 2 & \infty & \infty \\ 1 & \infty & 2 & \infty & 4 & 5 \\ 3 & 4 & \infty & 4 & \infty & 3 \\ 3 & \infty & \infty & 5 & 3 & \infty \end{bmatrix}.$$

The two algorithms have generated minimal spanning trees. They may also be used to generate *maximal* spanning trees, i.e., spanning trees which have *greatest* possible weight. To do this we create a new weighted graph with the same vertices and edges as the one given but where we replace each weight $w(e)$ by $M - w(e)$ where M is any number greater than the weight $w(e)$ of every edge e of the graph. Then any minimal spanning tree of this new weighted graph has the sum of its weights $M - w(e)$ at a minimum, i.e., the sum of the weights $w(e)$ are at a *maximum*, and so the corresponding spanning tree in the original weighted graph is a *maximal* spanning tree.

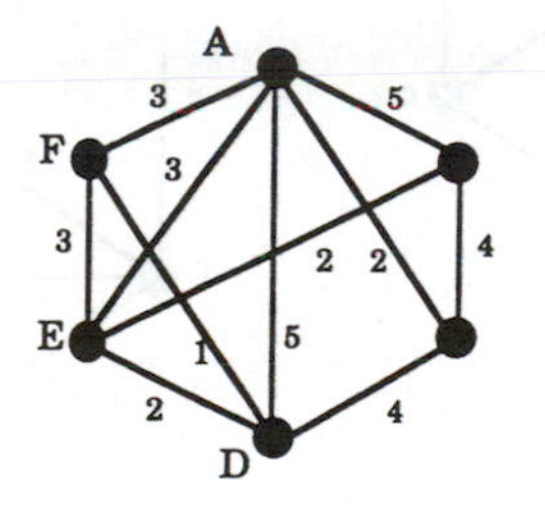

Figure 2.21: The weighted graph G'.

For example, in our graph G of Figure 2.17 take $M = 6$, since 6 is greater than each $w(e)$, to give the new weighted graph G' shown in Figure 2.21. By performing one of the two algorithms on G' we may find a minimal spanning tree for G', as for instance shown in Figure 2.22. We then convert this into a maximal spanning tree for G. (Again see Figure 2.22.) The maximal spanning tree here has weight $5+4+4+3+4=20$. Of course, every other maximal spanning tree also has this weight.

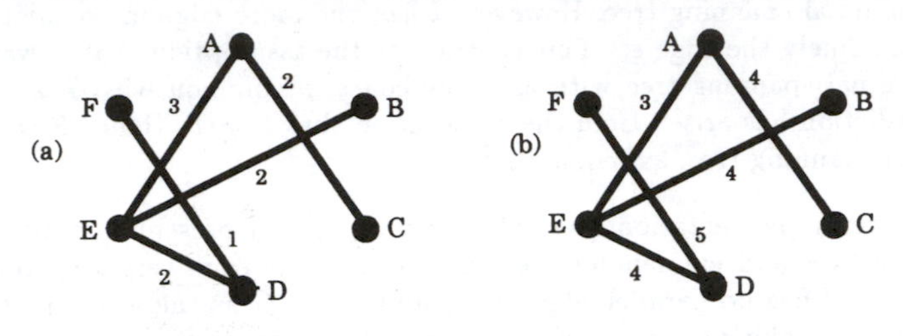

Figure 2.22: (a) A *minimal* spanning tree for the weighted graph G' of Figure 2.21 which gives (b) a *maximal* spanning tree for the weighted graph G of Figure 2.17.

Exercises for Section 2.4

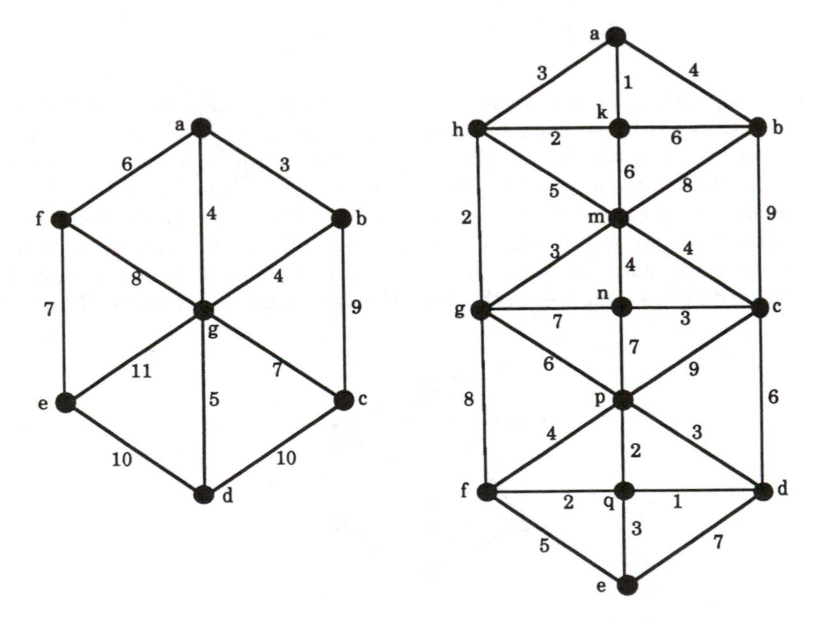

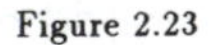
Figure 2.23

2.4.1 Find a minimal spanning tree for each of the connected weighted graphs of Figure 2.23 using both Kruskal's algorithm and Prim's algorithm.

2.4.2 Find a *maximal* spanning tree for each of graphs of Figure 2.23 using either Kruskal's algorithm or Prim's algorithm.

2.4.3 The following is a third algorithm for finding a minimal spanning tree of a connected weighted graph G with n vertices where each weight is non-negative. Simply delete one by one those edges of G with largest weight, provided each such deletion does not result in a disconnected graph, until there are just $n-1$ edges left. Then the resulting subgraph is a minimal spanning tree of G. Perform this algorithm on the graphs of Figure 2.23.

2.4.4 Prove that if G is a connected weighted graph in which no two edges have the same weight then G has a unique minimum spanning tree.

2.5 Shortest Path Problems

(1) *The Breadth First Search (BFS) technique.* Let G be a graph and let s, t be two specified vertices of G. We will now describe a method of finding a path from s to t, if there is any, which uses the least number of edges. Such a path, if it exists, is called a shortest path from s to t. The method assigns labels $0, 1, 2, \ldots$ to vertices of G and is called the Breadth First Search (BFS for short) technique. It is given by the following algorithm.

The Breadth First Search Algorithm

Step 1. Label vertex s with 0. Set $i = 0$.

Step 2. Find all unlabelled vertices in G which are adjacent to vertices labelled i. If there are no such vertices then t is not connected to s (by a path). If there are such vertices, label them $i + 1$.

Step 3. If t is labelled, go to Step 4. If not, increase i to $i + 1$ and go to Step 2.

Step 4. The length of a shortest path from s to t is $i + 1$. Stop.

For example, for the graph of Figure 2.24, first s is labelled 0. Then a and f are labelled 1. Then b, d and e are labelled 2. Then c and t are labelled 3. Since t is labelled 3, the length of a shortest path from s to t is 3.

For the graph of Figure 2.25, first s is labelled 0. Then a, b and c are labelled 1. Then d is labelled 2. But now there are no unlabelled vertices adjacent to d, the only vertex labelled 2. Hence, by Step 2, we can conclude that there is no path from s to t.

We now show that Step 4's statement is justified by proving that the Breadth First Search works.

Theorem 2.16 *A vertex v of the graph G is labelled the number $\lambda(v)$ by the BFS algorithm above if and only if the length of a shortest path from s to v is $\lambda(v)$. (In particular if t is labelled n then the shortest path from s to t has n edges.)*

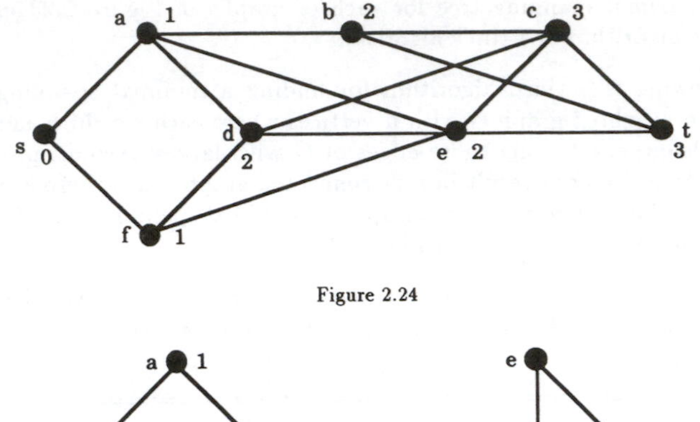

Figure 2.24

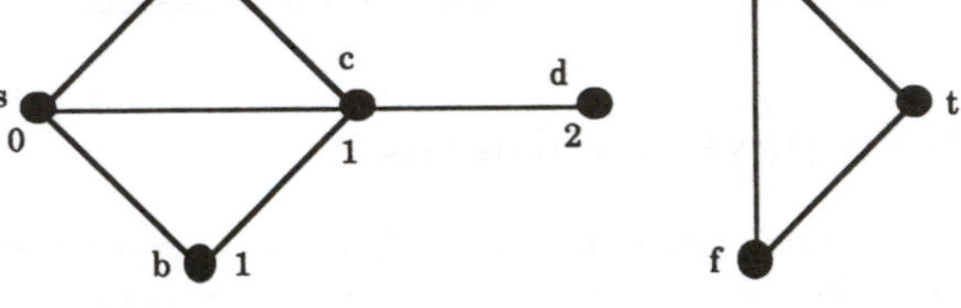

Figure 2.25

Proof The proof is by induction on the value i of $\lambda(v)$. First if $i = 0$ then v has label $\lambda(v) = 0$ and by the algorithm this occurs precisely when $v = s$. However the length of a shortest path from s to s is 0 and so in this case the label $\lambda(v)$ does indeed give the length of the shortest path from s to v.

Now assume that the statement is true for all values of i from 0 up to and including a fixed value k, i.e., if the vertex v of G has been labelled $\lambda(v)$ where $0 \leq \lambda(v) \leq k$ then the length of a shortest path from s to v is $\lambda(v)$ and conversely any vertex v having shortest path to s of length m where $0 \leq m \leq k$ has label $\lambda(v) = m$. Let u be a vertex of G which has label $\lambda(u) = k+1$. Then by the algorithm there is an edge e from a vertex v with $\lambda(v) = k$ to the vertex u. By our assumption there is a shortest path of length k from s to v and concatenating this with e gives a path of length $k+1$ from s to u. Moreover this path is the shortest possible since if there were one shorter then by our induction assumption u would have been given a label $\lambda(u)$ with $0 \leq \lambda(u) \leq k$. Thus if u is a vertex with $\lambda(u) = k+1$ then there is a shortest path of length $k+1$ from s to u.

Conversely, if there is a shortest path from s to a vertex w of length $k+1$ then the second last vertex v, say, of this path has a shortest path to s of length k and so, by our assumption, $\lambda(v) = k$. Since w is adjacent to v, the algorithm gives $\lambda(w) = k+1$. This completes the verification of the statement for $k+1$. Hence by induction the result is true for all values of $\lambda(v)$, as required. □

Once the algorithm has been performed successfully, i.e., Step 4 has been accomplished, we can use the following trace-back (or back-tracking) algorithm to find an actual shortest path from s to t. This algorithm uses the labels $\lambda(v)$ which were produced in the BFS algorithm. It produces a path $v_0, v_1, \ldots, v_{\lambda(t)}$ such that $v_0 = s$ and $v_{\lambda(t)} = t$.

The Back-tracking Algorithm for a Shortest Path

Step 1. Step 1. Set $i = \lambda(t)$ and assign $v_i = t$.

Step 2. Step 2. Find a vertex u adjacent to v_i and with $\lambda(u) = i-1$. Assign $v_{i-1} = u$.

Step 3. Step 3. If $i = 1$, stop. If not, decrease i to $i - 1$ and go to Step 2.

For example for the graph of Figure 2.24 we have $\lambda(t) = 3$ so we start with $i = 3$ and $v_3 = t$ (Step 1). We can then choose e adjacent to $v_3 = t$ with $\lambda(e) = 2$ and assign $v_2 = e$. Next we can choose f adjacent to $v_2 = e$ with $\lambda(f) = 1$ and assign $v_1 = f$. Finally we take s adjacent to f with $\lambda(s) = 0$ and assign $v_0 = s$. This gives the shortest path $v_0v_1v_2v_3 = s\ f\ e\ t$ from s to t. In general there may of course be many shortest paths from s to t and the previous algorithm finds just one of them. A simple extra labelling, assigning a label $\mu(v)$ to each vertex v which has been labelled in the BFS algorithm in a back-tracking manner, actually produces $\mu(s)$ which is the number of shortest paths from s to t. This is done by the following algorithm.

The Back-tracking Algorithm for the Number of Shortest Paths

Step 1. Set $i = \lambda(t)$ and $\mu(t) = 1$. All other vertices v for which $\lambda(v) = \lambda(t)$ are assigned $\mu(v) = 0$.

Step 2. For each vertex v which satisfies $\lambda(v) = i - 1$ compute the sum

$$\sum \mu(u)$$

over all u's which satisfy the following condition: $\lambda(u) = i$ and v is adjacent to u; if there are parallel edges, $\mu(u)$ is repeated in this summation as many times as there are parallel edges. For each such v, set $\mu(v)$ equal to this sum.

Step 3. If $i = 1$, stop. If not, decrease i to $i - 1$ and go to Step 2.

We illustrate the algorithm using the graph of Figure 2.24. We get:

Step 1. $i = 3$, $\mu(t) = 1$, $\mu(c) = 0$ (since $\lambda(c) = 3 = \lambda(t)$).

Step 2. The vertices v with $\lambda(v) = 2$ are b, d, e. Then $\mu(b) = \mu(t) + \mu(c)$ (since t, c are the vertices $u = 1 + 0$ with $\lambda(u) = 3$ and adjacent to b) $= 1$. Similarly $\mu(d) = \mu(c) = 0$ and $\mu(e) = \mu(t) + \mu(c) = 1$. In summary we have produced three new labels : $\mu(b) = 1$, $\mu(d) = 0$, $\mu(e) = 1$.

Step 3. Decrease $i = 3$ to 2 and go to Step 2.

Step 2. The vertices v with $\lambda(v) = 1$ are a, f. Then $\mu(a) = \mu(b) + \mu(d) + \mu(e) = 1 + 0 + 1 = 2$ and $\mu(f) = \mu(d) + \mu(e) = 0 + 1 = 1$.

Step 3. Decrease $i = 2$ to 1 and go to Step 2.

Step 2. The only vertex v with $\lambda(v) = i - 1 = 0$ is s. Then $\mu(s) = \mu(a) + \mu(f) = 2 + 1 = 3$.

Step 3. Since $i = 1$ we stop.

Our calculations give $\mu(s) = 3$ so there are 3 shortest paths from s to t. (In fact they are $s\ a\ e\ t$, $s\ a\ b\ t$ and $s\ f\ e\ t$.)

We omit the proof of the algorithm's validity. (The method is by induction, showing that for all vertices v for which $\lambda(v) < \lambda(t), \mu(v)$ is the number of paths of length $\lambda(t) - \mu(v)$ from v to t.)

(2) *Dijkstra's algorithm*

We now consider shortest path problems in weighted graphs. Given a path P from vertex s to vertex t in a weighted graph G we define the length of P to be the sum of the weights of its edges. (In fact this corresponds to the usual length of a path in an unweighted graph if we assign the weight 1 to each edge.) We wish to consider the problem of finding a shortest path from s to t, i.e., a path of least length. Since in general the weight of an edge can be negative it is possible to have paths of negative length. However the following algorithm due to Dijkstra [16] will be restricted to weighted graphs where the weight $w(e)$ of each edge e is non-negative, i.e., $w(e) \geq 0$.

As with BFS we use a labelling technique. Initially we set $\lambda(s) = 0$ and, for $v \neq s, \lambda(v)$ is as yet undetermined. We next label all neighbours v of s by $\lambda(v)$ where $\lambda(v)$ is the weight of the edge from s to v. Let u be the vertex among those v for which $\lambda(u)$ is minimum. Now find those neighbours w of u and, for those w not already labelled assign the label $\lambda(w) = \lambda(u) + w(e)$, $w(e)$ being the weight of the edge from u to w, while for those w already labelled $\lambda(w)$ change the label to $\lambda(u) + w(e)$ if this is smaller. Now find the w among these for which $\lambda(w)$ is a minimum ... this is getting too complicated! We now present the algorithm in stepwise form, using the convention that $\infty + x = \infty$ for any real number x, and $\infty + \infty = \infty$.

Dijkstra's Algorithm

Step 1. Set $\lambda(s) = 0$ and for all vertices $v \neq s, \lambda(v) = \infty$. Set $T = V$, the vertex set of G. (We will think of T as the set of vertices "uncoloured".)

Step 2. Let u be a vertex in T for which $\lambda(u)$ is minimum.

Step 3. If $u = t$, stop.

Step 4. For every edge $e = uv$ incident with u, if $v \in T$ and $\lambda(v) > \lambda(u) + w(e)$ change the value of $\lambda(v)$ to $\lambda(u) + w(e)$ (i.e., given an edge e from an "uncoloured" vertex v to u, change $\lambda(v)$ to $min(\lambda(v), \lambda(u) + w(e))$).

Step 5. Change T to $T - \{u\}$ and go to Step 2, (i.e., "colour" u and then go back to Step 2 to find an "uncoloured" vertex with minimum label).

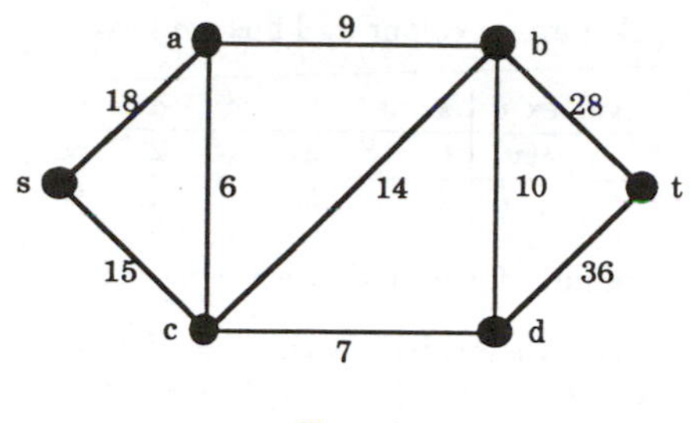

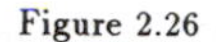
Figure 2.26

We illustrate this with the weighted graph of Figure 2.26.

Step 1. The initial labelling is given by:

vertex v	s	a	b	c	d	t
$\lambda(v)$	0	∞	∞	∞	∞	∞
T	$\{s,$	$a,$	$b,$	$c,$	$d,$	$t\}$

Step 2. $u = s$ has $\lambda(u)$ a minimum (with value 0).

Step 4. There are two edges incident with u, namely sa and sc. Both a and c are in T, i.e., they are not yet coloured. $\lambda(a) = \infty > 18 = 0 + 18 = \lambda(s) + w(sa)$ so $\lambda(a)$ becomes 18. Similarly $\lambda(c)$ becomes 15.

Step 5. T becomes $T - \{s\}$, i.e., we colour s. Thus we have

vertex v	s	a	b	c	d	t
$\lambda(v)$	0	18	∞	15	∞	∞
T	$\{$	$a,$	$b,$	$c,$	$d,$	$t\}$

Step 2. $u = c$ has $\lambda(u)$ a minimum and u in T (with $\lambda(u) = 15$).

Step 4. There are 3 edges cv incident with $u = c$ having v in T, namely ca, cb, cd. $\lambda(a) = 18 < 21 = 15 + 6 = \lambda(c) + w(ca)$ so $\lambda(a)$ remains as 18. $\lambda(b) = \infty > 29 = 15 + 14 = \lambda(c) + w(cb)$ so $\lambda(b)$ becomes 29. Similarly $\lambda(d)$ becomes $15 + 7 = 22$.

Step 5. T becomes $T - \{c\}$, i.e., we colour c. Thus we have

vertex v	s	a	b	c	d	t
$\lambda(v)$	0	18	29	15	22	∞
T	$\{$	$a,$	$b,$		$d,$	$t\}$

Step 2. $u = a$ has $\lambda(u)$ a minimum for u in T (with $\lambda(u) = 18$).

Step 4. There is only one edge av incident with $u = a$ having v in T, namely ab. $\lambda(b) = 29 > 27 = 18 + 9 = \lambda(a) + w(ab)$ so $\lambda(b)$ becomes 27.

Step 5. T becomes $T - \{a\}$, i.e., we colour a. Thus we have

vertex v	s	a	b	c	d	t
$\lambda(v)$	0	18	27	15	22	∞
T	$\{$		$b,$		$d,$	$t\}$

Step 2. $u = d$ has $\lambda(u)$ minimum for u in T (with $\lambda(u) = 22$).

Step 4. There are two edges dv incident with $u = d$ having v in T, namely db and dt. $\lambda(b) = 27 < 32 = 22 + 10 = \lambda(d) + w(db)$ so $\lambda(b)$ remains as 27. $\lambda(t) = \infty > 58 = 22 + 36 = \lambda(d) + w(dt)$ so $\lambda(t)$ becomes 58.

Step 5. T becomes $T - \{d\}$, i.e., we colour d. Thus we have

vertex v	s	a	b	c	d	t
$\lambda(v)$	0	18	27	15	22	58
T	$\{$		$b,$			$t\}$

Step 2. $u = b$ has $\lambda(u)$ minimum for u in T (with $\lambda(u) = 27$).

Step 4. There is only one edge bv for v in T, namely bt. $\lambda(t) = 58 > 55 = 27 + 28 = \lambda(b) + w(bt)$ so $\lambda(t)$ becomes 55.

Step 5. T becomes $T - \{b\}$, i.e., we colour b. Thus we have

vertex v	s	a	b	c	d	t
$\lambda(v)$	0	18	27	15	22	55
T	$\{$					$t\}$

Step 2. $u = t$, the only choice.

Step 3. Stop.

When the algorithm stops the values $\lambda(v)$ give the lengths of the shortest paths from the vertex s to each vertex v. Thus the lengths of these paths from s to a, b, c, d, t are 18, 27, 15, 22 and 55 respectively.

In the following table we summarise the above process as carried out on our example, recording the choices of u (in the Steps 2), the changes in the $\lambda(v)$ values (in the Steps 4), and the changes to T, i.e., the colouring of the vertices, (in the Steps 5). The values in bold type indicate the choices of u and the corresponding $\lambda(u)$ given by Step 2 of the algorithm.

	vertices v						T
	s	a	b	c	d	t	
	0	∞	∞	∞	∞	∞	$\{s, a, b, c, d, t\}$
stepwise		18	∞	**15**	∞	∞	$\{a, b, c, d, t\}$
$\lambda(v)$		**18**	29		22	∞	$\{a, b, d, t\}$
values			27		**22**	∞	$\{b, d, t\}$
			27			58	$\{b, t\}$
						55	$\{t\}$

We prove the validity of Dijkstra's algorithm in two stages. First

Theorem 2.17 *In Dijkstra's algorithm, if at some stage $\lambda(v)$ is finite (i.e., not ∞) for the vertex v then there is a path from s to v whose length is $\lambda(v)$.*

Proof If $v = s$ then $\lambda(v)$ is always 0 and the trivial path from s to itself has length 0. Thus the result is true if $v = s$. Now suppose $v \neq s$. Then, from the algorithm, since at Step 1 $\lambda(v) = \infty$ but now $\lambda(v)$ is finite, there must be some vertex u_1 given by Step 2 of the algorithm such that u_1 is adjacent to v, joined by the edge e_1 say, and the value $\lambda(v)$ is given by $\lambda(u_1) + w(e_1)$. Immediately after $\lambda(v)$ was given this value, Step 5 was applied to colour u_1. Also Steps 2 and 4 imply that once a vertex is coloured, i.e., no longer in T, then its label remains the same thereafter. Since $\lambda(u_1)$ is finite we can repeat the process to find the vertex u_2 adjacent to u_1, joined by the edge e_2 say, and with label $\lambda(u_1) = \lambda(u_2)+w(e_1)$. We keep on repeating this backward search until we get back to s. The vertices thus found form a path of length $\lambda(v)$ from v back to s since

$$\begin{aligned}
\lambda(v) &= \lambda(u_1) + w(e_1) \\
&= \lambda(u_2) + w(e_2) + w(e_1) \\
&= \lambda(u_3) + w(e_3) + w(e_2) + w(e_1) \\
&= \dots \\
&= \lambda(s) + w(e_n) + \cdots + w(e_1) \\
&= w(e_n) + \cdots + w(e_1),
\end{aligned}$$

the length of the path consisting of the n weighted edges $e_1, \dots, e_n$ produced in the process. We do have a path, i.e., no vertex u_i is repeated, since by the method of their choice, u_1 is coloured after u_2, u_2 is coloured after u_3, etc. □

For the second stage of the verification we let, for any vertex v, $\delta(v)$ denote the length of a shortest path from s to v , i.e., $\delta(v)$ is the distance from s to v , taking $\delta(v) = \infty$ if there is no path from s to v.

Theorem 2.18 *In Dijkstra's algorithm when a vertex u is chosen in Step 2 its label $\lambda(u)$ has value $\delta(u)$. (In particular, upon termination of the algorithm, $\lambda(t) = \delta(t)$.)*

Proof We use induction on the order in which the vertices are coloured, i.e., in which they leave T. The first vertex to be coloured is $v = s$ and since $\lambda(s) = 0 = \delta(s)$ the result is true in this case. Now suppose that u is a vertex different from s and that the result holds for all vertices coloured before u. If $\lambda(u) = \infty$ when u is chosen at Step 2 then, since $\lambda(s)$ is finite, there must be a vertex u' which is the first to be coloured having label $\lambda(u') = \infty$, and $u' \neq s$. Because of the minimum requirement in Step 2 when u' is chosen in Step 2 all remaining uncoloured vertices v have $\lambda(v) = \infty$. Moreover since u' is the first, all vertices v' coloured before u' have $\lambda(v')$ finite. This implies that there is no edge e from such a v' to either u' or any uncoloured v, because otherwise $\lambda(u')$ (or $\lambda(v)$) would be changed to the finite $\lambda(v') + w(e)$. It follows that there is no path from s to u since $\lambda(s)$ is finite while $\lambda(u) = \infty$. Thus $\delta(u) = \infty$, as required.

It remains to consider the case where $\lambda(u)$ is finite. By the previous theorem, $\lambda(u)$ is the length of some path from s to u. Thus $\lambda(u) \geq \delta(u)$. We have to show that $\lambda(u) > \delta(u)$ is impossible. Let a shortest path from s to u be given by $s = v_0 v_1 \dots v_k = u$. Then, denoting the edge from v_{i-1} to v_i by e_i (for $1 \leq i \leq k$), its length is

$$\sum_{i=1}^{k} w(e_i) = \delta(u).$$

Let v_j denote the last vertex on this path to be coloured before u. Then, by the induction hypothesis,

$$\lambda(v_j) = \delta(v_j) = \sum_{i=1}^{j} w(e_i).$$

If $v_{j+1} \neq u$ $(= v_k)$, then $\lambda(v_{j+1}) \leq \lambda(v_j) + w(e_{j+1})$ after v_j has been coloured (because of Step 4). Since labels only decrease if they change at all in step 4, when u is coloured $\lambda(v_{j+1})$ still satisfies this inequality. We have [1] $\lambda(v_{j+1}) \leq \lambda(v_j) + w(e_{j+1}) = \delta(v_j) + w(e_{j+1}) = \delta(v_{j+1}) \leq \delta(u)$. Thus if $\lambda(u) > \delta(u)$ we would get $\lambda(v_{j+1}) < \lambda(u)$ which, since v_{j+1} is not coloured before u, contradicts the minimality assumption for u in Step 2. Thus if $v_{j+1} \neq u$ we must have $\lambda(u) = \delta(u)$, as required. Finally if $v_{j+1} = u$ then as before $\lambda(u) = \lambda(v_{j+1}) \leq \lambda(v_j) + w(e_{j+1}) = \delta(v_j) + w(e_{j+1}) = \delta(u)$, giving again $\lambda(u) \leq \delta(u)$ and so $\lambda(u) = \delta(u)$ as required. □

The method of proof used in Theorem 2.17 actually produces a shortest path. It is a backtracking technique: starting at the final label assigned to t we go back along the temporary (previously assigned) labels for t until we get a change; we then move to the vertex which has caused this change and do a similar backtrack on its labels until we find a change, and so on. The vertices found in this way then give us a shortest path. We illustrate this using the table on page 74. Here the vertices of the shortest path are boxed.

	vertices v						T					
	s	a	b	c	d	t						
	[0]	∞	∞	∞	∞	∞	$\{s,$	$a,$	$b,$	$c,$	$d,$	$t\}$
stepwise		18	∞	**15**	∞	∞	$\{$	$a,$	$b,$	$c,$	$d,$	$t\}$
$\lambda(v)$		[18]	29		22	∞	$\{$	$a,$	$b,$		$d,$	$t\}$
values			27		**22**	∞	$\{$		$b,$		$d,$	$t\}$
			[27]			58	$\{$		$b,$			$t\}$
						[55]	$\{$					$t\}$

[1] It is at this stage in the proof we use our assumption that the weights of the edges are non-negative:

$$\delta(u) = \sum_{i=1}^{k} w(e_i) = w(e_1) + \cdots + w(e_{j+1}) + w(e_{j+2}) + \cdots + w(e_k) \geq w(e_1) + \cdots + w(e_{j+1})$$

since $w(e_{j+2}), \dots, w(e_k)$ are all non-negative.

In detail, $\lambda(t)$'s change from 58 to 55 was caused by b, $\lambda(b)$'s change from 29 to 27 was caused by a, $\lambda(a)$'s change from ∞ to 18 was caused by s, so the shortest path is $s\ a\ b\ t$. (Check that it has length 55 from Figure 2.26.)

Exercises for Section 2.5

2.5.1 Carry out the BFS algorithms on the graphs of Figure 2.27 to find the length of a shortest path from vertex a to vertex z, an example of such a shortest path, and the number of shortest paths from a to z.

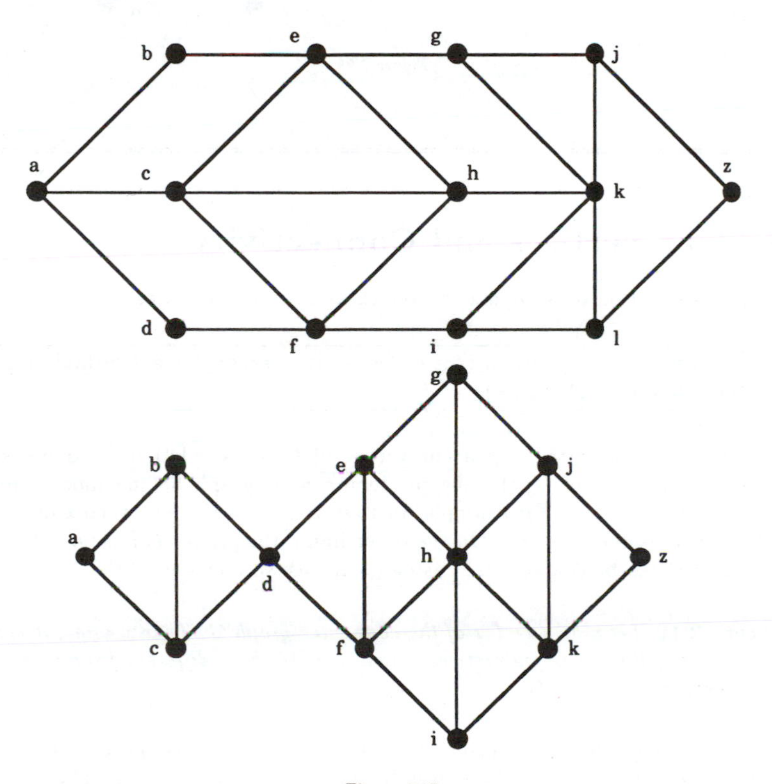

Figure 2.27

2.5.2 Use Dijkstra's algorithm on the connected weighted graphs of Figure 2.28 to find the length of shortest paths from the vertex a to each of the other vertices and to give examples of such paths. Set out your answer in tabular form as explained in the text.

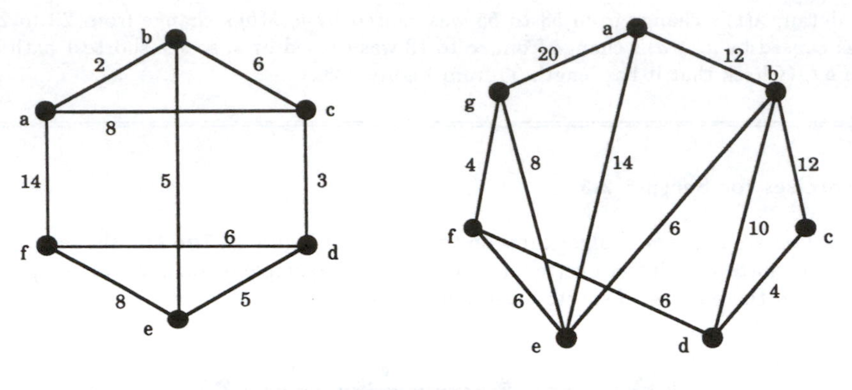

Figure 2.28

2.6 Cut Vertices and Connectivity

We begin this section by studying the vertex analogue of a bridge.

> A vertex v of a graph G is called a **cut vertex** (or **articulation point**) of G if $\omega(G - v) > \omega(G)$.

In other words, a vertex v is a cut vertex of G if its deletion disconnects some connected component of G, thereby producing a subgraph having more connected components than G has. For example, in Figure 2.29, v is a cut vertex of G_1 since $\omega(G_1) = 1$ while $\omega(G_1 - v) = 3$. On the other hand, the graph G_2 has no cut vertices.

Cut vertices can be characterised using paths, as we now see.

Theorem 2.19 *Let v be a vertex of the connected graph G. Then v is a cut vertex of G if and only if there are two vertices u and w of G, both different from v, such that v is on every u–w path in G.*

Proof First let v be a cut vertex of G. Then $G - v$ is disconnected and so there are vertices u and w of G which lie in different components of $G - v$. Thus, although there is a path in G from u to w, there is no such path in $G - v$. This implies that every path in G from u to w contains the vertex v, as required.

Conversely, suppose that u and w are two vertices of G, different from v, such that every path in G from u to w contains v. Then there can be no path from u to w in $G - v$. Thus $G - v$ is disconnected (with u and w lying in different components). Hence v is a cut vertex, as required. □

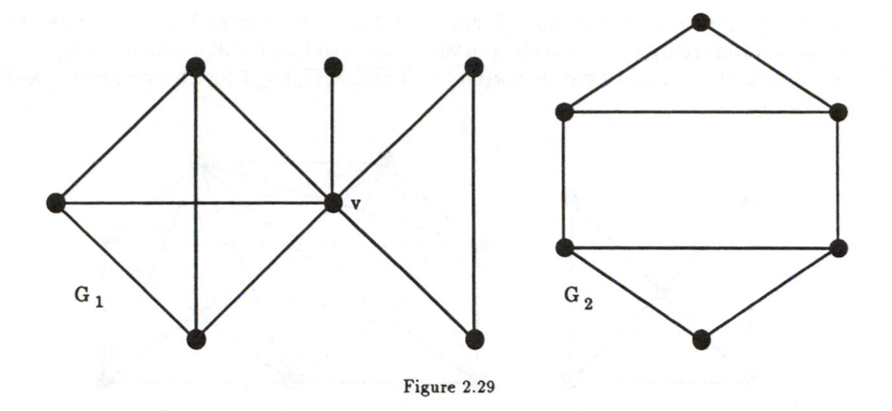

Figure 2.29

Clearly no vertex of a complete graph is a cut vertex. On the other hand, if P_n is the path of length n, where $n \geq 3$, then taking u and w to be the end vertices of P_n in Theorem 2.19, we see that every vertex v, apart from u and w, is a cut vertex. In other words, all but two vertices of P_n are cut vertices.

There is no graph in which every vertex is a cut vertex, as can be seen from our next result.

Theorem 2.20 *Let G be a graph with n vertices, where $n \geq 2$. Then G has at least two vertices which are not cut vertices.*

Proof Clearly we may suppose that G is a connected graph. We proceed by assuming the result is false for our G and so the proof will be complete if we derive a contradiction from this.

Thus we are assuming that there is at most one vertex in G which is not a cut vertex. Now let u, v be vertices in G such that the distance $d(u,v)$ between them is the greatest of distances between pairs of vertices in G, i.e., $d(u,v) = \text{diam}(G)$. (See Exercises 1.6.8, 1.6.9.) Since G is connected and has at least two vertices, $u \neq v$. Thus, by our assumption, one of these two vertices must be a cut vertex, say v. Then $G - v$ is disconnected and so there is a vertex w in G which does not belong to the same component as u does in $G - v$. This implies that every uw path in G contains the vertex v.

It follows from this that the shortest path in G from u to w contains the shortest path from u to v and this contradiction completes our proof. □

If a connected graph G has a cut vertex v then the connectedness of G is vulnerable at v. If G has no cut vertex then its connectedness is not vulnerable to a deletion of one of its vertices. We now put a measure on this vulnerability.

> Let G be a simple graph. The **(vertex) connectivity** of G, denoted by $\kappa(G)$, is the smallest number of vertices in G whose deletion from G leaves either a disconnected graph or K_1.

For example, since the graph G_2 of Figure 2.29 has no cut vertex but a disconnected subgraph is produced when the vertices u and v are deleted, it follows that $\kappa(G_2) = 2$. We leave it to the reader to see that $\kappa(G_3) = 3$ and $\kappa(G_4) = 4$ for the graphs G_3 and G_4 of Figure 2.30.

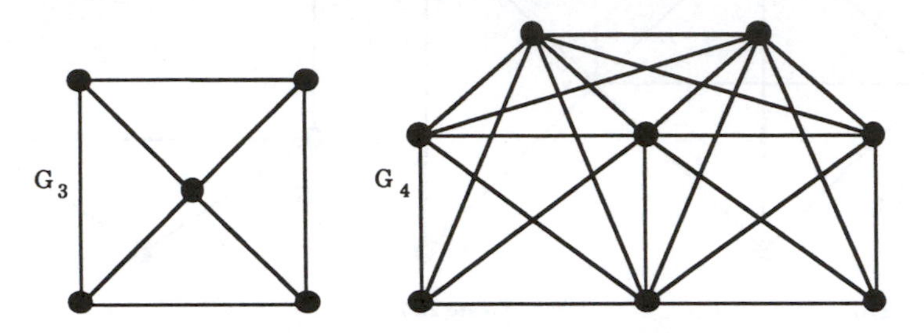

Figure 2.30: $\kappa(G_3) = 3$ and $\kappa(G_4) = 4$.

For $n \geq 2$, the deletion of any vertex from K_n results in K_{n-1} and in general the deletion of t vertices (where $t < n$) results in K_{n-t}. This shows that $\kappa(K_n) = n - 1$.

It is also easy to see that a connected graph G has $\kappa(G) = 1$ if and only if either $G = K_2$ or G has a cut vertex. Moreover $\kappa(G) = 0$ if and only if either $G = K_1$ or G is disconnected.

A simple graph G is called n-**connected** (where $n \geq 1$) if $\kappa(G) \geq n$.

It follows that G is 1-connected if and only if G is connected and has at least two vertices. Moreover G is 2-connected if and only if G is connected with at least three vertices but no cut vertices. We finish this chapter with a characterisation of 2-connected graphs due to Whitney [64]. First we need a definition.

Let u and v be two vertices of a graph G. A collection $\{P_{(1)}, \ldots, P_{(n)}\}$ of $u - v$ paths is said to be **internally disjoint** if, given any distinct pair $P_{(i)}$ and $P_{(j)}$ in the collection, u and v are the only vertices $P_{(i)}$ and $P_{(j)}$ have in common.

Theorem 2.21 (Whitney, 1932) *Let G be a simple graph with at least three vertices. Then G is 2-connected if and only if for each pair of distinct vertices u and v of G there are two internally disjoint $u - v$ paths in G.*

Proof Suppose first that any pair of distinct vertices is connected by a pair of internally disjoint paths. Then clearly G is connected so it remains to prove that G has no cut vertices. Assume, to the contrary, that v is a cut vertex of G. Then, by Theorem 2.19, there are two vertices u and w of G, both different from v, such that v is on every $u - w$ path in G. However, by the hypothesis there are two internally disjoint $u - w$ paths in G and at most one of these can then pass through v. Thus v is *not* on every $u - w$ path, a contradiction. Hence G has no cut vertices.

Now conversely suppose that G is 2-connected. Let u and v be a pair of distinct vertices of G. We use induction on $d(u, v)$, the distance between u and v (see Exercise 1.6.8), to show that there is a pair of internally disjoint $u - v$ paths.

First, if $d(u, v) = 1$ then u and v are joined by an edge, say e. It follows from Exercise 2.6.2 that e is not a bridge, since G has no cut vertices. Hence, there is a $u - v$ path P different and so internally disjoint from the $u - v$ path Q given by the single edge e.

We now assume that $d(u, v) = k \geq 2$ and that if x and y are any pair of vertices with $d(x, y) < k$ then there are two internally disjoint $x - y$ paths. Let P be a path of length k from u to v and let w be the second last vertex of P. Then $d(u, w) = k - 1$ and there are two internally disjoint $u - w$ paths, say Q_1 and Q_2.

Since G is 2-connected, w is not a cut vertex, i.e., $G - w$ is connected and so there is a $u - v$ path P' which does not pass through w. Let x be the last vertex of P' which is also a vertex of either Q_1 or Q_2 (such a vertex exists since u is common to all three paths). See Figure 2.31 for an illustration of this.

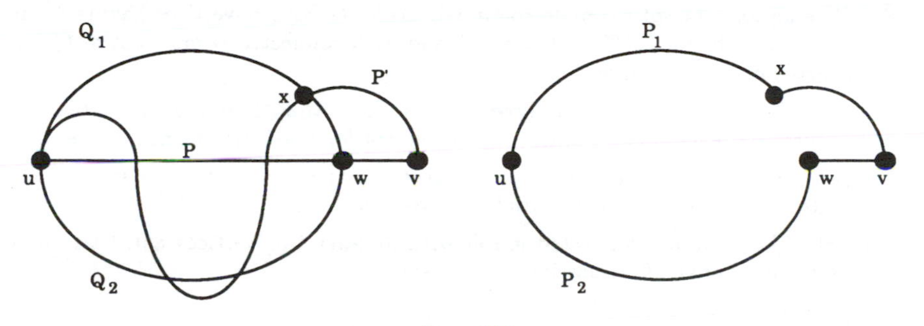

Figure 2.31

Suppose that $x \in Q_1$. Let P_1 be the $u - v$ path given by the $u - x$ section of Q_1 followed by the $x - v$ section of P'. Let P_2 be the $u - v$ path given by the path Q_2 followed by the edge wv. Then P_1 and P_2 are internally disjoint, by the definition of Q_1, Q_2, x and P'. (See Figure 2.31.) The proof now follows by induction. □

Corollary 2.22 *Let u and v be two vertices of the 2-connected graph G. Then there is a cycle C of G passing through both u and v.*

Proof Let P_1 and P_2 be two $u - v$ internally disjoint $u - v$ paths, as guaranteed by Theorem 2.21. Then $P_1 \cup P_2$ gives a cycle C containing both u and v, as required.

Exercises for Section 2.6

2.6.1 Prove that a vertex v of a tree T is a cut vertex if and only if $d(v) > 1$.

2.6.2 Let G be a connected graph with at least three vertices. Prove that if G has a bridge then G has a cut vertex.

2.6.3 Find $\kappa(G)$ for the graphs G of Figure 2.32. If $\kappa(G) = 1$, identify the cut vertices of G.

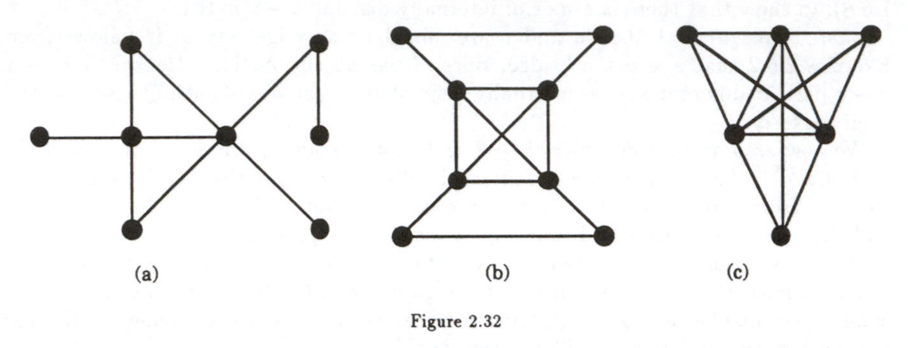

Figure 2.32

2.6.4 Give an example of a simple connected graph G with n vertices having a cut vertex v such that $\omega(G - v) = n - 1$ and each connected component of $G - v$ consists of an isolated vertex.

2.6.5 Let T be a tree with at least three vertices. Prove that there is a cut vertex v of T such that every vertex adjacent to v, except for possibly one, has degree 1.

2.6.6 Let v be a cut vertex of the simple connected graph G. Prove that v is *not* a cut vertex of its complement $\overline{G}$. (Hint: see Exercise 1.6.12.)

2.6.7 Let G be a simple connected graph with at least two vertices and let v be a vertex in G of smallest possible degree, say k.

(a) Prove that $\kappa(G) \leq k$.

(b) Prove that $\kappa(G) \leq 2e/n$, where e is the number of edges and n is the number of vertices in G.

2.6.8 Let $v_1, v_2, \ldots, v_n$ be n distinct vertices of the n-connected graph G and form the supergraph H of G by introduction of a new vertex v, not in G, which is adjacent to each of $v_1, v_2, \ldots, v_n$. Prove that H is n-connected.

2.6.9 Let G be an n-connected graph and let H be the join $G + K_1$. Prove that H is $(n+1)$-connected.

Chapter 3

Euler Tours and Hamiltonian Cycles

3.1 Euler Tours

Recall, from Section 1.6, that a trail in a graph G is a walk in G in which the edges are distinct, i.e., no edge of G appears in the trail more than once.

> A trail in G is called an **Euler trail** if it includes every edge of G.

Thus a trail is Euler if each edge of G is in the trail exactly once.

> A **tour** of G is a closed walk of G which includes every edge of G at least once.
>
> An **Euler tour** of G is a tour which includes each edge of G *exactly* once.

Thus an Euler tour is just a closed Euler trail.

> A graph G is called **Eulerian** or **Euler** if it has an Euler tour.

For example, the graphs G_1 and G_2 of Figure 3.1 have an Euler trail and an Euler tour respectively.

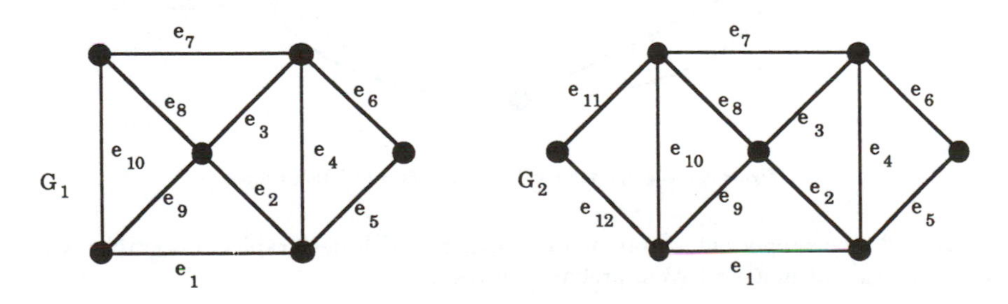

Figure 3.1: G_1 has an Euler trail. G_2 is an Euler graph.

In G_1 an Euler trail, from u to v is given by the sequence of edges $e_1e_2e_3\ldots e_9e_{10}$, while in G_2 an Euler tour from u to u is given by $e_1e_2e_3\ldots e_{11}e_{12}$. We will see below that G_1 has no Euler tour. Euler trails and tours are, historically, the most famous walks in graph theory. Graphs and graph theory probably began in the early 18th century when the Swiss mathematician, Leonhard Euler, considered the problem of the seven Königsberg bridges. To describe this problem we use Figure 3.2 which gives a simplified map of the Prussian city of Königsberg, as it appeared in the 18th century, showing its site on the banks of the river Pregel:

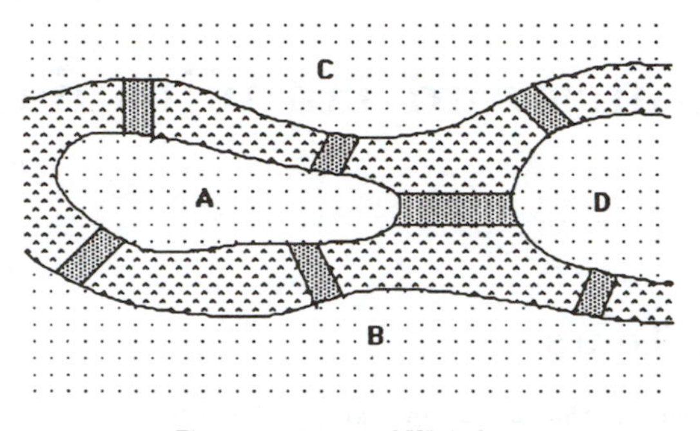

Figure 3.2: A map of Königsberg.

The river was crossed by seven bridges, which connected two islands in the river, shown in Figure 3.2 by A and D, with each other and with the opposite banks B and C. It is said that the townsfolk of Königsberg amused themselves by trying to find a route that crossed each bridge just once. Euler considered this problem by using the graph of Figure 3.3, where each edge represents one of the seven bridges. He then showed [21] the impossibility of such a route by in effect showing that the graph has no Euler trail.

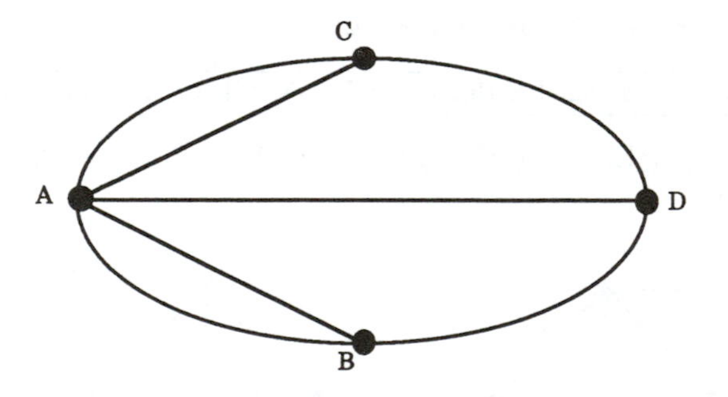

Figure 3.3: A graph representing the bridges of Königsberg.

We will give simple but useful characterizations of Euler graphs and graphs with Euler trails. But first we have a preliminary result.

Theorem 3.1 *Let G be a graph in which the degree of every vertex is at least two. Then G contains a cycle.*

Proof If G is not simple then it contains a cycle since any loop is a cycle of length 1 while a pair of parallel edges gives a cycle of length 2. We now suppose that G is simple. Let v_0 be any vertex of G. Since $d(v_0) \geq 2$, we can choose an edge e_1 with one end v_0 and the other v_1, say. Since $d(v_1) \geq 2$ we can choose an edge e_2 with one end v_1 and the other v_2, say, different from v_0. We repeat this process so that (see Figure 3.4) at the $(i+1)$th stage we have an edge e_i incident with v_i and v_{i+1} and $v_{1+1} \neq v_{i-1}$.

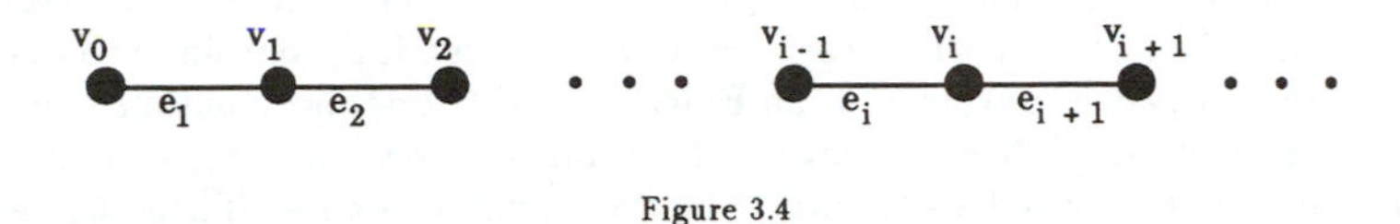

Figure 3.4

Since G has only finitely many vertices, we must eventually choose a vertex which has been chosen before. If v_k is the first such vertex then the walk between the first two occurrences of v_k is a cycle (since the internal vertices of this walk are distinct and also different from v_k as v_k is the first vertex to be repeated).(See Figure 3.5.) □

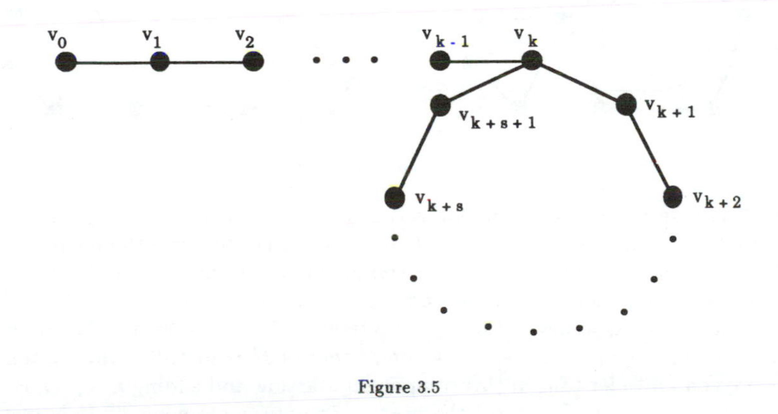

Figure 3.5

Now for our characterization of Euler graphs:

Theorem 3.2 *A connected graph G is Euler if and only if the degree of every vertex is even.*

Proof Suppose that G is Euler. Let C be an Euler tour in G, starting and ending at the vertex u. Then if v is a vertex of G different from u, v must be a vertex on the tour C since u is connected to v (G being connected) and C involves every edge of G. Moreover each time v is met on the tour C, it is entered and left by different edges (because each edge of G only occurs once in C). Thus each occurrence of v in C represents a contribution of 2 to its degree. Thus $d(v)$ is even. Finally, since C begins and ends with u, the first and last edges of C contribute 2 to the degree of u and any internal occurrence of u on C will (as above) also contribute 2 to $d(u)$. Hence $d(u)$ is also even. Thus the degree of every vertex of G is even, as required.

Conversely, suppose that G is connected and that every vertex is even. We use induction on the number of edges of G to show that G is Euler. Firstly, if there are no edges then, since G is connected, G must consist of a single vertex u, with degree 0. Then the trivial trail $C = u$ involves all the edges of G — since there are none! Thus G, in this case, is Euler.

Now suppose that G does have edges. Then, since G is connected, no vertex of G can have degree 0. Thus, since each vertex is even, each vertex must have degree at least two. Then, by Theorem 3.1, G contains a cycle, C say. If C contains every edge of G then we are finished since then C is an Euler tour. If C does not contain every edge of G then we delete from G every edge in C to form a new (possibly disconnected) graph H which has fewer edges than G, but the same vertex set. Then the vertices of H are still even since any vertex which has had a change in degree has had two distinct incident edges removed from it (because these edges come from the cycle C). (See Figure 3.6.)

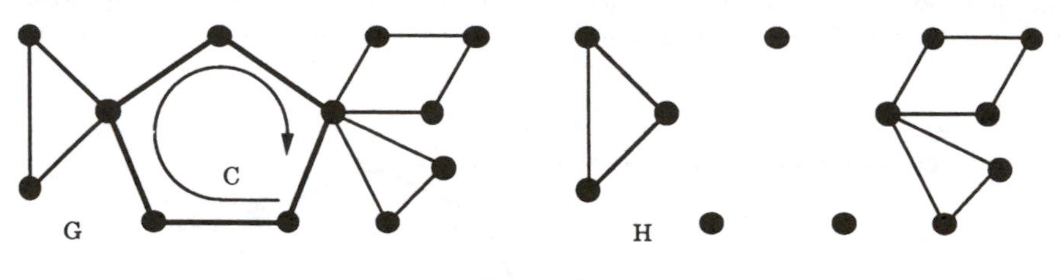

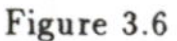
Figure 3.6

Now, assuming the induction hypothesis that the result holds for all graphs with less edges than G, each component of H must be Euler (because the degree of each vertex of H is even). Moreover, since the components were formed by the deletion of C's edges, each component has at least one vertex in common with C. We now obtain an Euler tour for G as follows: start at any vertex of C and go round the edges of C until a vertex belonging to a nonempty component of H is met. Say this vertex is v_1 and then go on an Euler tour in this component starting and ending at v_1. Once back at v_1, resume going round C until the next nonempty component of H is met and repeat the process of going on an Euler tour in this new component. We repeat this procedure eventually going all the way round C back to our starting point, making an Euler tour in each nonempty component of H on the way. We illustrate this in Figure 3.7. Then, since the edges of G are just those of H together with those of C, we have completed an Euler tour of G. By induction the proof is complete. □

Although Euler proved in 1736 the necessity part of Theorem 3.2, i.e., that a connected graph is Euler only if all its vertices are even, surprisingly it was not until 1873 that the sufficiency part was established, by Hierholzer [34]. We refer the reader to Biggs, Lloyd and Wilson [6] for an interesting account of the history of the Königsberg bridges problem (and for a general history of graph theory).

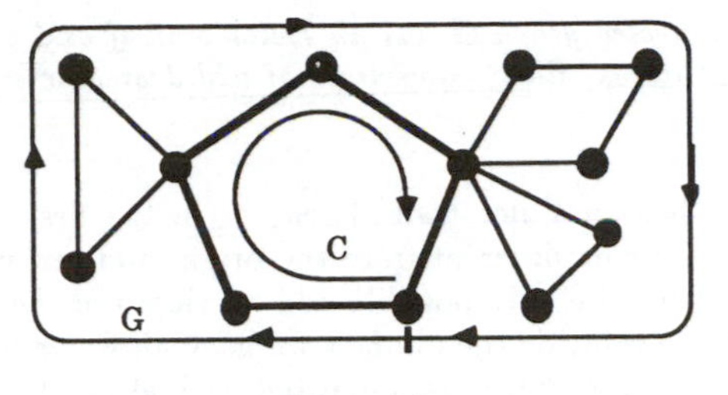

Figure 3.7

We now outline an alternative argument for the second part of the proof of Theorem 3.2, that each vertex of an Euler graph is even. This is a recent simple proof due to Fowler [25].

Thus let G be a connected graph in which every vertex is even. As in the proof of Theorem 3.2, we use induction on the number of edges of G to show that G has an Euler tour. If G has less than three *vertices* then it is an easy matter to see that G has an Euler tour. Hence we may assume that G has at least three vertices. Suppose that G has q edges and, so that we may use induction, that any connected graph H with less than q edges in which every vertex is even is Euler. Since G is connected and has at least three vertices, there exists a vertex v in G joined by two edges to vertices x and y (so that v is distinct from both x and y but x and y are not necessarily distinct). Delete both edges vx and vy from G. Now form a new graph H from this edge deleted subgraph by (i) inserting a new edge xy if $x \neq y$ or (ii) inserting a loop at x if $x = y$. Then it is easy to see that every vertex of H is even and that H has $q - 1$ edges. Thus, by our assumption, if H is connected it must be an Euler graph, say with Euler tour T'. In this case, an Euler tour T for G is then obtained from T' by replacing x, xy and y by x, xv, vy and y.

If H is not connected then it has exactly two connected components, say H_1 and H_2. Moreover, x and y must belong to one component, say H_1, and v to the other, H_2. Our induction assumption provides Euler tours T_1 and T_2 for H_1 and H_2 respectively. We can now construct an Euler tour T for G from T_1 and T_2 by replacing x, xy and y in T_1 by x, xv, T_2, vy and y.

Using the first proof of Theorem 3.2 we can prove another characterisation of Euler graphs, as follows.

Theorem 3.3 *A connected graph G is Euler if and only if G has cycles $C_{(1)}, \ldots, C_{(n)}$ such that every edge of G belongs to exactly one cycle $C_{(i)}$, i.e., G is the union of edge disjoint cycles.*

Proof We leave the proof of this as Exercise 3.1.8. □

Using Theorem 3.2 it is a simple matter to give a characterisation of graphs which have Euler trails:

Theorem 3.4 *A connected graph G has an Euler trail if and only if it has at most two odd vertices, i.e., it has either no vertices of odd degree or exactly two vertices of odd degree.*

Proof Suppose G has an Euler trail. Then, as in the first part of the proof of Theorem 3.2, if v is a vertex different from the origin and terminus of the trail, the degree of v is even. Thus the only possible odd vertices are the origin and terminus of the trail (and if these coincide then in fact we have an Euler tour and every vertex is even). Conversely, suppose that G is connected with at most two odd vertices. If G has no odd vertices then, by Theorem 3.2, G is Euler and so has an Euler trail. This leaves us to treat the case where G has two odd vertices, u and v, say. (G can not have just one odd vertex by Corollary 1.2.)

Now let $G+e$ denote the graph obtained from G by adding in a new edge, joining u to v. Then e increases the degrees of u and v by 1 and so in $G+e$ every vertex is even. Hence, by Theorem 3.2, $G+e$ has an Euler tour say $C = v_0e_1v_1e_2\ldots e_nv_n$, involving each edge of $G+e$ exactly once. We may suppose that $e_1 = e, v_0 = u, v_1 = v$ and so $v_n = u$. Then, deleting e from this tour, gives the Euler trail $v_1e_2\ldots e_nv_n$, from v to u, in G since it involves each edge of G exactly once. □

Theorems 3.2 and 3.4 can now be applied to the graphs G_2 and G_1 of Figure 3.1 to show that they have an Euler trail, tour respectively. Also the Theorems show that the graph of Figure 3.3, of the Königsberg bridge problem, has no Euler trail since every vertex has odd degree. Euler graphs and, more generally, graphs with Euler trails often appear in books on recreational mathematics or children's puzzle books. For example, a puzzle might ask whether a given diagram can be drawn without lifting one's pencil from the paper and without repeating any lines. In effect this is asking if the corresponding graph has an Euler trail. For example, the first and second diagrams of Figure 3.8 can be drawn in this way, but the third can not.

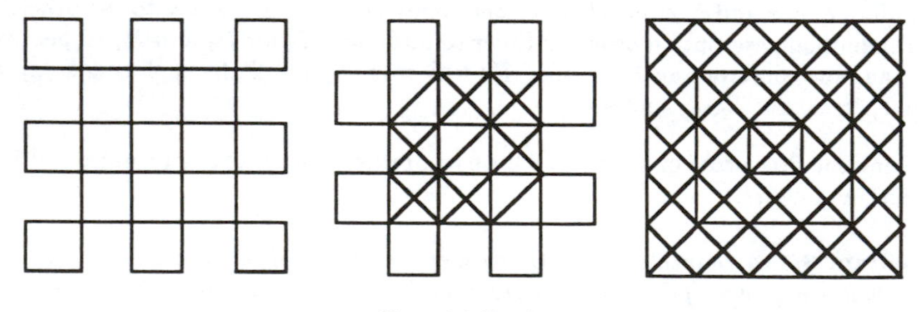

Figure 3.8: Puzzles

The first part of the proof of Theorem 3.4 shows that if a graph G has an Euler trail which is not a tour then the trail must start at one of the odd vertices and end at the other. (This is useful for example when one wants to draw the second diagram above without lifting one's pencil.) We now give an algorithm which constructs an Euler tour in an Euler graph.

Fleury's algorithm

Step 1. Choose any vertex v_0 in the Euler graph G and set $W_0 = v_0$.

Step 2. If the trail $W_i = v_0 e_1 v_1 \ldots e_i v_i$ has been chosen (so that $e_1, \ldots, e_i$ are all different), choose an edge e_{i+1} different from $e_1, \ldots, e_i$ such that

(i) e_{i+1} is incident with v_i and

(ii) unless there is no alternative, e_{i+1} is not a bridge of the edge-deleted subgraph $G - \{e_1, \ldots, e_i\}$.

Step 3. Stop if W_i contains every edge of G. Otherwise repeat Step 2.

We illustrate Fleury's algorithm with the Euler graph G of Figure 3.9.

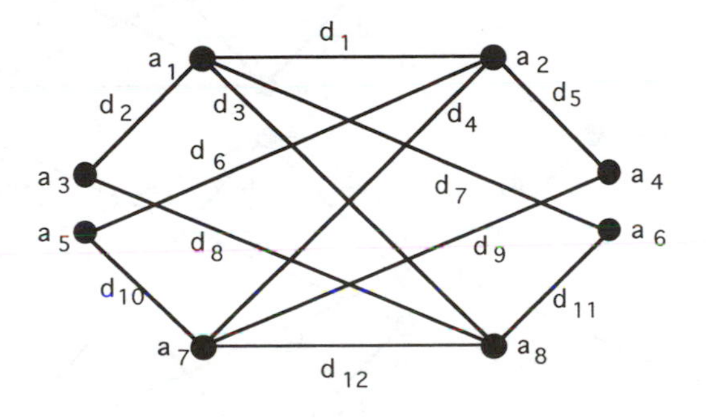

Figure 3.9

Step 1. Choose $v_0 = a_1$. $W_0 = a_1$.

Step 2. Choose edge d_1 for e_1. Figure 3.10 shows W_1.

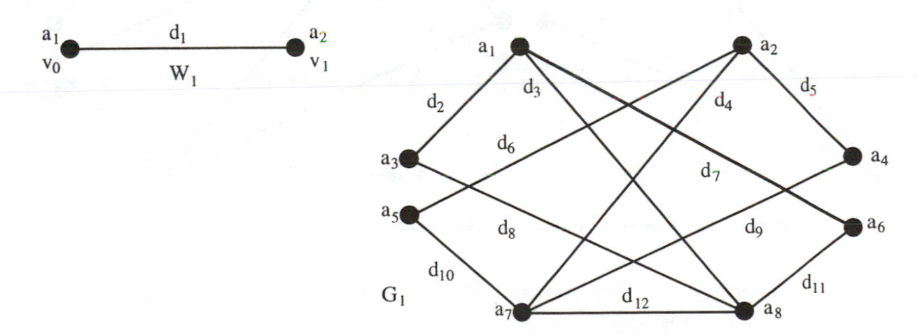

Figure 3.10: The first stage of the Euler tour construction.

Step 2. Choose edge d_5 for e_2. Figure 3.11 shows W_2 and G_2.

Step 2. Choose $e_3 = d_9$. (Although d_9 is a bridge there is no alternative.) Figure 3.11 shows W_3 and G_3.

Step 2. Choose $e_4 = d_4$. (We do not choose d_{12} since it is a bridge in G_3, whereas d_4 is not a bridge.) Figure 3.11 shows W_4 and G_4.

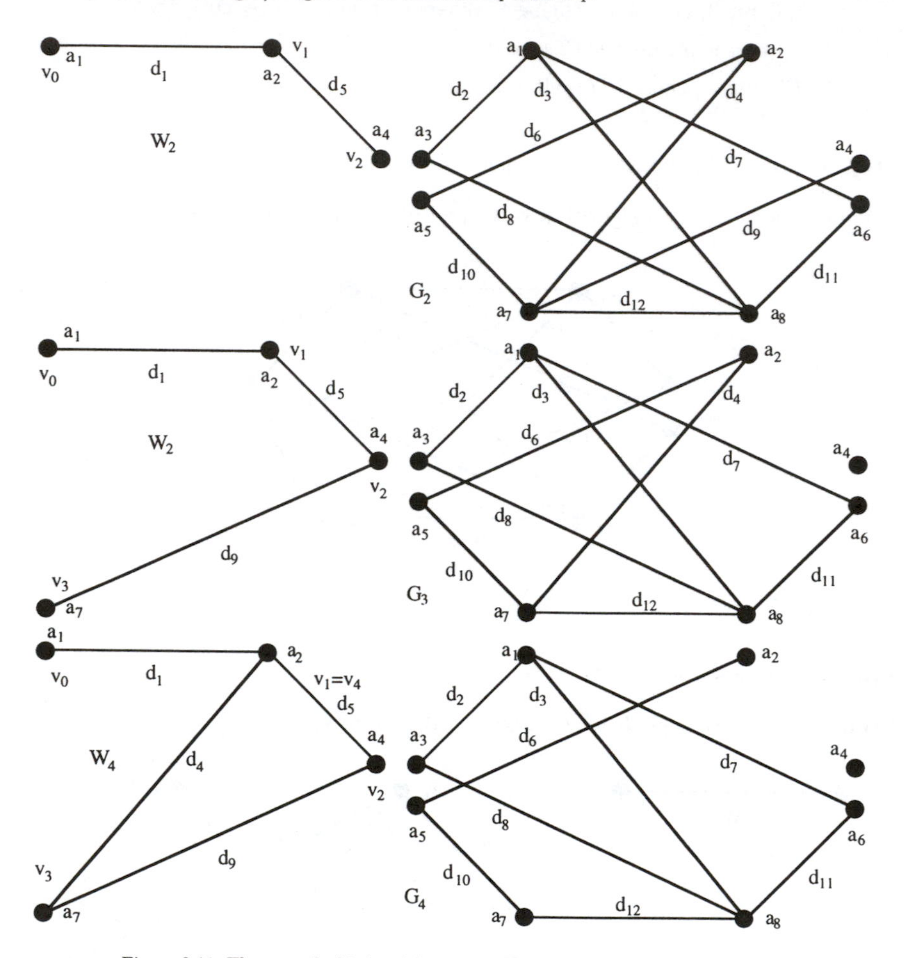

Figure 3.11: The second, third and fourth stages of the Euler tour construction.

Step 2. Choose $e_5 = d_6$ (no choice). (We leave the reader to draw the remaining walks W_i and subgraphs G_i.)

Step 2. Choose $e_6 = d_{10}$ (no choice).

Step 2. Choose $e_7 = d_{12}$ (no choice).

Step 2. Choose $e_8 = d_8$.

Step 2. Choose $e_9 = d_2$ (no choice).

Step 2. Choose $e_{10} = d_3$.

Step 2. Choose $e_{11} = d_{11}$ (although a bridge in G_{10}, no choice).

Step 2. Choose $e_{12} = d_4$ (although a bridge in G_{11}, no choice).

Step 3. Stop.

We have produced the Euler tour:

$$a_1\, d_1\, a_2\, d_5\, a_4\, d_9\, a_7\, d_4\, a_2\, d_6\, a_5\, d_{10}\, a_7\, d_{12}\, a_8\, d_8\, a_3\, d_2\, a_1\, d_3\, a_8\, d_{11}\, a_6\, d_7\, a_1.$$

We now prove that Fleury's algorithm actually does produce an Euler tour.

Theorem 3.5 *Fleury's algorithm produces an Euler tour in an Euler graph G.*

Proof Suppose that the trail $W_i = v_0 e_1 v_1 \ldots e_i v_i$ has been chosen by the algorithm and denote the edge-deleted subgraph $G - \{e_1, \ldots, e_i\}$ by G_i. Each time a vertex v occurs on the trail W_i which is different from both v_0 and v_i, v has one trail edge going in and one going out, i.e., W_i supplies two to the degree of v each time v occurs. Thus the degree of v in G_i is still even (although possibly 0). Similarly if $v_0 = v_i$ then there is an even number of edges in W_i incident with v_i, namely e_1, e_i and 2 more for each of the other occurrences of v_i. Thus if $v_0 = v_i$ then $d(v_i)$ in G_i is still even. However if $v_0 \neq v_i$ then there is an odd number of edges in W_i incident with v_i, namely e_i and two more for every other occurrence, and so $d(v_i)$ is odd in G_i.

In this latter case, when $v_0 \neq v_i$, if $d(v_i) = 1$ in G_i, i.e., if there is only one edge, e say, in G_i, incident with v_i, then, by Step 2 of the algorithm, we have no option but to choose this edge e to extend W_i to W_{i+1}. In doing so we remove the only edge still incident with v_i to leave v_i isolated. In particular, e is a bridge. (See Figure 3.12.)

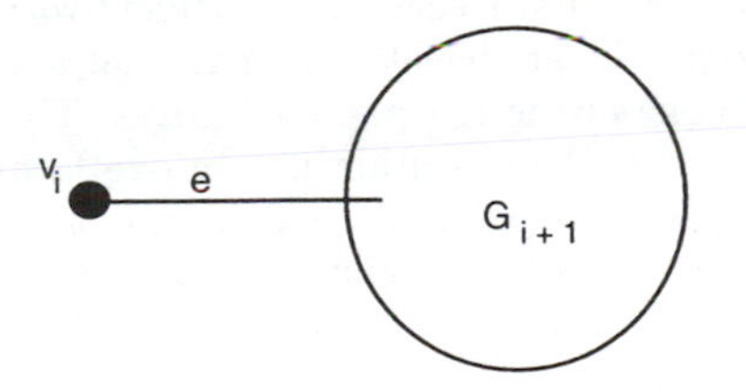

Figure 3.12: $d(v_i) = 1$ in G_i.

If, on the other hand, $d(v_i) \neq 1$ in G_i, then, as we now show, we may choose an edge incident with v_i in G_i which is not a bridge. To prove this it suffices to show that if there are at least two edges in G_i incident with v_i then at most one of these edges is a bridge. Suppose, to the contrary, that e' and f' are two bridges in G_i joining v_i to vertices u, v respectively. Then deleting both e' and f' from G_i drops the degree

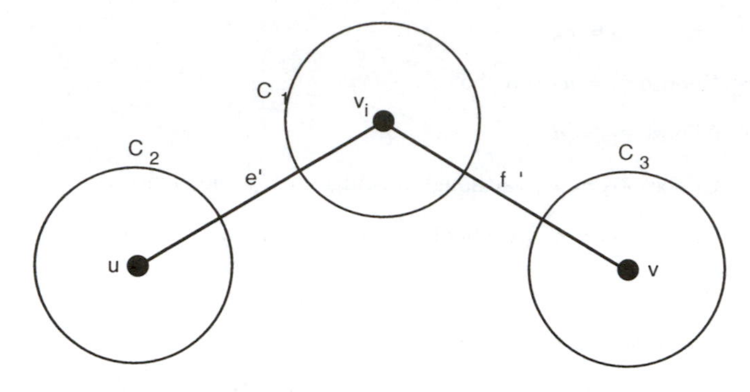

Figure 3.13: e' and f' are bridges in G_i.

of v_i down by 2 so that it is still odd, drops the degree of both u and v by 1, and creates a deleted subgraph $G_i - \{e', f'\}$ with at least three components, v_i being in one component C_1, say, u and v being in two others, say C_2 and C_3. (See Figure 3.13.)

Now the first paragraph of the proof says that the only odd vertices in G_i are v_0 and v_i. Thus if v_0 is different from both u and v, $G_i - \{e', f'\}$ has exactly four odd vertices and so at least one of the above components has exactly one odd vertex — impossible by Corollary 1.2. If on the other hand $v_0 = u$ (or similarly v) then $G_i - \{e', f'\}$ has two odd vertices, namely v_i and v (respectively u). In particular the component C_1 has only one odd vertex, namely v_i — again impossible. This completes the proof that if there is more than one edge incident with v_i then one of these is not a bridge in G_i. Moreover we have shown that in Step 2 of the algorithm, when $v_0 \neq v_i$, if we are forced to choose a bridge e_{i+1} to form the extended trail $W_{i+1} = v_0e_1v_1 \ldots v_ie_{i+1}v_{i+1}$ then v_i is an isolated vertex in G_{i+1}. In fact, in the process of extending W_i to W_{i+1}, when we disconnect G_i (i.e., choose a bridge) then one of the new components formed is always an isolated vertex.

In the case where $v_0 = v_i$ in W_i, as the first paragraph indicates, each vertex of G_i is of even degree. Thus, if there are still edges in G_i incident with $v_0 = v_i$, then there must be at least two such edges. An argument similar to that above, illustrated by Figure 3.13, then shows that none of these edges are bridges. This shows that throughout the entire implementation of the algorithm if a bridge has to be chosen then doing so leaves behind an isolated vertex and Step 2 can no longer be implemented only if the trail has used all the edges of the graph and has returned to its initial vertex. In other words, the algorithm finishes with an Euler tour of G. □

Exercises for Section 3.1

3.1.1 Determine which of the graphs in Figure 3.14 have (a) Euler trails and (b) Euler tours. For those that have, use Fleury's algorithm to produce such a trail or tour.

3.1.2 (a) Which of the following graphs are Euler?

i. the complete graph K_n, for $n \geq 3$

ii. the n-cube Q_n, for $n \geq 3$ (see Exercise 1.4.7),

iii. the wheel W_n, for $n \geq 4$ (see Exercise 1.6.4).

(b) Which graphs of (a) have Euler trails?

(c) For which $m, n \geq 1$ does the complete bipartite graph $K_{m,n}$ have an Euler tour (or Euler trail)?

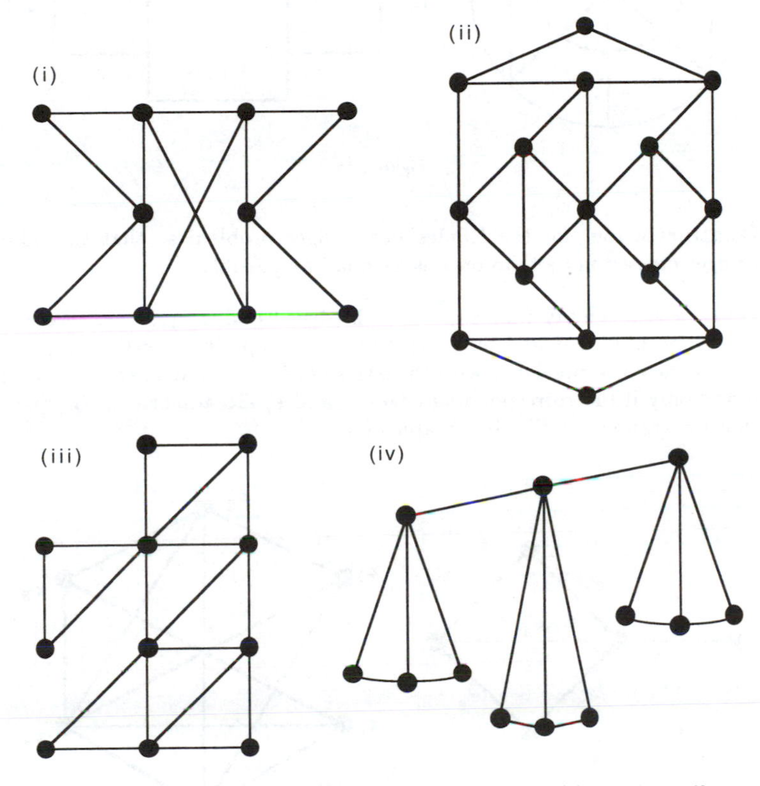

Figure 3.14: Which of these graphs have (a) an Euler tour, (b) an Euler trail?

3.1.3 Let G be a connected graph in which there are exactly four odd vertices. Prove that there are two trails in G such that each edge of G belongs to exactly one of these trails. (Hint: consider the proof of Theorem 3.4.)

3.1.4 Let G be a connected graph in which there are exactly $2k$ odd vertices, where $k \geq 1$. Prove that there are k open trails in G such that each edge of G belongs to exactly one of these trails. Prove also that it is not possible to find $k-1$ open trails with this property.

3.1.5 Using Exercise 3.1.4 determine the minimum number of times you must lift your pencil in order to draw each diagram of Figure 3.15 without repeating a line.

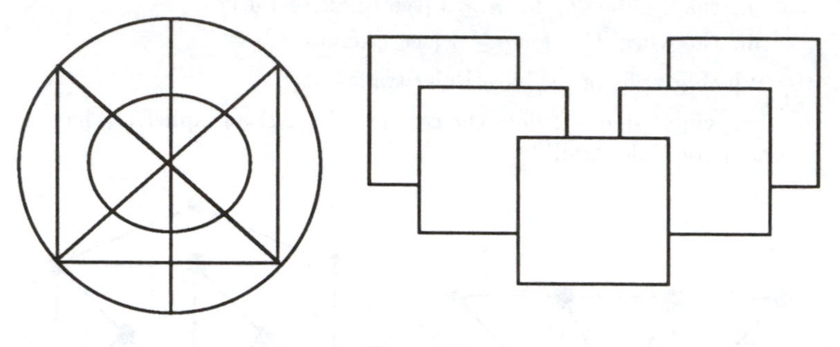

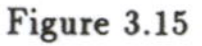

Figure 3.15

3.1.6 Construct a tour for the Königsberg bridges problem so that the number of bridges crossed more than once is as small as possible.

3.1.7 Let G be a nonempty simple graph with edges listed as $e_1, \ldots, e_n$. We define the line graph of G, denoted by $L(G)$, to be the graph with vertices $x_1, \ldots, x_n$ (in one-to-one correspondence with the edges of G) such that x_i and x_j are adjacent if and only if the corresponding edges e_i and e_j are adjacent in G. Figure 3.16 shows a graph G and its line graph $L(G)$.

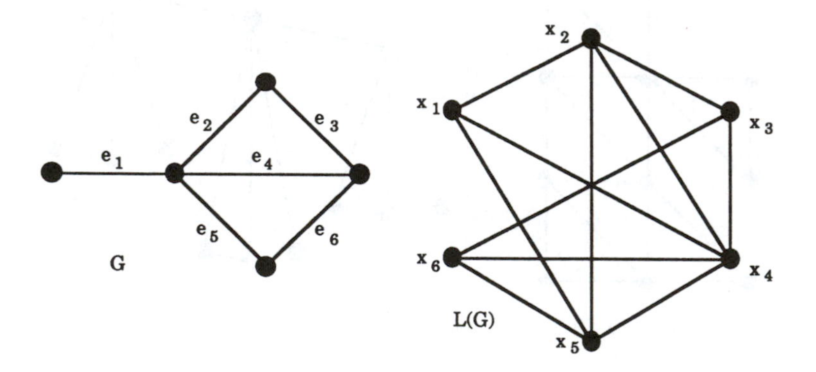

Figure 3.16: A graph G and its line graph $L(G)$.

(a) Prove that if G is Euler then so is $L(G)$.

(b) For each $n \geq 3$ find the line graph of the complete bipartite graph $K_{1,n}$.

(c) Give an example of a simple graph G such that $L(G)$ is Euler but G is not.

3.1.8 Prove Theorem 3.3, i.e., that a connected graph G is Euler if and only if G has cycles $C_{(1)}, \ldots, C_{(n)}$ such that every edge of G belongs to exactly one cycle $C_{(i)}$, i.e., G is the union of edge disjoint cycles. (Hint: try to modify the proof of Theorem 3.2.)

3.1.9 An Euler graph G is called **randomly traceable** from a vertex v if every trail in G starting at v can be extended to an Euler tour, starting and ending at v.

(a) Show that the Euler graph of Figure 3.17 is randomly traceable from the vertex v indicated, but not from any other vertex.

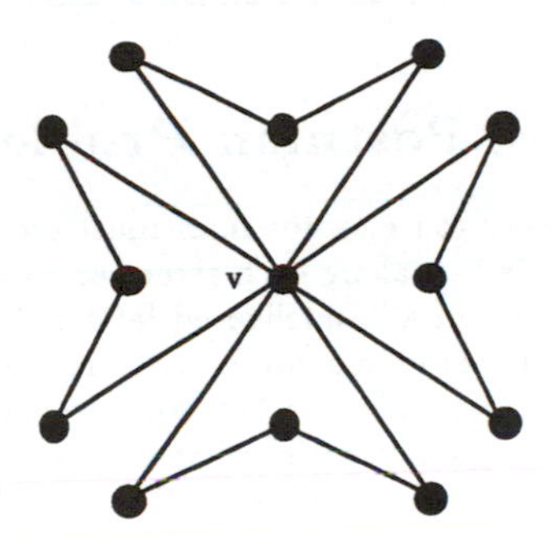

Figure 3.17: Show that this graph is randomly traceable from v but not from any other vertex.

(b) Give an example of an Euler graph which is not randomly traceable from any of its vertices.

(c) Give an example of an Euler graph which is randomly traceable from all of its vertices.

(d) Prove that the Euler graph G is randomly traceable from its vertex v if and only if v lies on every cycle of G.

3.1.10 Here is another algorithm, due to Hierholzer, which also produces an Euler tour in an Euler graph G. Its method is to start with any closed trail in G and "attach" to it, systematically, "detour" trails until all edges of G are used up. (In what follows, $E(H)$ denotes the set of edges of a subgraph H.)

Hierholzer's algorithm

Step 1. Choose any vertex v in G and choose any closed trail W_0 in G starting and ending at v. Set $i = 0$. (i is a counter.)

Step 2. If $E(W_i) = E(G)$ stop, since then W_i is an Euler tour of G.

Otherwise choose a vertex v_i on W_i which is incident with an edge in G which is not in W_i.

Now choose a closed trial W_i^* in the subgraph $G - E(W_i)$, starting at the vertex v_i. (W_i^* is a "detour" trail.)

Step 3. Let W_{i+1} be the closed trail consisting of the edges of both W_i and W_i^* obtained by starting at the vertex v, traversing the trail W_i until v_i is reached, then traversing the closed trail W_i^* and, on returning to v_i, completing the rest of the trail W_i.

Now set $i = i + 1$ and return to Step 2.

Use Hierholzer's algorithm to produce an Euler tour for the graph of Figure 3.9 starting with $W_0 = a_1 a_2 a_6 a_1$.

3.2 The Chinese Postman Problem

Before starting on his delivery route, a postman must pick up his letters at the post office, then he must deliver letters along each street on his route, and finally he must return to the post office to return all undelivered letters. Wishing to conserve energy, every postman would like to cover his route with as little walking as possible. In nongraph terms, the postman problem is how to cover all the streets in the route and return back to the starting point with as little travelling as possible. It is usually referred to as the Chinese Postman Problem (CPP for short) because the first work published on the problem was by a Chinese mathematician, Kuan, in 1962 [39].

In graph-theoretical terms the problem is dealt with as follows. We construct a weighted graph G where each edge represents a street in the postman's route, each vertex represents a junction of streets and the weight assigned to each edge represents the length of the street between junctions. If we define the weight of a tour $v_0 e_1 v_1 e_2 \ldots e_n v_0$ in G to be the sum of the weights of its edges, i.e., $w(e_1) + w(e_2) + \cdots w(e_n)$, the Chinese postman problem is just that of finding a tour of minimum weight in a weighted connected graph with non-negative weights. (Recall that a tour involves each edge of G at least once.)

If the weighted graph G so constructed is Euler then any Euler tour of G is a tour of minimum weight since it involves each edge of G once and only once. Thus in practice we may use Fleury's algorithm to produce such a tour, i.e., a solution to the CPP.

If G is not Euler then any tour in G has to involve some edges more than once. For example, in the weighted graph of Figure 3.18, an optimal tour, i.e., a minimal weighted tour is given by

$$v_1\ v_2\ v_3\ v_6\ v_2\ v_3\ v_6\ v_7\ v_5\ v_4\ v_7\ v_5\ v_1\ v_4\ v_3\ v_1$$

where the edges v_2v_3, v_3v_6 and v_5v_7 are each used twice. Of course, at this stage we have no easy way of telling that this tour is optimal.

In order to look at a particular case of the non-Eulerian situation we introduce the process of duplication of an edge.

> An edge e is said to be **duplicated** when its ends are joined by a new edge with the same weight $w(e)$ as e.

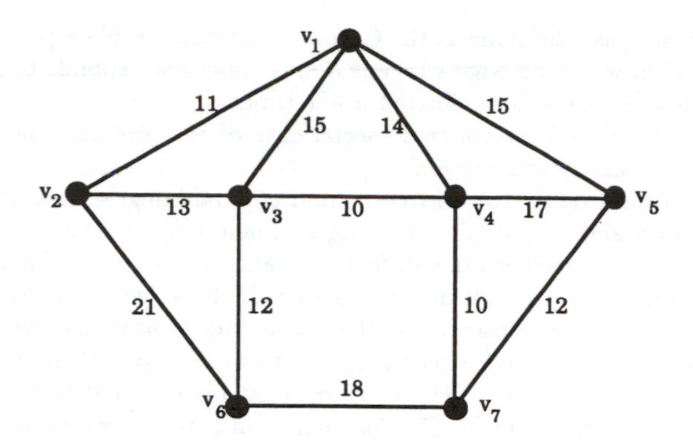

Figure 3.18: A postman's delivery area.

For example, by duplicating the edges v_4v_6, v_6v_8,v_8v_7 and v_7v_5 in the graph G of Figure 3.19, we produce the supergraph G^* of G:

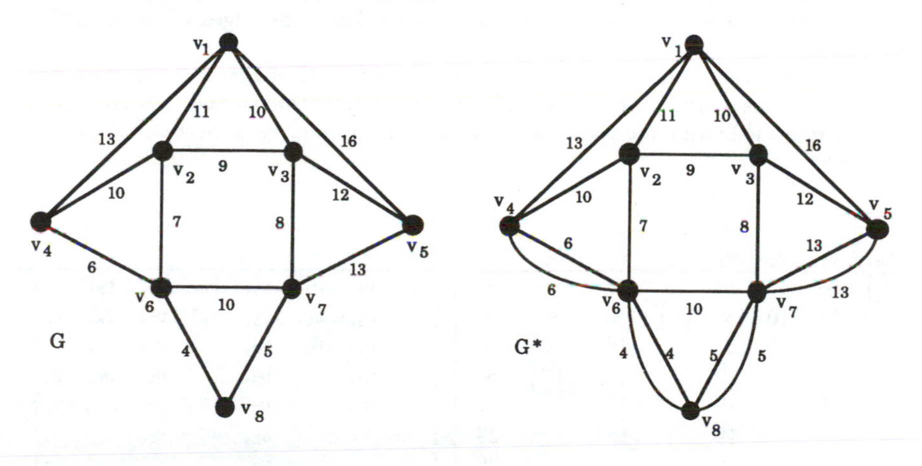

Figure 3.19: A graph G with an Euler supergraph G^* obtained by duplicating edges.

The Chinese Postman Problem may now be rephrased as follows.

Given a connected weighted graph G with non-negative weights,

(i) find, by duplicating edges if necessary, an Euler weighted supergraph G^* of G such that the sum of the weights of the duplicated edges is as small as possible, i.e.,

$$\sum_{e \in E(G^*) - E(G)} w(e)$$

is at a minimum where $E(G^*) - E(G)$ is the set of edges in G^* but not in G, and

(ii) find an Euler tour in G^*.

To see that this is just the same as the Chinese Postman Problem just note that an optimal tour of G in which an edge e is repeated n times corresponds to an (optimal) tour of G^* where the edge e is duplicated $n-1$ times.

At this stage we will only consider a special case of the problem, namely when G *has exactly two vertices of odd degree.*

Suppose that G has exactly two vertices, u and v, of odd degree. Let G^* be an Euler supergraph of G obtained by duplicating edges. Then, since u and v now have even degree in G^*, the duplicated edges will form a walk from u to v. (There must be a duplicated edge out of u to give u its even degree. If this edge, e_1 say, goes to vertex v_1 then the degree of v_1 increases by 1. If $v_1 = v$ then v now has even degree and the edge e_1 gives a walk (in fact a path) from u to v. If $v_1 \neq v$ then the edge e_1 has changed v_1's even degree in G to odd so there must be a second edge, e_2 say, going out of v_1 to give v_1 even degree in G^*. Continuing in this way we eventually reach v by a sequence of duplicated edges since, in order to make v have even degree in G^*, v must have a duplicated edge incident.)

Clearly to satisfy criterion (i), i.e., to optimise G^*, the length of such a walk of duplicated edges from u to v in G^* should be as small as possible. Thus in the case where G has exactly two vertices, u and v, of odd degree, to solve (i) we just have to find a shortest path in G from u to v and then duplicate the edges which form this path to get the Euler supergraph G^*.

To illustrate the complete procedure we use the graph G of Figure 3.19, which has two odd vertices v_4 and v_5. First we apply Dijkstra's algorithm to find a shortest path from v_4 to v_5. Following the presentation of the algorithm given in Section 2.5, we get the table:

vertices v								
v_4,	v_1	v_2	v_3	v_6	v_7	v_8	v_5	T
$\boxed{0}$	∞	∞	∞	∞	∞	∞	∞	$\{v_4, v_1, v_2, v_3, v_6, v_7, v_8, v_5\}$
	13	10	∞	$\boxed{6}$	∞	∞	∞	$\{v_1, v_2, v_3, v_6, v_7, v_8, v_5\}$
	13	**10**	∞		16	10	∞	$\{v_1, v_2, v_3, v_7, v_8, v_5\}$
	13		19		16	$\boxed{10}$	∞	$\{v_1, v_3, v_7, v_8, v_5\}$
	13		19		15		∞	$\{v_1, v_3, v_7, v_5\}$
			19		$\boxed{15}$		29	$\{v_3, v_7, v_5\}$
			19				29	$\{v_3, v_5\}$
							$\boxed{29}$	$\{v_5\}$

Since $v_4\ v_6\ v_8\ v_7\ v_5$ is a shortest path from v_4 to v_5, (as indicated in the table), duplicating each of the edges in this path gives the Euler weighted supergraph G^* shown in Figure 3.19.

It is now left to carry out (ii), i.e., find an Euler tour in G^*. We use Fleury's algorithm starting at vertex v_1 (the Post Office). Then an Euler tour is given by

$$v_1\ v_4\ v_6\ v_8\ v_7\ v_5\ v_1\ v_2\ v_4\ v_6\ v_8\ v_7\ v_5\ v_3\ v_2\ v_6\ v_7\ v_3\ v_1.$$

It has length 162. Fleury's algorithm can of course provide several different Chinese postman tours apart from this one.

Exercise for Section 3.2

3.2.1 Solve the Chinese Postman Problem for the graphs of Figure 3.20.

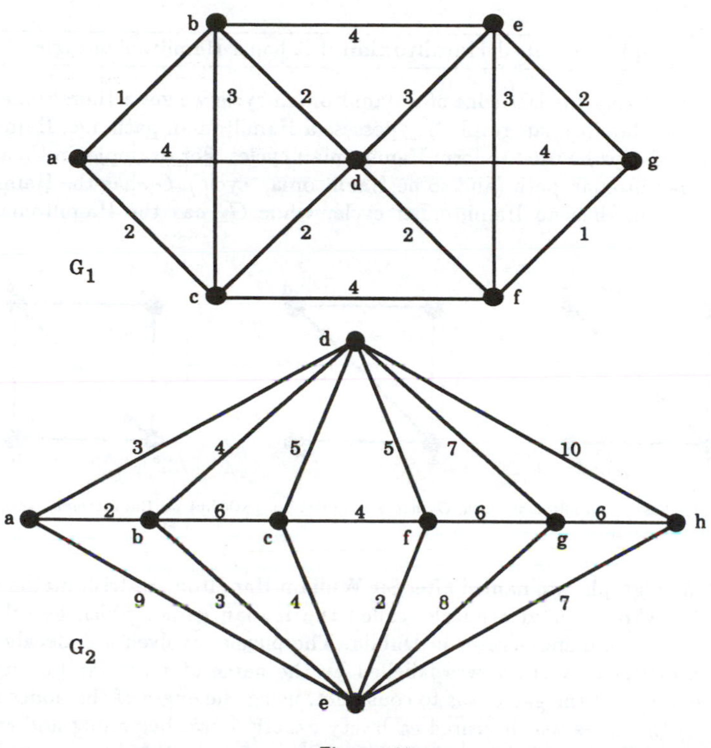

Figure 3.20

3.3 Hamiltonian Graphs

> A **Hamiltonian path** in a graph G is a path which contains every vertex of G.

Since, by definition (see Section 1.6), no vertex of a path is repeated, this means that a Hamiltonian path in G contains every vertex of G once and only once.

A **Hamiltonian cycle** (or **Hamiltonian circuit**) in a graph G is a cycle which contains every vertex of G.

Since, by definition (again see Section 1.6), no vertex of a cycle is repeated apart from the final vertex being the same as the first vertex, this means that a Hamiltonian cycle in G with initial vertex v contains every other vertex of G precisely once and then ends back at v.

A graph G is called **Hamiltonian** if it has a Hamiltonian cycle.

By simply deleting the last edge of a Hamiltonian cycle we get a Hamiltonian path. However a non-Hamiltonian graph may possess a Hamiltonian path, i.e., Hamiltonian paths cannot always be used to form Hamiltonian cycles. For example, in Figure 3.21, G_1 has no Hamiltonian path (and so no Hamiltonian cycle), G_2 has the Hamiltonian path $a\ b\ c\ d$ but has no Hamiltonian cycle, while G_3 has the Hamiltonian cycle $a\,b\,d\,c\,a$.

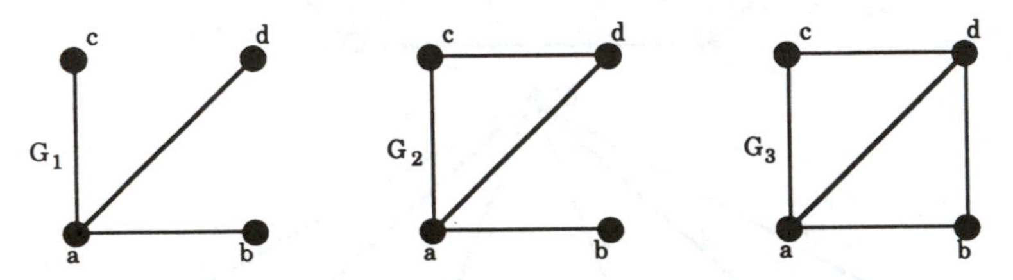

Figure 3.21: G_1 has no Hamiltonian path, G_2 has a Hamiltonian path but no Hamiltonian cycle, while G_3 has a Hamiltonian cycle.

Hamiltonian graphs are named after Sir William Hamilton, an Irish mathematician (1805–1865), who invented a puzzle, called the Icosian game, which he sold for 25 guineas to a games manufacturer in Dublin. The puzzle involved a dodecahedron on which each of the 20 vertices was labelled by the name of some capital city in the world. The object of the game was to construct, using the edges of the dodecahedron, a tour of all the cities which visited each city exactly once, beginning and ending at the same city. In other words, one had essentially to form a Hamiltonian cycle in the graph corresponding to the dodecahedron. We show such a cycle, using bolder lines, in Figure 3.22.

Clearly the n-cycle C_n with n distinct vertices (and n edges) is Hamiltonian. Moreover, given any Hamiltonian graph G, then, if G' is a supergraph of G obtained by adding in new edges between vertices of G, G' will also be Hamiltonian, since any Hamiltonian cycle in G will also be a Hamiltonian cycle in G'. In particular since K_n, the complete graph on n vertices, is such a supergraph of an n-cycle, K_n is Hamiltonian.

A graph G will be Hamiltonian if and only if its underlying simple graph is Hamiltonian since if G is Hamiltonian then any Hamiltonian cycle in G will remain a Hamiltonian cycle in the underlying simple graph of G (provided we delete the appropriate parallel edges). Conversely, if the underlying simple graph of a graph G

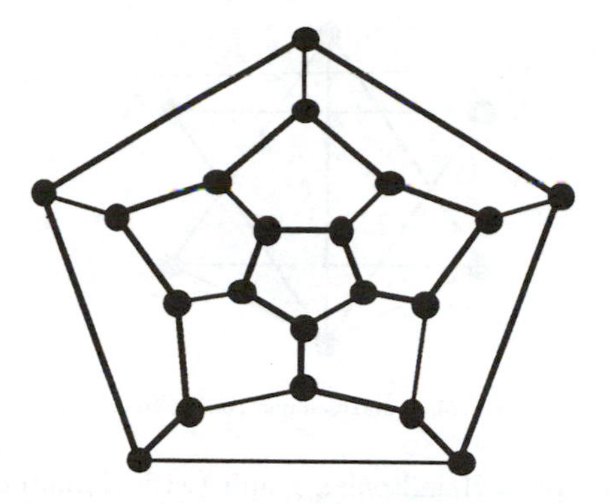

Figure 3.22: A Hamiltonian cycle in the graph of the dodecahedron.

is Hamiltonian then G will also be, because of the remarks of the previous paragraph. For this reason one usually only considers the Hamiltonian property for simple graphs.

Given a simple graph G with n vertices, since G is a subgraph of the complete graph K_n, we can construct step-by-step simple supergraphs of G to eventually get K_n, simply by adding in an extra edge at each step between two vertices that are not already adjacent. We illustrate this in Figure 3.23.

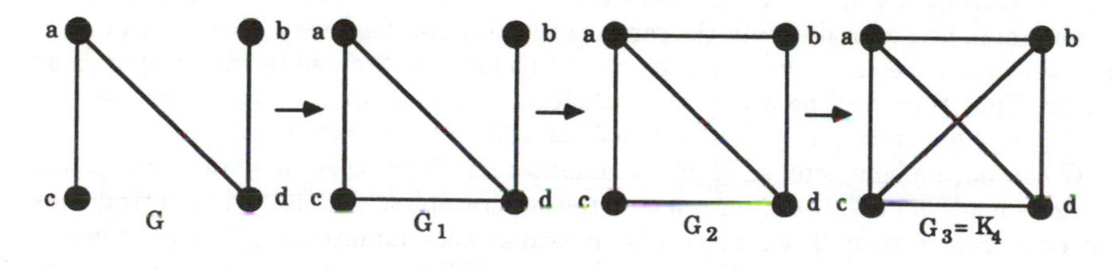

Figure 3.23: A build-up to K_4.

If, moreover, we start with a graph G that is not Hamiltonian, then, since the final outcome of the procedure is the Hamiltonian graph K_n, at some stage during the procedure we change from a non-Hamiltonian graph to a Hamiltonian graph. For example, the non-Hamiltonian graph G_1 above is followed by the Hamiltonian graph G_2. Notice that since supergraphs of Hamiltonian graphs are Hamiltonian, once a Hamiltonian supergraph is reached in the procedure, all the subsequent supergraphs are Hamiltonian. This discussion leads to the following definition.

> A simple graph G is called **maximal non-Hamiltonian** if it is not Hamiltonian but the addition to it of any edge connecting two non-adjacent vertices forms a Hamiltonian graph.

For example, G_1 of Figure 3.21 is maximal non-Hamiltonian since the addition of the edge ab gives the Hamiltonian G_2 as shown, while the other possibility, the addition

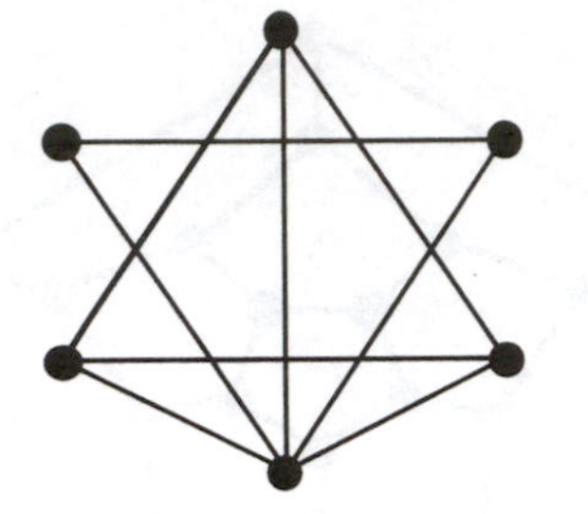

Figure 3.24: A maximal non-Hamiltonian graph.

of the edge bd, also gives a Hamiltonian graph (with Hamiltonian cycle $b\ d\ a\ c\ b$). Similarly the graph of Figure 3.24 is also maximal non-Hamiltonian:

Because of the stepwise procedure described above any non-Hamiltonian graph with n vertices will be a subgraph of a maximal non-Hamiltonian graph with n vertices. We use this to prove the following theorem, due to Dirac [18].

Theorem 3.6 (Dirac, 1952) *If G is a simple graph with n vertices, where $n \geq 3$, and the degree $d(v) \geq n/2$ for every vertex v of G, then G is Hamiltonian.*

Proof We suppose that the result is false. Then, for some value $n \geq 3$, there is a non-Hamiltonian graph in which every vertex has degree at least $n/2$. Any spanning supergraph, i.e., with precisely the same vertex set, also has every vertex with degree at least $n/2$, since any proper supergraph of this form is obtained by introducing more edges. Thus there will be a maximal non-Hamiltonian graph G with n vertices and $d(v) \geq n/2$ for every v in G. Using this G we obtain a contradiction.

G can not be complete, since K_n is Hamiltonian. Thus there are two nonadjacent vertices u and v in G. Let $G+uv$ denote the supergraph of G obtained by introducing an edge from u to v. Then, since G is maximal non-Hamiltonian, $G + uv$ must be Hamiltonian. Also, if C is a Hamiltonian cycle of $G + uv$, then it must contain the edge uv (since otherwise it would be a Hamiltonian cycle in G). Thus, choosing such a C, we may write $C = v_1v_2 \ldots v_nv_1$ where $v_1 = u, v_n = v$ (and the edge v_nv_1 is just vu, i.e., uv). Now let

$$S = \{v_i \in C : \text{ there is an edge from } u \text{ to } v_{i+1} \text{ in } G\} \text{ and}$$

$$T = \{v_j \in C : \text{ there is an edge from } v \text{ to } v_j \text{ in } G\}.$$

Then $v_n \notin T$, since otherwise there would be an edge from v to $v_n = v$, i.e., a loop, impossible because G is simple. Also $v_n \notin S$ (interpreting v_{n+1} as v_1), since otherwise we would again get a loop, this time from u to $v_1 = u$. Thus $v_n \notin S \cup T$. Then, letting $|S|$, $|T|$ and $|S \cup T|$ denote the number of elements in S, T, and $S \cup T$ respectively, we get

$$|S \cup T| < n \tag{3.1}$$

Also, for every edge incident with u there corresponds precisely one vertex v_i in S. Thus

$$|S| = d(u) \tag{3.2}$$

Similarly

$$|T| = d(v) \tag{3.3}$$

Moreover, if v_k is a vertex belonging to both S and T, then there is an edge e joining u to v_{k+1} and an edge f joining v to v_k. This would give

$$C' = v_1\ v_{k+1}\ v_{k+2}\ \ldots\ v_n\ v_k\ v_{k-1}\ \ldots\ v_2\ v_1.$$

as a Hamiltonian cycle in G, (see Figure 3.25), a contradiction since G is non-Hamiltonian.

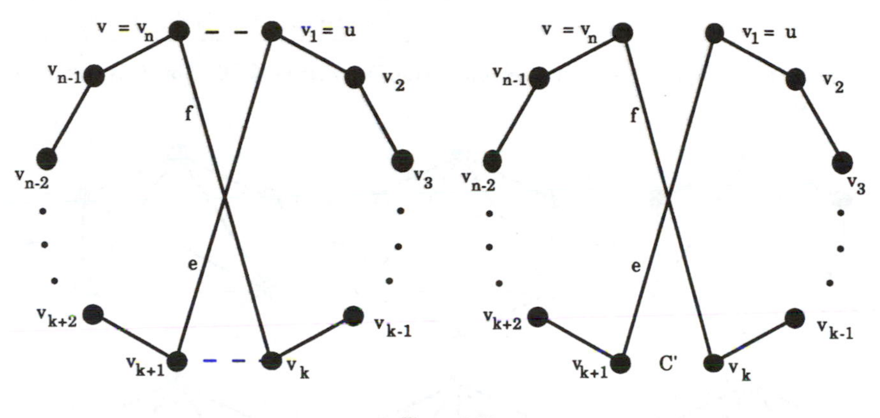

Figure 3.25

This shows that there is no vertex v_k in $S \cap T$, i.e., $S \cap T = \emptyset$. Thus $|S \cup T| = |S| + |T|$. Hence, by (3.1), (3.2) and (3.3) above,

$$d(u) + d(v) = |S| + |T| = |S \cup T| < n. \tag{3.4}$$

This is impossible since in G, $d(u) \geq n/2$ and $d(v) \geq n/2$, and so $d(u) + d(v) \geq n$. This contradiction tells us that we have wrongly assumed the result to be false. □

We now use the ideas of the above proof to present some results on Hamiltonian graphs by Bondy and Chvatal [8].

Theorem 3.7 *Let G be a simple graph with n vertices and let u and v be non-adjacent vertices in G such that*

$$d(u) + d(v) \geq n.$$

Let $G + uv$ denote the supergraph of G obtained by joining u and v by an edge. Then G is Hamiltonian if and only if $G + uv$ is Hamiltonian.

Proof Suppose that G is Hamiltonian. Then, as noted earlier, the supergraph $G+uv$ must also be Hamiltonian.

Conversely, suppose that $G+uv$ is Hamiltonian. Then, if G is not Hamiltonian, just as in the proof of Theorem 3.6 we obtain the inequality $d(u) + d(v) < n$. However, by hypothesis, $d(u) + d(v) \geq n$. Hence G must be Hamiltonian also, as required. □

Motivated by Theorem 3.7 we now define what we mean by the closure $c(G)$ of a simple graph G.

> Let G be a simple graph. If there are two nonadjacent vertices u_1 and v_1 in G such that $d(u_1) + d(v_1) \geq n$ in G, join u_1 and v_1 by an edge to form the supergraph G_1. Then, if there are two nonadjacent vertices u_2 and v_2 such that $d(u_2) + d(v_2) \geq n$ in G_1, join u_2 and v_2 by an edge to form the supergraph G_2. Continue in this way, recursively joining pairs of nonadjacent vertices whose degree sum is at least n until no such pair remains. The final supergraph thus obtained is called the **closure** of G and is denoted by $c(G)$.

We give an example of the closure operation in Figure 3.26. For this example, $c(G) = K_7$.

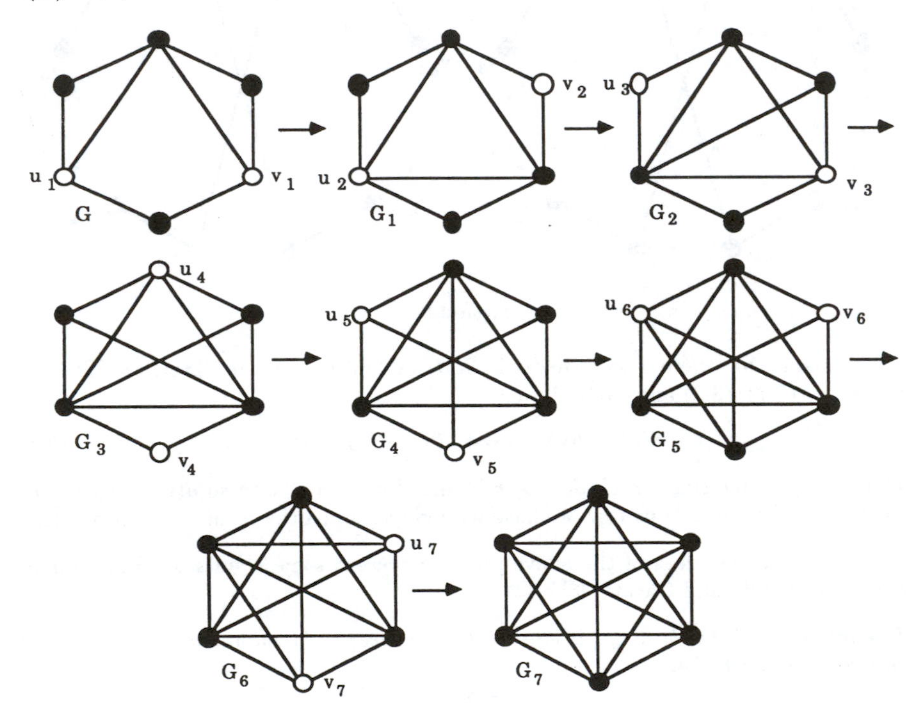

Figure 3.26: Here the closure operation joins the pairs of vertices shown in white and the closure is reached after seven such joins.

On the other hand, for the graph G on 7 vertices of Figure 3.27, $d(u) + d(v) < 7$ for any pair u, v of nonadjacent vertices in G and so the closure operation does not get off the ground, i.e., $c(G) = G$.

Notice that in the example of Figure 3,26 there were often various choices available of pairs of nonadjacent vertices u, v with $d(u) + d(v) \geq n$. Thus the closure procedure

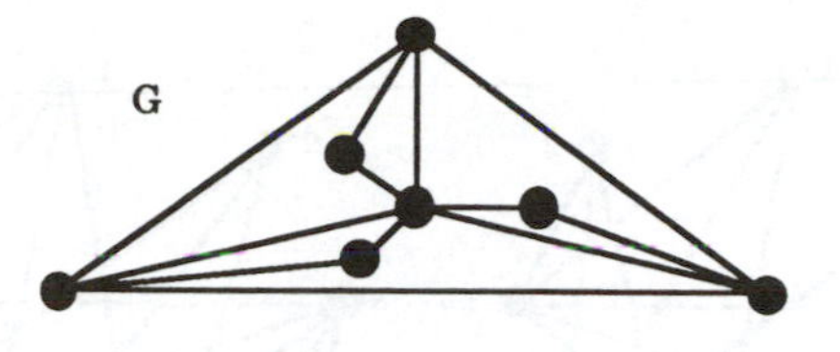

Figure 3.27: $c(G) = G$.

could have been carried out in several different ways. The question arises as to whether each different way gives the same result, i.e., do we always end with the same $c(G)$? Yes, we do — we omit the details here. (The interested reader is referred to Lemma 4.4.2 of Bondy and Murty [7].)

The importance of $c(G)$ is given in the following result:

Theorem 3.8 (Bondy and Chvatal, 1976) *A simple graph G is Hamiltonian if and only if its closure $c(G)$ is Hamiltonian.*

Proof Since $c(G)$ is a supergraph of G, if G is Hamiltonian then $c(G)$ must be Hamiltonian.

Conversely, suppose that $c(G)$ is Hamiltonian. Let G, G_1,$G_2, \ldots, G_{k-1}$, $G_k = c(G)$ be the sequence of graphs obtained by performing the closure procedure on G. Since $c(G) = G_k$ is obtained from G_{k-1} by setting $G_k = G_{k-1} + uv$, where u, v is a pair of nonadjacent vertices in G_{k-1} with $d(u)+d(v) \geq n$, it follows by Theorem 3.7 that G_{k-1} is Hamiltonian. Similarly G_{k-2}, so $G_{k-3}, \ldots$, so G_1, and so G must be Hamiltonian, as required. □

Corollary 3.9 *Let G be a simple graph on n vertices, with $n \geq 3$. If $c(G)$ is complete, i.e., if $c(G) = K_n$, then G is Hamiltonian.*

Proof This is immediate from the Theorem since any complete graph is Hamiltonian. □

For example, for the graph G of Figure 3.26 we got $c(G)$ as the complete graph K_7 and so, by the Corollary, it follows that G is Hamiltonian.

Unfortunately, the closure operation is not always helpful in determining if a graph is Hamiltonian. For example, $c(G) = G$ for the graph G of Figure 3.27 so here the operation provides no additional information.

Although the closure operation tells us that the graph G of Figure 3.26 is Hamiltonian, this is obvious from the drawing of G. For this reason we look at a less obvious example G in Figure 3.28.

We perform the closure operation on G in Figures 3.28 and 3.29, illustrating the nonadjacent pair of vertices u, v with $d(u) + d(v) \geq n = 9$ we use at each step by white dots. Since the final graph G_{15} in our closure construction is the complete graph K_9, we may conclude from the Corollary that our initial graph G is Hamiltonian.

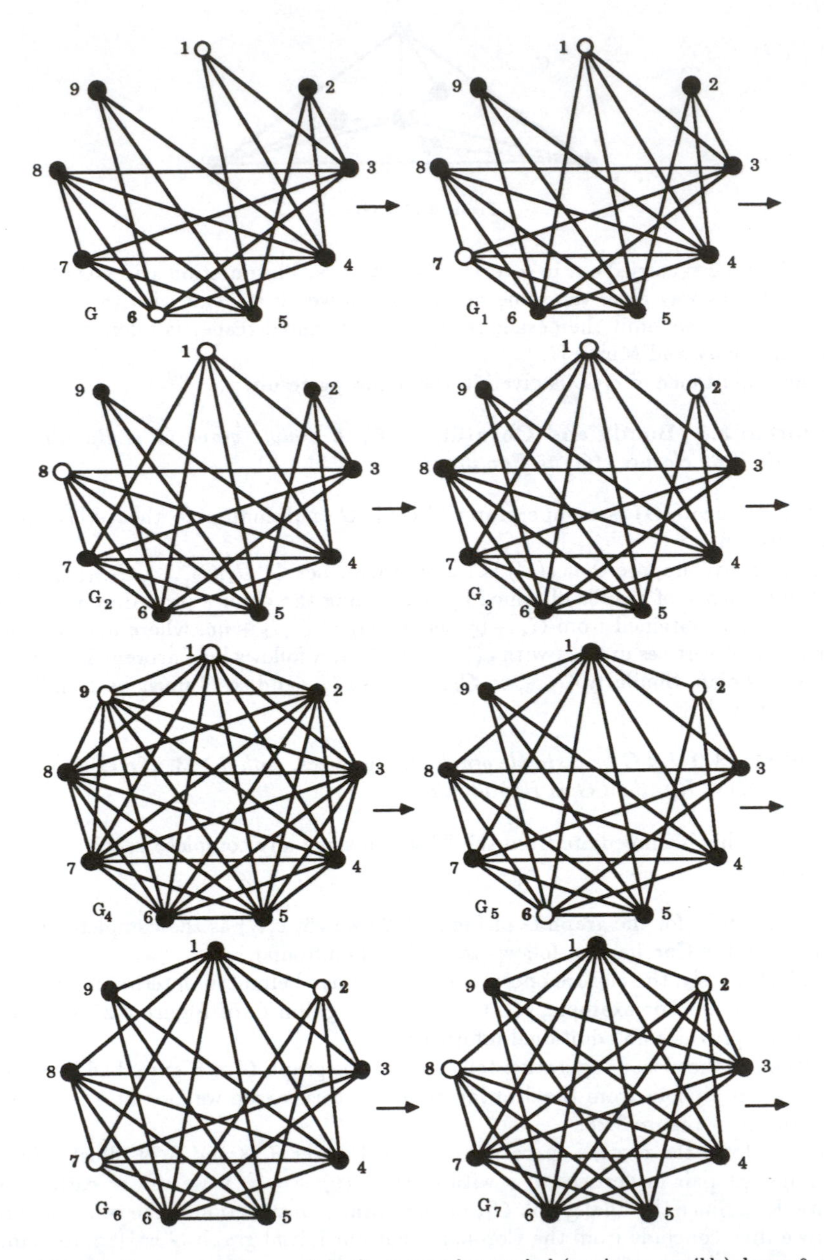

Figure 3.28: The closure operation on G. In G_5 vertex 1 has reached (maximum possible) degree 8, while vertex 6 reaches degree 8 in G_6.

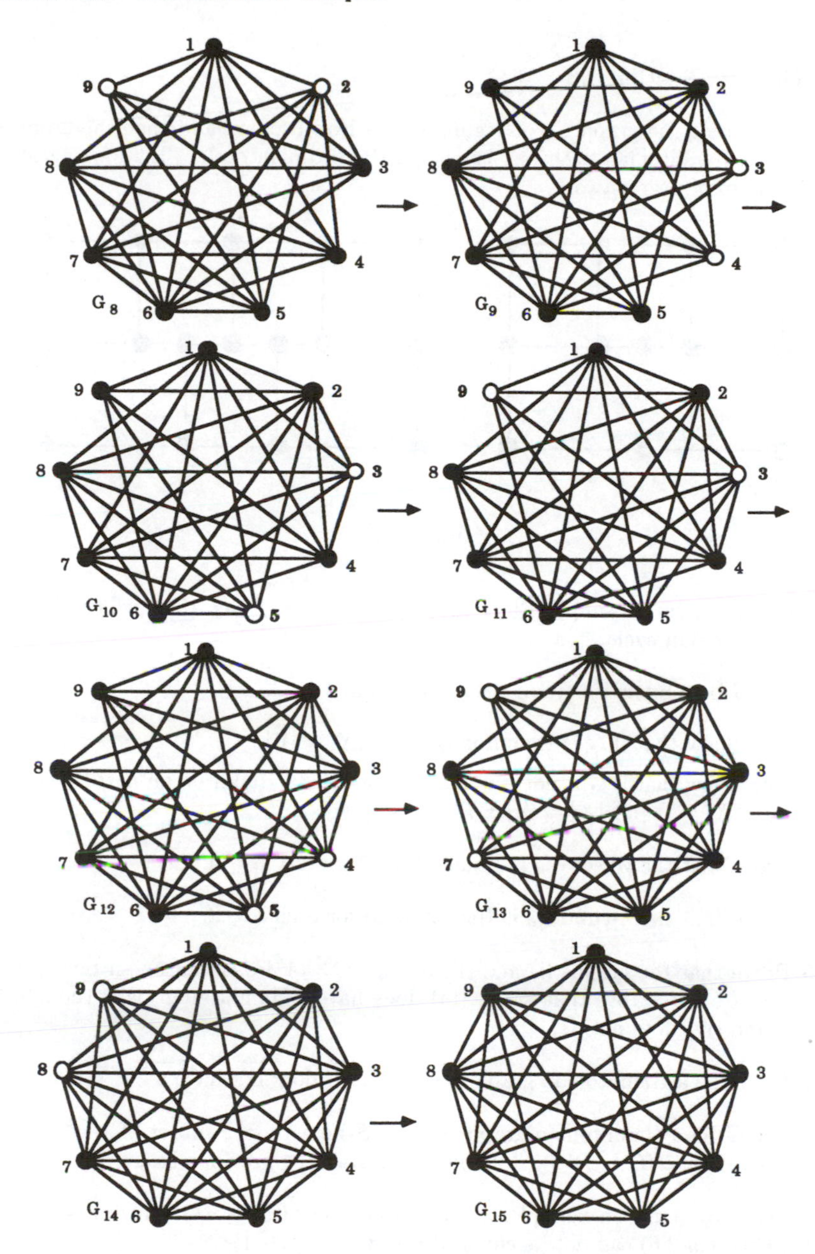

Figure 3.29: The closure operation continued. Vertices 2 and 3 reach degree 8 in G_9 and G_{12} respectively, vertices 4 and 5 reach degree 8 in G_{13}, vertex 7 reaches degree 8 in G_{14} and, finally, vertices 8 and 9 reach degree 8 in G_{15}.

Exercises for Section 3.3

3.3.1 Show that the graph G_1 of Figure 3.30 is Hamiltonian and that the graph G_2 has a Hamiltonian path but has not a Hamiltonian cycle. (Try looking at the vertices of degree two.)

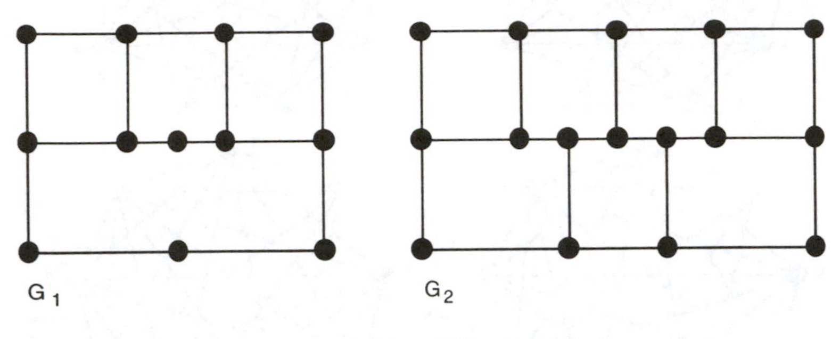

Figure 3.30

3.3.2 Characterise all simple Euler graphs having an Euler tour which is also a Hamiltonian cycle.

3.3.3 Let G be a bipartite graph with bipartition $V = X \cup Y$.

(a) Show that if G is Hamiltonian then $|X| = |Y|$.

(b) Show that if G is not Hamiltonian but has a Hamiltonian path then either $|X| = |Y|$ or $|X| = |Y| \pm 1$.

3.3.4 Prove that the wheel W_n is Hamiltonian for every $n \geq 4$.

3.3.5 Prove that the n-cube Q_n is Hamiltonian for each $n \geq 2$.

3.3.6 Prove that the graphs (i) and (ii) of Figure 3.14 are Hamiltonian but (iii) and (iv) are not. Prove that graph (iii) does have a Hamiltonian path though, but graph (iv) does not.

3.3.7 Let G be a Hamiltonian graph. Show that G does not have a cut vertex.

3.3.8 Let G be a Hamiltonian graph and let S be a proper subset of vertices of G. Prove that $\omega(G - S) \leq |S|$. (This generalises the previous Exercise.)

3.3.9 Show, by giving an example, that the condition "$d(v) \geq n/2$" in Dirac's Theorem (Theorem 3.6) can not be changed to "$d(v) \geq (n-1)/2$".

3.3.10 Let H be an Euler graph and let $G = L(H)$, the line graph of H. (See Exercise 3.1.7.) Prove that G is Hamiltonian.

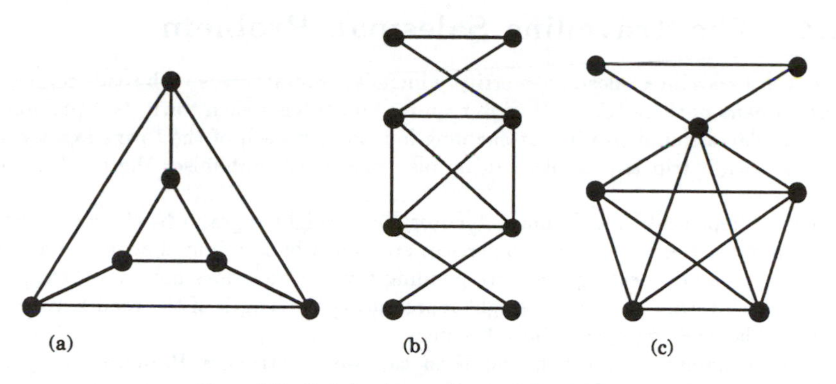

Figure 3.31: Find the closure of each of these graphs.

3.3.11 Find the closure $c(G)$ for each of the graphs of Figure 3.31. Which of these graphs are Hamiltonian?

3.3.12 Prove that, for each $n \geq 1$, the complete tripartite graph $K_{n,2n,3n}$ is Hamiltonian but $K_{n,2n,3n+1}$ is not Hamiltonian. (See Exercise 1.6.13 for the definition of $K_{r,s,t}$.)

3.3.13 In a Hamiltonian graph G two Hamiltonian cycles C and C' are considered to be the same if C is a cyclic rotation of C' or a cyclic rotation of the reverse of C'. Thus, for example, if $C = v_1 v_2 v_3 v_4 v_1$ is a Hamiltonian cycle in G we consider it to be the same as the Hamiltonian cycles $C_{(1)} = v_2 v_3 v_4 v_1 v_2$, $C_{(2)} = v_3 v_4 v_1 v_2 v_3$, $C_{(3)} = v_4 v_1 v_2 v_3 v_4$, $C_{(4)} = v_1 v_4 v_3 v_2 v_1$, $C_{(5)} = v_4 v_3 v_2 v_1 v_4$, $C_{(6)} = v_3 v_2 v_1 v_4 v_3$ and $C_{(7)} = v_2 v_1 v_4 v_3 v_2$.

(a) Prove that the complete graph K_n has $(n-1)!/2$ different Hamiltonian cycles.

(b) How many different Hamiltonian cycles does $K_{n,n}$ have?

3.3.14 There are n guests at a dinner party, where $n \geq 4$. Any two of these guests know, between them, all the other $n - 2$. Prove that the guests can be seated round a circular table so that each one is sitting between two people they know.

3.3.15 Let G be a simple k-regular graph with $2k - 1$ vertices. Prove that G is Hamiltonian.

3.3.16 Let G_1 and G_2 be two simple graphs and let G denote their join $G_1 + G_2$. (See Exercise 1.5.5.) Prove that if G_1 is Hamiltonian and has at least as many vertices as G_2 has then G is also Hamiltonian.

3.4 The Travelling Salesman Problem

Suppose a travelling salesman's territory includes several towns with roads connecting certain pairs of these towns. His job requires him to visit each town. Is it possible for him to plan a round trip by car enabling him to visit each of the towns exactly once and, if such a trip is possible, can he plan one which minimises the total distance travelled?

We can represent the salesman's territory by a weighted graph G where the vertices correspond to the towns and two vertices are joined by a weighted edge if and only if there is a road connecting the corresponding towns which does not pass through any of the other towns, the edge's weight representing the length of the road between the towns. The question posed above becomes:

Is G a Hamiltonian graph and if so can we construct a Hamiltonian cycle of minimum weight (length)?

This problem is known as the Travelling Salesman Problem and we shall refer to a Hamiltonian cycle of minimum weight as an **optimal circuit**. There are two difficulties that arise with the problem:

(1) It is sometimes difficult to determine if a graph is Hamiltonian (since there is no easy characterization of Hamiltonian graphs such as Theorem 3.2 for Euler graphs).

(2) Given a weighted graph G which is Hamiltonian there is no easy or efficient algorithm for finding an optimal circuit in G, in general.

Here we look at a special case, namely when *the weighted graph is complete*, i.e., it is simple and every pair of distinct vertices is joined by an edge. Even with this special case we will have to be content with looking at two algorithms which produce "reasonably good" solutions, i.e., they produce Hamiltonian circuits which are quite short in length but there is no guarantee that they are the shortest possible.

In the first algorithm, known as the **two-optimal method**, we begin by choosing a Hamiltonian cycle C of the complete weighted graph G. Thereafter we perform a sequence of modifications to C in the hope of finding a cycle of smaller weight.

In more detail, let $C = v_1\ v_2 \ldots v_n\ v_1$ be the initial Hamiltonian cycle in our complete graph G. Then, for every pair of numbers i, j such that $1 < i+1 < j \leq n$, we can form a new Hamiltonian cycle C_{ij} from C given by

$$C_{ij} = v_1\ v_2 \ldots v_i\ v_j\ v_{j-1} \ldots v_{i+1}\ v_{j+1}\ v_{j+2} \ldots v_n\ v_1$$

obtained by deleting the edges v_iv_{i+1} and v_jv_{j+1} and adding the edges v_iv_j and $v_{i+1}v_{j+1}$, as shown in Figure 3.32. (If $j = n$ then C_{ij} is to be interpreted as $v_1v_2 \ldots v_iv_nv_{n-1} \ldots v_{i+1}v_1$.)

Now if the sum of the weights of the two new edges is less than that of the edges they replaced this new circuit C_{ij} is an improvement on C, i.e., if

$$w(v_iv_j) + w(v_{i+1}v_{j+1}) < w(v_iv_{i+1}) + w(v_jw_{j+1})$$

then C_{ij} is of smaller length than C. In this case we replace C by C_{ij} and perform a similar comparison on C_{ij}. We repeat the procedure until we reach a cycle which

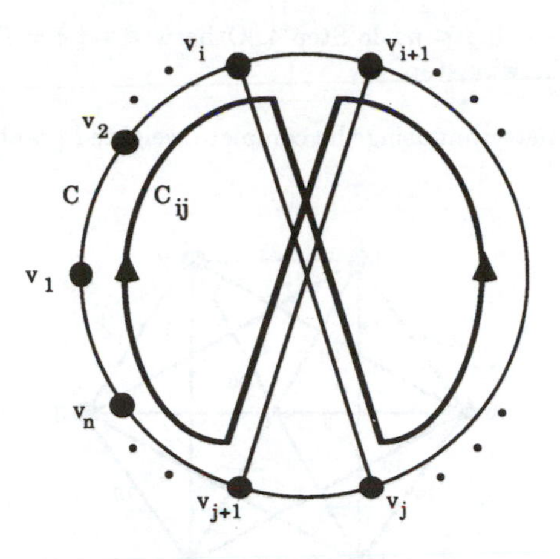

Figure 3.32: The Hamiltonian cycle C_{ij} obtained from C.

cannot be improved upon by using the same technique. This process is called the *two-optimal* method since at each stage we are choosing the optimal from two pairs of edges.

We carry out the procedure systematically, first by taking $i = 1$ and looking at the j values $3, 4, \dots, n$ in turn, then taking $i = 2$ and looking at the j values $4, 5, \dots, n$ in turn, and so on until we reach $i = n - 2$, the final step having $i = n - 2$, $j = n$. This leads us to the following two-optimal algorithm:

The Two-Optimal Algorithm

Step 1. Let $C = v_1\ v_2 \dots v_n\ v_1$ be any Hamiltonian cycle of the weighted graph G and let w be the weight of C, i.e.,

$$w = w(v_1v_2) + w(v_2v_3) + \cdots + w(v_{n-1}v_n) + w(v_nv_1).$$

Step 2. Set $i = 1$.

Step 3. Set $j = i + 2$.

Step 4. Let C_{ij} be the Hamiltonian cycle

$$C_{ij} = v_1\ v_2 \dots v_i\ v_j\ v_{j-1} \dots v_{i+1}\ v_{j+1}\ v_{j+2} \dots v_n\ v_1$$

and let w_{ij} denote the weight of C_{ij}, so that

$$w_{ij} = w - w(v_iv_{i+1}) - w(v_jv_{j+1}) + w(v_iv_j) + w(v_{i+1}v_{j+1}).$$

If $w_{ij} < w$, i.e., if $w(v_iv_j)+w(v_{i+1}v_{j+1}) < w(v_iv_{i+1})+w(v_jw_{j+1})$, then replace C by C_{ij} and w by w_{ij}, i.e., set $C = C_{ij}$, $w = w_{ij}$, and return to Step 1, taking the sequence of vertices $v_1v_2 \dots v_nv_1$ as given by our new C.

Step 5. Set $j = j+1$. If $j \leq n$, do Step 4. Otherwise set $i = i+1$. If $i \leq n-2$, do Step 3. Otherwise stop.

We illustrate the algorithm using the complete weighted graph G on $n = 6$ vertices shown in Figure 3.33.

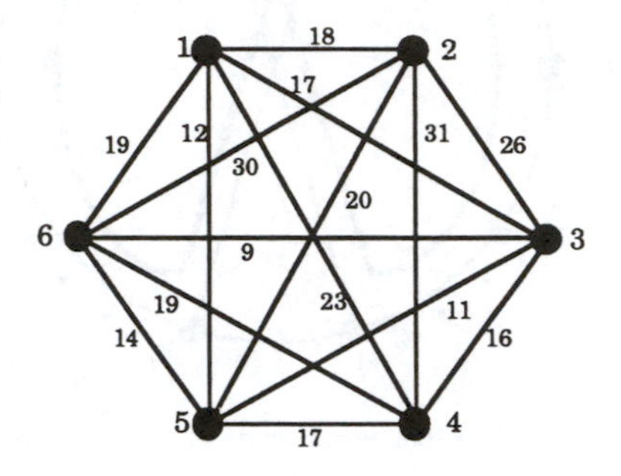

Figure 3.33

Step 1. Let $C = v_1\, v_2\, v_3\, v_4\, v_5\, v_6\, v_1$ be 1 2 3 4 5 6 1, so that
$$w = 18 + 26 + 16 + 17 + 14 + 19 = 110.$$

Step 2. Set $i = 1$.

Step 3. Set $j = i + 2 = 1 + 2 = 3$.

Step 4. Set $C_{13} = v_1\, v_3\, v_2\, v_4\, v_5\, v_6\, v_1 = 1\,3\,2\,4\,5\,6\,1$, so that
$$w_{13} = \underline{17} + 26 + \underline{31} + 17 + 14 + 19 = 124.$$
(Here, and in the following, the weights underlined are those of the two new edges.) Since $w_{13} > w$, we move on to

Step 5. Increase j to 4.

Step 4. Set $C_{14} = v_1\, v_4\, v_3\, v_2\, v_5\, v_6\, v_1 = 1\,4\,3\,2\,5\,6\,1$, so that
$$w14 = \underline{23} + 16 + 26 + \underline{20} + 14 + 19 = 118.$$
Since $w_{14} > w$, we move on to

Step 5. Increase j to 5.

Step 4. Set $C_{15} = v_1\, v_5\, v_4\, v_3\, v_2\, v_6\, v_1 = 1\,5\,4\,3\,2\,6\,1$, so that
$$w_{15} = \underline{12} + 17 + 16 + 26 + \underline{30} + 19 = 120.$$
Since $w_{15} > w$, we move on to

Step 5. Increase j to 6.

Step 4. Set $C_{16} = v_1\, v_6\, v_5\, v_4\, v_3\, v_2\, v_1 = 1\,6\,5\,4\,3\,2\,1$; this is just the reverse of the cycle C so that $w_{16} = 80$. Since $w_{16} = w$, we move on to

Step 5. Increase j to 7. Since $j > n$, we increase i to 2. Since $i \leq 4 = n - 2$, we move on to

Step 3. Set $j = i + 2 = 2 + 2 = 4$.

Step 4. Set $C_{24} = v_1\, v_2\, v_4\, v_3\, v_5\, v_6\, v_1 = 1\,2\,4\,3\,5\,6\,1$, so that
$$w_{24} = 18 + \underline{31} + 16 + \underline{11} + 14 + 19 = 109.$$
Since $w_{24} < w$, we replace C by C_{24} and w by 109 and move on to

Step 1. Now $C = 1\,2\,4\,3\,5\,6\,1$, as shown in Figure 3.34, and $w = 109$. We denote C by $v_1\, v_2\, v_3\, v_4\, v_5\, v_6\, v_1$, (not the same v's as before).

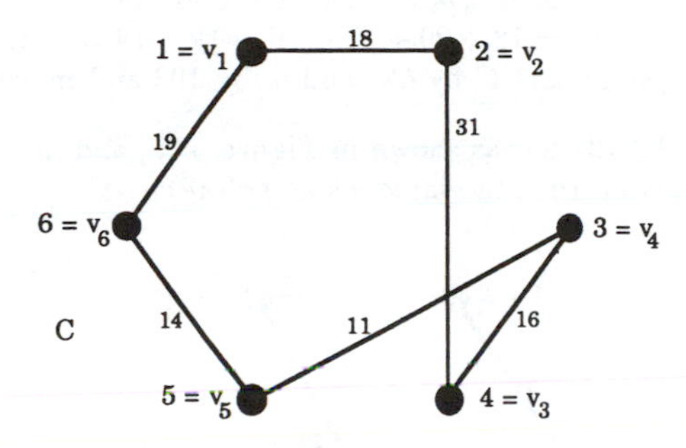

Figure 3.34: The new improved Hamiltonian cycle C.

Step 2. Set $i = 1$.

Step 3. Set $j = i + 2 = 1 + 2 = 3$.

Step 4. Set $C_{13} = v_1\, v_3\, v_2\, v_4\, v_5\, v_6\, v_1 = 1\,4\,2\,3\,5\,6\,1$, so that
$$w_{13} = \underline{23} + 31 + \underline{26} + 11 + 14 + 19 = 124.$$
Since $w_{13} > w$, we move on to

Step 5. Increase j to 4.

Step 4. Set $C_{14} = v_1\, v_4\, v_3\, v_2\, v_5\, v_6\, v_1 = 1\,3\,4\,2\,5\,6 1$, so that
$$w_{14} = \underline{17} + 16 + 31 + \underline{20} + 14 + 19 = 117.$$
Since $w_{14} > w$, we move on to

Step 5. Increase j to 5.

Step 4. Set $C_{15} = v_1\, v_5\, v_4\, v_3\, v_2\, v_6\, v_1 = 1\,5\,3\,4\,2\,6\,1$, so that
$$w_{15} = \underline{12} + 11 + 16 + 31 + \underline{30} + 19 = 119.$$
Since $w_{15} > w$, we move on to

Step 5. Increase j to 6.

Step 4. Set $C_{16} = v_1\, v_6\, v_5\, v_4\, v_3\, v_2\, v_1 = 1\,6\,5\,3\,4\,2\,1$; this is just the reverse of the cycle C so that $w_{16} = 79$. Since $w_{16} = w$, we move on to

Step 5. Increase j to 7. Since $j > 6 = n$, we increase i to 2. Since $i \leq 4 = n - 2$, we move on to

Step 3. Set $j = i + 2 = 2 + 2 = 4$.

Step 4. Set $C_{24} = v_1\, v_2\, v_4\, v_3\, v_5\, v_6\, v_1 = 1\,2\,3\,4\,5\,6\,1$, so that
$$w_{24} = 18 + \underline{26} + 16 + \underline{17} + 14 + 19 = 110.$$
Since $w_{24} > w$, we move on to

Step 5. Increase j to 5.

Step 4. Set $C_{25} = v_1\, v_2\, v_5\, v_4\, v_3\, v_6\, v_1 = 1\,2\,5\,3\,4\,6\,1$, so that
$$w_{25} = 18 + \underline{20} + 11 + 16 + \underline{19} + 19 = 103.$$
Since $w_{25} < w$, we replace C by C_{25} and w by 103 and move on to

Step 1. Now $C = 1\,2\,5\,3\,4\,6\,1$, as shown in Figure 3.35, and $w = 103$. We denote C by $v_1\, v_2\, v_3\, v_4\, v_5\, v_6\, v_1$, (not the same v's as before).

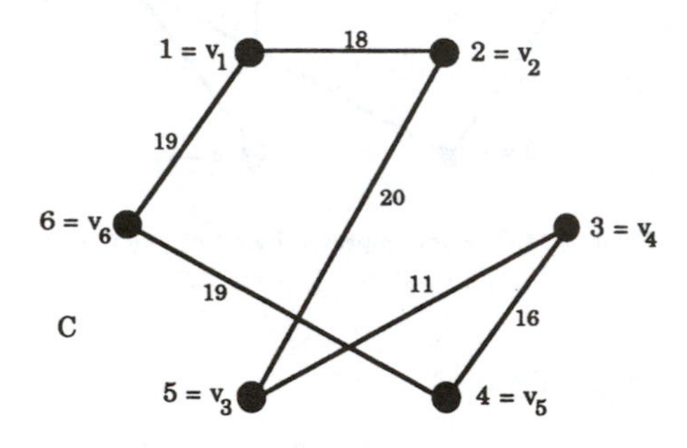

Figure 3.35: The new improved Hamiltonian cycle C.

Set $i = 1$.

Set $j = i + 2 = 1 + 2 = 3$.

Set $C_{13} = v_1\, v_3\, v_2\, v_4\, v_5\, v_6\, v_1 = 1\,5\,2\,3\,4\,6\,1$, so that
$$w_{13} = \underline{12} + 20 + \underline{26} + 16 + 19 + 19 = 112.$$

We omit most of the remaining details. The reader can check that the next change occurs with $i = 3$ and $j = 5$, so that

$$C_{ij} = C_{35} = v_1\, v_2\, v_3\, v_4\, v_6\, v_1 = 1\,2\,5\,4\,3\,6\,1$$

and $w_{ij} = w_{35} = 18 + 20 + 17 + 16 + 9 + 19 = 99$, giving the new cycle C as shown in Figure 3.36.

Moreover if we continue the process, no further improvement is obtained.

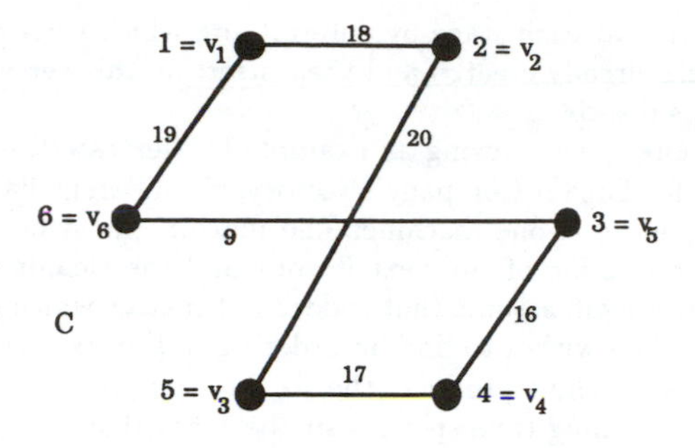

Figure 3.36: The optimal Hamiltonian cycle C.

We now indicate how to get some idea of how good this solution is. Given any optimal circuit C in G, then for any vertex v of G, $C - v$ will be a spanning tree of $G - v$. Now if T is an optimal spanning tree of $G - v$ and if e and f are two edges incident with v such that $w(e) + w(f)$ is as small as possible then certainly $w(T) + w(e) + w(f)$ is not greater than $w(C)$ i.e. $w(T) + w(e) + w(f) < w(C)$. Now Kruskal's algorithm gives us a spanning tree and applying it to our graph $G - v$ with v as vertex 2 we get the following optimal tree T.

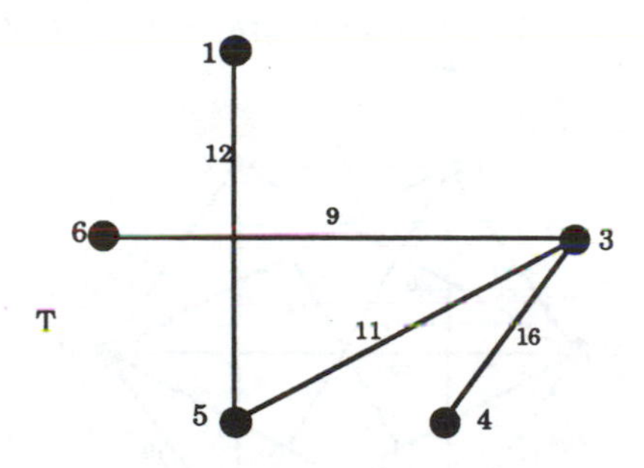

Figure 3.37: A minimal spanning tree T, of weight 48, for the graph $G - 2$.

On examination, the two edges incident with vertex 2 with smallest weights are those joining 2 to 1 and 5 with weights 13 and 15 respectively. Thus these are chosen as our e and f and we get

$$w(T) + w(e) + w(f) = 48 + 18 + 20 = 86.$$

Hence any optimal cycle C has weight $w(C) > 86$. So our solution, which has weight 99, seems reasonable — note that it may indeed be the best possible.

We now consider a different method, again finding a reasonably good solution, but with no guarantee that it is the best. It is called the *closest insertion algorithm*. The general idea of this method is to gradually build up a sequence of cycles in the graph which involve more and more vertices, until all the vertices are used up, and to involve one more vertex at each stage by determining which vertex, as yet unchosen, is nearest to the cycle already created and then inserting this vertex into the cycle in as economical way as possible.

We will now be more specific, using an example to illustrate the method.

In the Real Cool Ice Cream Company's factory, six different flavours of ice cream are produced in sequence on one machine. The machine must be cleaned after each flavour before the production of the next flavour and the cleaning time depends on the two flavours (but we will assume that it does not matter which of the two flavours is mixed first). Real Cool wishes to find an ordering of the six flavours, starting with banana, so that, if the machine produces the six flavours (once and) only once in this order, then the total cleaning time spent is smallest possible.

The cleaning times (in minutes) are given in the following table:

flavour	banana	chocolate	mint	raspberry	strawberry	vanilla
banana	0	10	11	11	6	10
chocolate	10	0	17	15	15	20
mint	11	17	0	10	15	19
raspberry	11	15	10	0	15	20
strawberry	6	15	15	15	0	11
vanilla	10	20	19	20	11	0

This table produces the complete weighted graph of Figure 3.38, where B denotes "banana", C denotes "chocolate", etc.

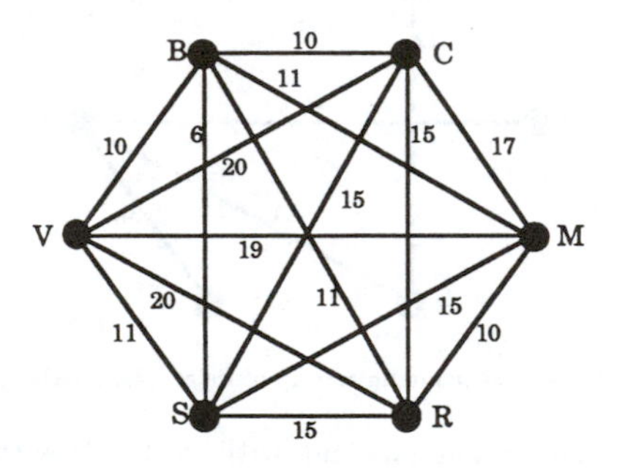

Figure 3.38

We now describe the closest insertion algorithm. The description uses the idea of the distance of a vertex v from a walk W. This is defined to be

$$d(v, W) = \min\{d(v, u) : u \text{ is a vertex of } W\}.$$

The vertex v, not in W, is then said to be **closest** to W if $d(v, W) \leq d(x, W)$ for any other vertex x not in W.

We emphasise the way in which a new vertex is inserted into a cycle by using bold type.

The Closest Insertion Algorithm

Step 1. Choose any vertex v_1 as a starting vertex.

Step 2. From among the $n-1$ vertices not chosen so far, find one, say v_2, which is closest to v_1. Let W_2 denote the walk $v_1 \boldsymbol{v_2} v_1$.

Step 3. From among the $n-2$ vertices not chosen so far, find one, say v_3, which is closest to the walk $W_2 = v_1\,v_2\,v_1$.
Let W_3 be the walk $v_1 v_2 \boldsymbol{v_3} v_1$ (so that W_3 is in fact a cycle).

Step 4. From among the $n-3$ vertices not chosen so far, find one, say v_4, which is closest to the walk W_3.
Determine which of the walks (cycles) $v_1\,v_2\,v_3\,\boldsymbol{v_4}\,v_1$, $v_1\,v_2\,\boldsymbol{v_4}\,v_3\,v_1$, $v_1\,\boldsymbol{v_4}\,v_2\,v_3\,v_1$ is the shortest.
Let W_4 denote the shortest one and relabel it, if necessary, as $v_1\,v_2\,v_3\,v_4\,v_1$.

Step 5. From among the $n-4$ vertices not chosen so far, find one, say v_5 which is closest to the walk W_4.
Determine which of the cycles $v_1\,v_2\,v_3\,v_4\,\boldsymbol{v_5}\,v_1$, $v_1\,v_2\,v_3\,\boldsymbol{v_5}\,v_4\,v_1$, $v_1\,v_2\,\boldsymbol{v_5}\,v_3\,v_4\,v_1$, $v_1\,\boldsymbol{v_5}\,v_2\,v_3\,v_4\,v_1$ is shortest.
Let W_5 denote the shortest one and relabel it, if necessary , as $v_1\,v_2\,v_3\,v_4\,v_5\,v_1$.
Continue in this way to eventually arrive at

Step n-1. From the 2 vertices not already chosen, find one, say v_{n-1} which is closest to the walk (cycle) $W_{n-2} = v_1\,v_2 \ldots v_{n-2}\,v_1$.
Determine which of the cycles $v_1\,v_2 \ldots v_{n-2}\,\boldsymbol{v_{n-1}}\,v_1$, $v_1\,v_2 \ldots v_{n-3}\,\boldsymbol{v_{n-1}}\,v_{n-2}\,v_1$, $\ldots$, $v_1\,\boldsymbol{v_{n-1}}\,v_2\,v_3 \ldots v_{n-2}\,v_1$ is shortest.
Let W_{n-1} denote the shortest of these cycles and relabel it, if necessary, as $v_1\,v_2 \ldots v_{n-2}\,v_{n-1}\,v_1$.

Step n. Denote the remaining unchosen vertex by v_n and determine which of the cycles $v_1\,v_2 \ldots v_{n-1}\,\boldsymbol{v_n}\,v_1$, $v_1\,v_2 \ldots v_{n-2}\,\boldsymbol{v_n}\,v_{n-1}\,v_1$, $\ldots$, $v_1\,\boldsymbol{v_n}\,v_2\,v_3 \ldots v_{n-1}\,v_1$ is shortest.
Let W_n denote the shortest of these cycles.

Conclusion: W_n is a Hamiltonian cycle of G which is, at least approximately, optimal.

For our example above we have

Step 1. Let $v_1 = B$.

Step 2. $v_2 = S$ is closest to B so $W_2 = v_1\,v_2\,v_1 = BSB$.

Step 3. $v_3 = V$ is closest to W_2. (It has distance 10 from B; but there is actually a choice here since C is another candidate.)
W_3 is the cycle $v_1\,v_2\,v_3\,v_1 = BSVB$.

Step 4. $v_4 = C$ is closest to W_3 (It has distance 10 from B.)
We find the lengths of the cycles $v_1\,v_2\,v_3\,\boldsymbol{v_4}\,v_1$, $v_1\,v_2\,\boldsymbol{v_4}\,v_3\,v_1$ and $v_1\,\boldsymbol{v_4}\,v_3\,v_2\,v_1$:
$v_1\,v_2\,v_3\,v_4\,v_1 = BSVCB$ has length $6 + 11 + 20 + 10 = 47$,
$v_1\,v_2\,v_4\,v_3\,v_1 = BSCVB$ has length $6 + 15 + 20 + 10 = 51$,
$v_1\,v_4\,v_2\,v_3\,v_1 = BCSVB$ has length $10 + 15 + 11 + 10 = 46$.
We let W_4 denote the shortest of these, namely $BCSVB$, and relabel with $BCSVB = v_1\,v_2\,v_3\,v_4\,v_1$.

Step 5. $v_5 = R$ is closest to W_4. (It has distance 11 from B; but again there is a choice here: M is also distance 11 from W_4.)
We find the lengths of the cycles $v_1\,v_2\,v_3\,v_4\,\boldsymbol{v_5}\,v_1$, $v_1\,v_2\,v_3\,\boldsymbol{v_5}\,v_4\,v_1$, $v_1\,v_2\,\boldsymbol{v_5}\,v_3\,v_4\,v_1$ and $v_1\,\boldsymbol{v_5}\,v_2\,v_3\,v_4\,v_1$:
$v_1\,v_2\,v_3\,v_4\,v_5\,v_1 = BCSVRB$ has length $10 + 15 + 11 + 20 + 11 = 67$,
$v_1\,v_2\,v_3\,v_5\,v_4\,v_1 = BCSRVB$ has length $10 + 15 + 15 + 20 + 10 = 70$,
$v_1\,v_2\,v_5\,v_3\,v_4\,v_1 = BCRSVB$ has length $10 + 15 + 15 + 11 + 10 = 61$,
$v_1\,v_5\,v_2\,v_3\,v_4\,v_1 = BRCSVB$ has length $11 + 15 + 15 + 11 + 10 = 62$.
We let W_5 denote the shortest of these, namely $BCRSVB$, and relabel with $BCRSVB = v_1\,v_2\,v_3\,v_4\,v_5\,v_1$.

Step n ($n = 6$). Let $v_6 = M$, the last remaining vertex.
We find the lengths of the cycles $v_1\,v_2\,v_3\,v_4\,v_5\,\boldsymbol{v_6}\,v_1$, $v_1\,v_2\,v_3\,v_4\,\boldsymbol{v_6}\,v_5\,v_1$, $v_1\,v_2\,v_3\,\boldsymbol{v_6}\,v_4\,v_5\,v_1$, $v_1\,v_2\,\boldsymbol{v_6}\,v_3\,v_4\,v_5\,v_1$ and $v_1\,\boldsymbol{v_6}\,v_2\,v_3\,v_4\,v_5\,v_1$:
$v_1\,v_2\,v_3\,v_4\,v_5\,v_6\,v_1 = BCRSVMB$ has length $10 + 15 + 15 + 11 + 19 + 11 = 81$,
$v_1\,v_2\,v_3\,v_4\,v_6\,v_5\,v_1 = BCRSMVB$ has length $10 + 15 + 15 + 15 + 19 + 10 = 84$,
$v_1\,v_2\,v_3\,v_6\,v_4\,v_5\,v_1 = BCRMSVB$ has length $10 + 15 + 10 + 15 + 11 + 10 = 71$,
$v_1\,v_2\,v_6\,v_3\,v_4\,v_5\,v_1 = BCMRSVB$ has length $10 + 17 + 10 + 15 + 11 + 10 = 73$,
$v_1\,v_6\,v_2\,v_3\,v_4\,v_5\,v_1 = BMCRSVB$ has length $11 + 17 + 15 + 15 + 11 + 10 = 79$.

Then the shortest cycle is $W_6 = BCRMSVB$ and it has length 71.

We can therefore conclude that a reasonably efficient cleaning-time cycle for Real Cool's ice cream machine is

banana-chocolate-raspberry-mint-strawberry-vanilla-banana,

the cycle taking 71 minutes to complete.

Exercises for Section 3.4

3.4.1 Six computer programs, $P_1, \ldots, P_6$, have to be run in sequence on a mainframe machine. Each program needs its own resources such as a part of the main memory, a compiler and drives, and changing from one set of resources to another uses up valuable time. The following matrix $C = (c_{ij})$ records the times c_{ij} used in converting from the resources for program P_i to those for program P_j.

$$C = \begin{bmatrix} 0 & 30 & 90 & 10 & 30 & 10 \\ 30 & 0 & 80 & 40 & 20 & 20 \\ 90 & 80 & 0 & 40 & 20 & 30 \\ 10 & 40 & 40 & 0 & 60 & 10 \\ 30 & 20 & 20 & 60 & 0 & 90 \\ 10 & 20 & 30 & 10 & 90 & 0 \end{bmatrix}.$$

Note that $c_{ij} = c_{ji}$ for each i and j. Using both the two optimal method and the closest insertion method for the Travelling Salesman Problem, find an ordering of the programs which, if they are run in this order, should result in a comparatively small total conversion time.

3.4.2 Carry out the two optimal method and the closest insertion method for the Travelling Salesman Problem for the complete weighted graph of Figure 3.39.

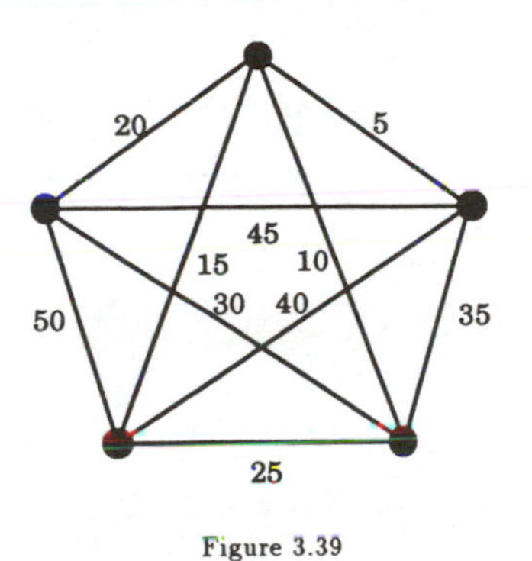

Figure 3.39

Chapter 4

Matchings

4.1 Matchings and Augmenting Paths

Let G be a graph with vertex set $V = V(G)$ and edge set $E = E(G)$. *We will assume throughout this chapter that G has no loops.*

> A subset M of E, i.e ubset M of edges of G, is called a **matching** in G if no two of the edges in M are adjacent, in other words, if for any two edges e and f in M the two end vertices of e are both different from the two end vertices of f.

More briefly, M is a matching if no two edges in M have an end vertex in common. For example, in the graph G of Figure 4.1 the sets $M_1 = \{e_1, e_2\}$ and $M_2 = \{e_1, e_3, e_4\}$ are both matchings.

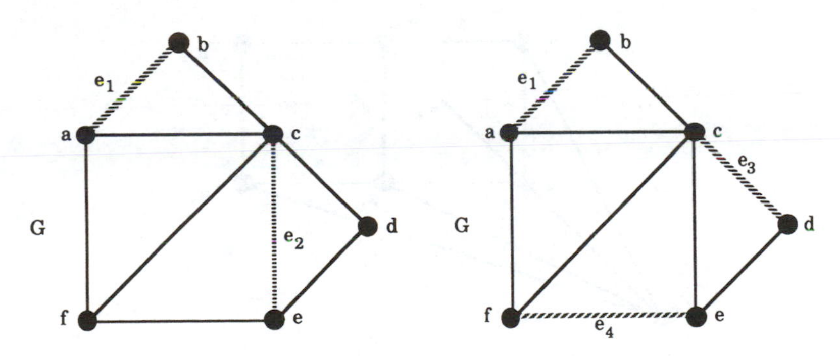

Figure 4.1: A graph G with two matchings.

> If the vertex v of the graph G is the end vertex of some edge in the matching M then v is said to be M-**saturated** and we say that M **saturates** v. Otherwise v is M-**unsaturated**.

Thus, in the example of Figure 4.1, a, b, c and e are all M_1-saturated while f and d are both M_1-unsaturated; every vertex of G is M_2-saturated.

If M is a matching in G such that every vertex of G is M-saturated then M is called a **perfect matching** .

A matching M in G is called **maximum** if G has no matching M' with a greater number of edges than M has.

Thus the matching $M_2 = \{e_1, e_3, e_4\}$ of Figure 4.1 is a perfect matching.

Clearly any perfect matching is a maximum matching. The matchings of the graphs in Figure 4.2 shown by the shaded lines are both maximum but not perfect.

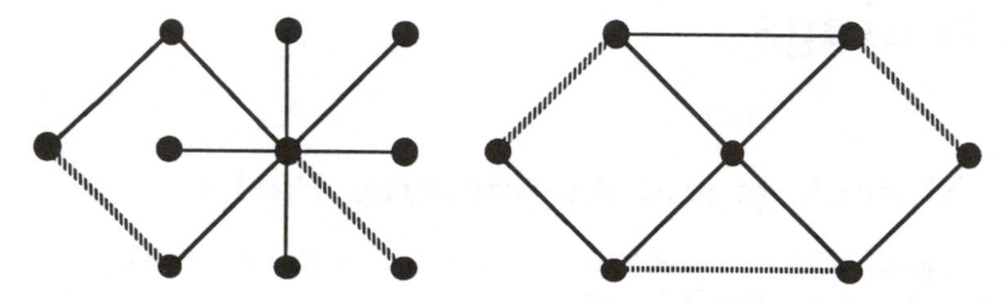

Figure 4.2: Two maximum matchings which are not perfect.

Let M be a matching in G and let $E = E(G)$ be the edge set of G. An M-**alternating path** in G is a path whose edges are alternately in M and $E - M$, i.e., alternately in M and not in M.

For example, if in the graph G_1 of Figure 4.3 we consider the matching $M = \{e_3, e_5, e_6\}$, then the path $v_1\ v_4\ v_2\ v_5\ v_6\ v_3$ is M-alternating.

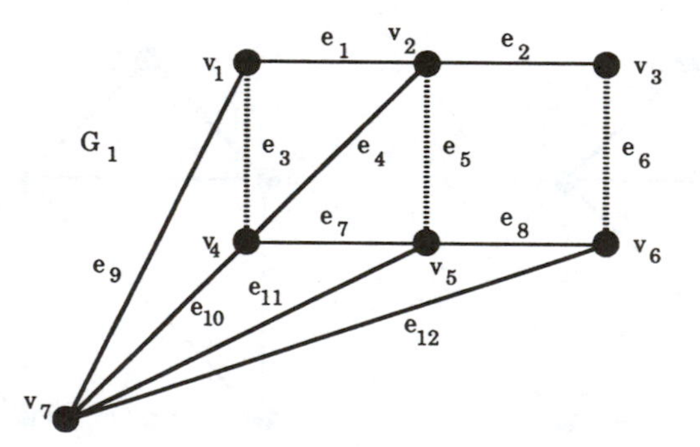

Figure 4.3: $v_1\ v_4\ v_2\ v_5\ v_6\ v_3$ is an M-alternating path.

An M-alternating path whose origin and terminus are both M-unsaturated is called an M-**augmenting path**.

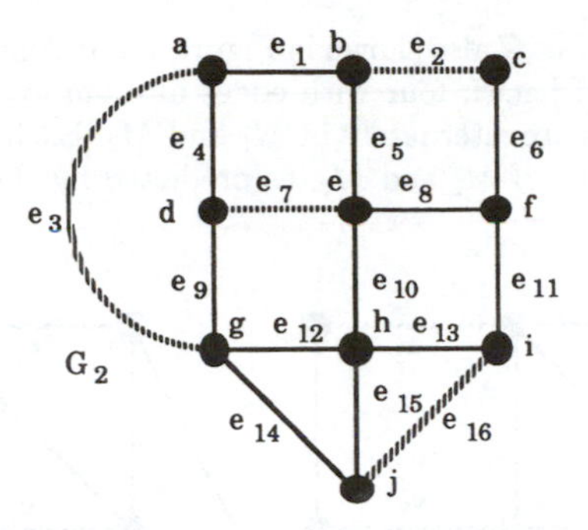

Figure 4.4: $f\ c\ b\ e\ d\ a\ g\ h$ is an augmenting path.

Thus, in the graph G_2 of Figure 4.4 with matching $M = \{e_2, e_3, e_7, e_{16}\}$, the path $f\ c\ b\ e\ d\ a\ g\ h$ is M-augmenting.

Theorem 4.1 *Let M_1 and M_2 be two matchings in a simple graph G. Let H be the subgraph of G induced by the set of edges*

$$M_1 \triangle M_2 = (M_1 - M_2) \cup (M_2 - M_1),$$

i.e., by the symmetric difference of the two matchings.[1] *Then each connected component of H is of one of the following two types:*

(1) a cycle of even length whose edges are alternately in M_1 and M_2,

(2) a path whose edges are alternately in M_1 and M_2 and whose end vertices are unsaturated in one of the two matchings.

Proof Let v be any vertex of the subgraph H. Then either

(i) v is an end vertex of an edge in $M_1 - M_2$ and also of an edge in $M_2 - M_1$ or

(ii) v is an end vertex of an edge in one of $M_1 - M_2$ and $M_2 - M_1$ but not both.

In either case, since M_1 is a matching, there is at most one edge in M_1 with v as one of its end points and, similarly, there is at most one edge in M_2 with v as one of its end points. Thus, in case (i), v has degree 2 in H while, in case (ii), v has degree 1 in H. Hence every vertex of H has either degree 1 or degree 2. It follows from Exercise 4.1.12 that the components of H are as described in the statement of the theorem. □

We illustrate the Theorem using the graph G_1 of Figure 4.3, shown again in Figure 4.5. Let M_1 be the matching $\{e_2, e_8, e_{10}\}$ and take M_2 to be the matching $\{e_1, e_7, e_{12}\}$ as shown in Figure 4.5. Then

$$M_1 \triangle M_2 = \{e_2, e_8, e_{10}\} \cup \{e_1, e_7, e_{12}\} = \{e_1, e_2, e_7, e_8, e_{10}, e_{12}\}.$$

[1] Given two sets A and B, their **symmetric difference**, denoted by $A \triangle B$, is the set of all elements which belong to either A or B but *not* to both A and B.

This gives the subgraph H of G also shown in Figure 4.5. It consists of two components, one of which is a cycle of length four with edges alternately in M_1 and M_2 and the other a path whose edges are alternately in M_1 and M_2. Each end vertex of this path is saturated by exactly one of M_1 and M_2, as predicted by the Theorem.

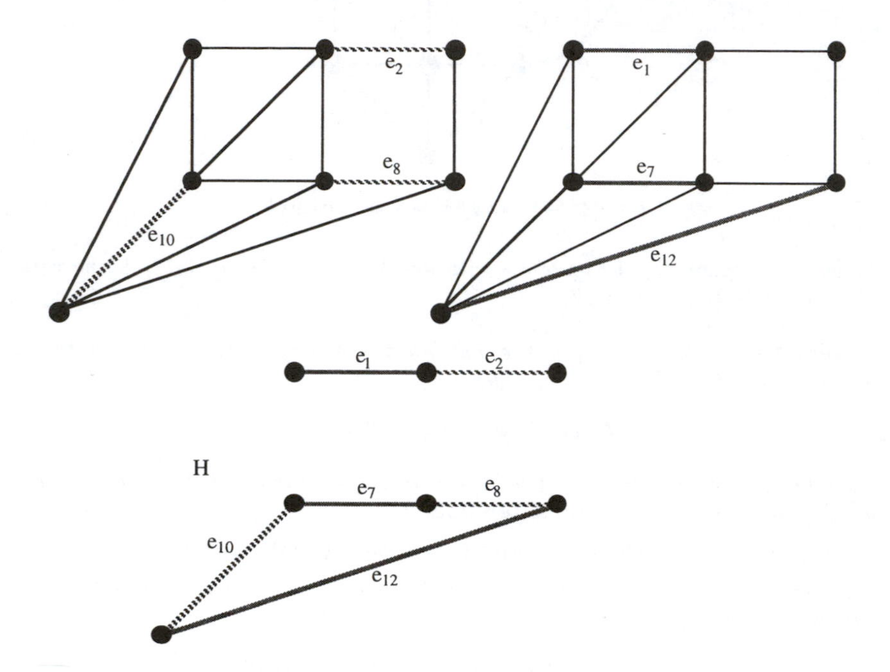

Figure 4.5: Two matchings M_1 and M_2 and the subgraph H induced by their symmetric difference.

Using Theorem 4.1, we prove the following important result of Berge [5] which characterizes maximum matchings in terms of augmenting paths.

Theorem 4.2 (Berge, 1957) *A matching M in a graph G is a maximum matching if and only if G contains no M-augmenting path.*

The proof actually involves a technique which can be used to find a maximum matching. For this reason we first describe the technique.

Let M be any matching in the graph G. We will refer to edges in M as **dark** edges and the other edges of G as **light** edges. Let P be an M-alternating path in G, so that the edges of P are alternately dark and light. If we further assume that P is M-augmenting, i.e., that the origin and terminus of P are not M-saturated, then the first and last edges of P must be light. Because of the alternation, the sequence of edges in P must be of the form

$$\text{light, dark, light, } \ldots, \text{ dark, light,}$$

and so P has an odd number of edges, say $2m+1$ of them, m of which are dark and $m+1$ light.

Also any edge of M which is not in P is not incident with any vertex of P because the origin and terminus of P are not saturated by M while the other vertices of P are already saturated by M. Thus if we let M' be the set of edges consisting of all the dark edges which are not in P and also the light edges which are in P, i.e., obtained by replacing in M the m dark edges in P by the $m+1$ light edges in P, then M' is a new matching having one more edge than M.

This operation which transforms the given matching M into a bigger matching M' using an M-augmenting path P is called a **transfer along the augmenting path P**.

For example, using the graph G_2 of Figure 4.4 with the matching $M = \{e_2, e_3, e_7, e_{16}\}$ and augmenting path $P = f\ c\ b\ e\ d\ a\ g\ h$, the transfer along P gives the matching $M' = \{e_4, e_5, e_6, e_{12}, e_{16}\}$. We show this in Figure 4.6.

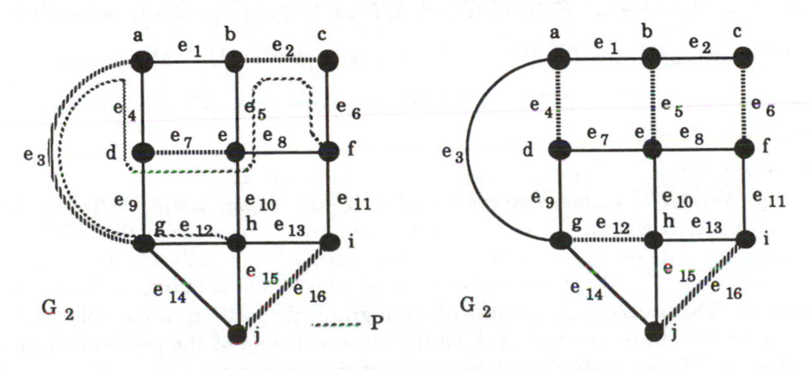

Figure 4.6: A transfer along an augmenting path.

We now give the proof of Theorem 4.2.

Proof Let M be a maximum matching in G. If there is an M-augmenting path P in G then, as shown above, we can transfer along P to produce a new matching M' in G which has one more edge than M has. This is impossible since M is maximum. Thus G has no M-augmenting path.

Conversely suppose that M is a matching in G such that there is no M-augmenting path in G. We wish to show that M is a maximum matching. Let M' be any maximum matching of G. We wish then to show that

$$|M| = |M'|,$$

where $|\ |$ denotes "the number of elements in". Let H be the subgraph of G induced by the set of edges

$$M \triangle M' = (M - M') \cup (M' - M),$$

i.e., by those edges which are in M or M' but not both. Then, by Theorem 4.1, the connected components of H are either

(i) cycles of even length whose edges are alternately in M and M' or
(ii) paths whose edges are alternately in M and M' and whose end vertices are unsaturated in one of the two matchings.

Now if there is a path as in (ii) which is of *odd* length then the origin and terminus of this path would be either both unsaturated in M or both unsaturated in M', so such a path would either be M-augmenting or M'-augmenting. We are denied the first possibility by our initial assumption on M while we are denied the second because M' is maximum and so, by the first part of the proof, has no M'-augmenting path. Thus the components of H are either paths of even length or cycles of even length and so each involves the same number of edges from M as from M'. Thus

$$|M - M'| = |M' - M|.$$

Now given any two subsets A and B of a set X, we have $|A| = |A - B| + |A \cap B|$. Taking X as the edge set $E(G)$ of G and A, B as M, M' respectively, we get that

$$|M| = |M - M'| + |M \cap M'| \quad \text{and similarly} \quad |M'| = |M - M'| + |M \cap M'|.$$

Since, from above, $|M - M'| = |M' - M|$, it follows that $|M| = |M'|$. Hence M is maximum, as required. □

We now mention two practical applications of matching.

During World War II, many aeroplane pilots from occupied countries fled to Britain to enlist in the Royal Air Force. In certain squadrons, each plane sent aloft by the RAF required a pilot and a navigator whose navigational skills and language skills were compatible. The RAF was interested in sending as many planes aloft at one time as possible. The problem can be described in graph-theoretic terms as follows.

Let G be the graph in which each vertex represents one of the pilots or navigators available and where an edge joins two vertices if these represent a pilot and a navigator who are compatible. Then any matching of this graph represents a possible set of planes that can be sent aloft simultaneously. Thus the RAF was interested in finding a *maximum matching* in G.

A real estate agent has for sale a variety of homes and a number of prospective buyers. Each prospective buyer is interested in possibly more than one of the available homes for sale. The agent can estimate fairly accurately just how much each buyer would pay for each of the homes he is interested in. Since he makes a certain percentage commission on each transaction, he is interested in maximizing the total dollar volume of his sales. How can he accomplish this?

Let G be the graph where buyers and homes are each represented by a vertex. Join a buyer vertex to a house vertex by an edge if the buyer would be willing to buy the home. Thus G is a bipartite graph with bipartition $V(G) = B \cup H$ where B is the set of vertices representing the buyers and H is the set of vertices representing the homes. Place a weight on each edge equal to the commission the agent would receive for the corresponding transaction. Then the agent can maximize his earnings by finding a matching with the greatest total weight in G, i.e., a *maximum weight matching*.

We will not be able to consider in detail matching problems in general graphs. Instead we concentrate on matching in bipartite graphs. (The two examples above are bipartite matching problems.)

Exercises for Section 4.1

4.1.1 For the graphs of Figure 4.7, with matchings M shown by the shaded lines, find

(a) an M-alternating path which is not M-augmenting,

(b) an M-augmenting path if one exists, and, if so, use it to obtain a bigger matching.

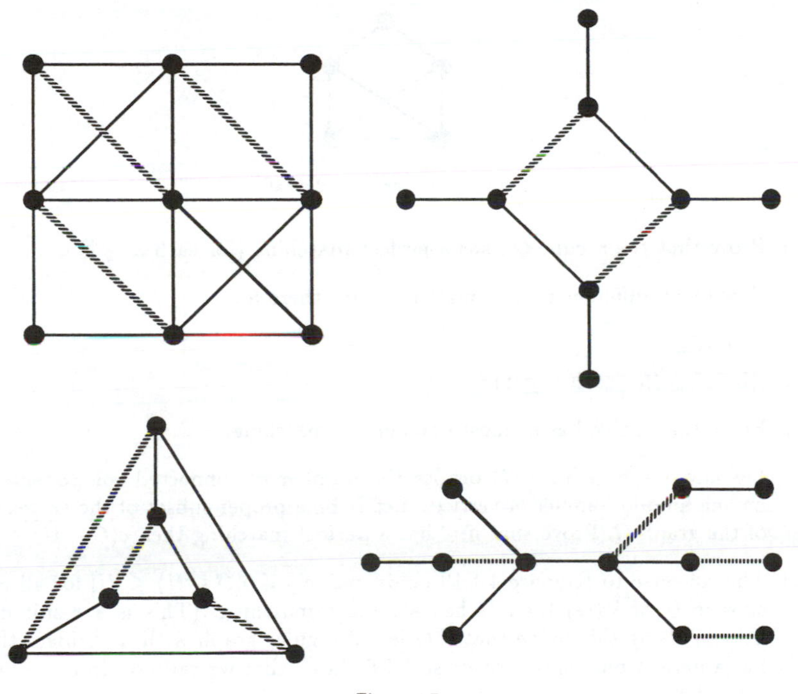

Figure 4.7

4.1.2 For which values of $n \geq 4$ does the wheel W_n have a perfect matching?

4.1.3 For which values of $n \geq 2$ does K_n have a perfect matching?

4.1.4 Which complete tripartite graphs of the form $K_{n,n,n}$ have perfect matchings?

4.1.5 Prove that a 2-regular graph G has a perfect matching if and only if each component of G is an even cycle.

4.1.6 Prove that the 3-regular graph of Figure 4.8, which we dub the **triple flyswat**, does not have a perfect matching but does have a matching with seven edges.

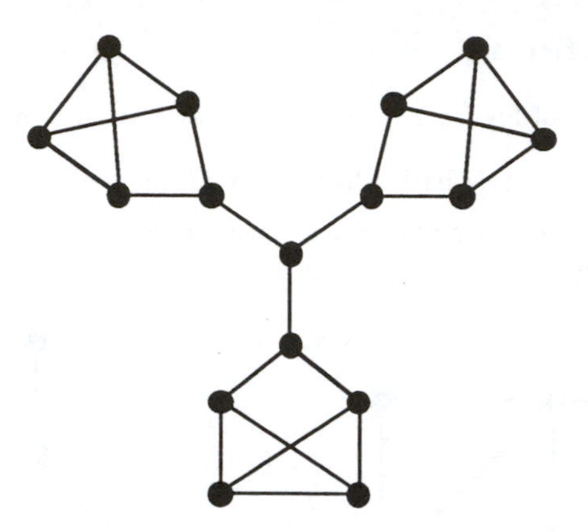

Figure 4.8: The triple flyswat.

4.1.7 Prove that the n-cube Q_n has a perfect matching (for each $n \geq 2$).

4.1.8 How many different perfect matchings are there for

(a) K_{2n},

(b) $K_{n,n}$ (for each $n \geq 1$).

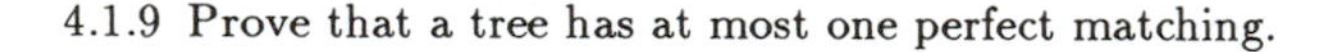

4.1.9 Prove that a tree has at most one perfect matching.

4.1.10 For any graph G let $o(G)$ denote the number of connected components of G having an odd number of vertices. Let U be a proper subset of the vertex set V of the graph G. Prove that if G has a perfect matching then $o(G - U) \leq |U|$.

4.1.11 The converse to Exercise 4.1.10 holds, namely if $o(G - U) \leq |U|$ for all proper subsets U of $V(G)$ then G has a perfect matching. (This is a result due to Tutte.) Using this prove that if G is a 3-regular graph with no bridges then G has a perfect matching. (Exercise 4.1.6 shows that we can not drop the bridges condition!)

4.1.12 Show that if G is a connected graph in which every vertex has degree either 1 or 2 then G is either a path or a cycle. (This fills in the details missing at the end of the proof of Theorem 4.1.)

4.2 The Marriage Problem

The following question is known as the **marriage problem**.

If we have a finite set of boys each of whom has several girlfriends, under what conditions can we marry off the boys in such a way that each boy marries one of his girlfriends? (We assume that no girl (boy) can marry more than one boy (girl)!)

The problem can be posed in graph-theoretical terms as follows:

Construct a bipartite graph G with bipartition $V(G) = X \cup Y$ where $X = \{x_1, \ldots, x_n\}$ represents the set of n boys and $Y = \{y_1, \ldots, y_m\}$ represents their girlfriends, m in all. An edge joins vertex x_i to vertex y_j if and only if y_j is a girlfriend of x_i. The marriage problem is then equivalent to finding conditions for the existence of a matching in G which saturates every vertex of X.

For example, if there are four boys, x_1, x_2, x_3, x_4, and five girls, y_1, y_2, y_3, y_4, y_5, and the relationships are given by

boy	girlfriends
x_1	y_2, y_3, y_4
x_2	y_2, y_4
x_3	y_1, y_4, y_5
x_4	y_1

we get the bipartite graph of Figure 4.9.

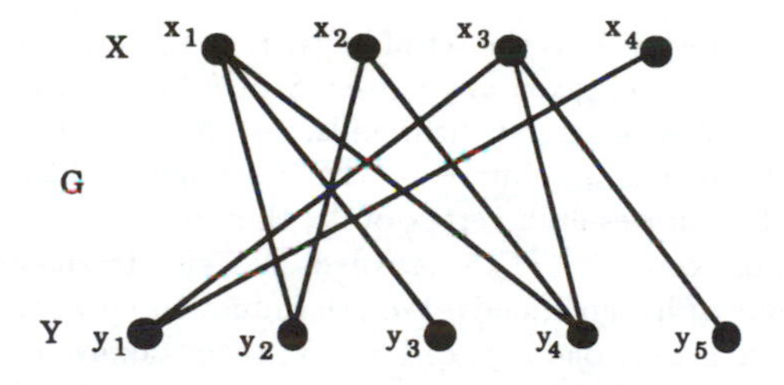

Figure 4.9: The bipartite graph for the girlfriends/boyfriends example.

A particular solution in this example is for x_3 to marry y_4, x_4 to marry y_1, x_1 to marry y_3, and x_2 to marry y_2, corresponding to the matching shown in Figure 4.10.

A restatement of the general problem is:

Let G be a bipartite graph with bipartition $V(G) = X \cup Y$. Find necessary and sufficient conditions for the existence of a matching in G which saturates every vertex in X.

We will give the solution to the problem as first presented by the Cambridge mathematician Philip Hall in 1935 [32]. First a definition:

For any set S of vertices in G, the **neighbour set** of S in G, denoted by $N(S)$, is defined to be the set of all vertices adjacent to vertices in S.

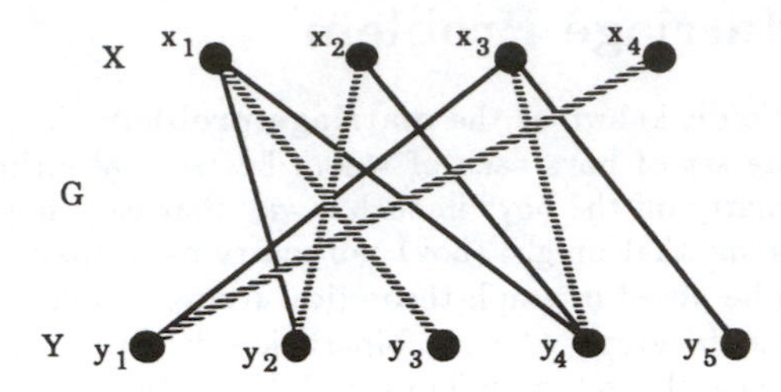

Figure 4.10: A maximum matching for the bipartite graph of Figure 4.9.

Thus, in the graph G of Figure 4.9, if S is a subset of the set of boys then $N(S)$ denotes the set of all their girlfriends.

Theorem 4.3 (Hall's Marriage Theorem, 1935) *Let G be a bipartite graph with bipartition $V = X \cup Y$. Then G contains a matching that saturates every vertex in X if and only if*

$$|N(S)| \geq |S| \quad \textit{for every subset } S \textit{ of } X, \qquad (*) \tag{4.1}$$

(where here $|\ |$ denotes "the number of elements in"). Thus the marriage problem with n boys has a solution if and only if for every $k, 1 \leq k \leq n$, every set of k boys has collectively at least k girlfriends.

Proof We refer to the set X as the set of boys, Y as the set of girls and let $n = |X|$. Then, for any k, $1 \leq k \leq n$, given any subset S of k boys, to marry them all off to girlfriends they must have at least k girlfriends, i.e. $|N(S)| \geq k$, and so $|N(S)| \geq |S|$. This shows that if there is a solution to the marriage problem, i.e., if there is a matching in G that saturates each vertex of X, then condition (*) must be satisfied.

Conversely, suppose condition (*) is satisfied. We wish to show that every boy can be married off to one of his girlfriends. We use induction on n, the number of boys.

If $n = 1$, i.e., there is only one boy, call him x, then, taking $S = \{x\}$, (*) says that he has at least one girlfriend. Thus we can easily marry him off. This proves the case for $n = 1$.

Now suppose that the result is true for all sets of boys of size t where $t = 1, 2, \ldots, n-1$. We wish to show that the result is true for a set of n boys. We consider two cases.

Case (i). Suppose first that, for any k with $1 \leq k < n$, every set of k boys together have (between them) at least $k+1$ girlfriends. This means that for any subset S of X with $|S| < n$ we have

$$|N(S)| \geq |S| + 1,$$

i.e., condition (*) is satisfied "with one girl left over". Now take any boy and marry him to one of his girlfriends, leaving $n-1$ boys still to be married off. In this process, given any subset consisting of k boys from the remaining $n-1$, we have possibly lost one of their girlfriends, namely the one married to the first boy. Thus we may have lost "the girl left over", but condition (*) is still satisfied for the subsets of the $n-1$

boys. Hence, by the induction assumption, we can marry each of these $n-1$ boys off to one of their girlfriends. Thus, in this case, we have married off each of the n boys.

Case (ii). Suppose now that there is a set of k boys (from the n), with $k < n$, who collectively know exactly k girls. (All this is saying is that (*) holds but Case (i) above does not.) Then, since $k < n$, our induction hypothesis allows us to marry off these k boys, leaving $n-k$ boys still unmarried. But any collection of h boys from these $n-k$ (with $1 \leq h \leq n-k$) must have at least h girl friends from among the remaining girls, since otherwise these h boys together with the above collection of k boys ($h+k$ boys in all) would have less than $h+k$ girlfriends, contrary to condition (*). It follows that conditon (*) also applies to the $n-k$ boys left and so by our induction hypothesis we can marry off each of them also to a girlfriend. Thus each of the n boys has been married off successfully.

This shows that assuming the result true for $1, \ldots, n-1$ we can prove that it is true for n. Hence, by mathematical induction, the result is true for all values of n. □

Corollary 4.4 *Let G be a k-regular bipartite graph with $k > 0$. Then G has a perfect matching.*

Proof Let G have bipartition $V = X \cup Y$. There are $|X|$ vertices in X and each of these vertices has k edges (all going to vertices in Y) incident with it. Thus there are $k \times |X|$ edges going from X to Y. Similarly since each of the $|Y|$ vertices in Y has k edges incident with it there are $k \times |Y|$ edges going from Y to X. By the bipartite nature of G each edge goes from X to Y (the same as from Y to X) and so G has $k|X| = k|Y|$ edges. Since $k > 0$ and $k|X| = k|Y|$, cancelling k gives

$$|X| = |Y|.$$

Now let S be a subset of X. Let E_1 denote the set of edges incident with vertices in S. Then, by the k-regularity of G,

$$|E_1| = k|S|. \tag{4.2}$$

Let E_2 denote the set of edges incident with vertices in $N(S)$. Then, since $N(S)$ is the set of vertices which are joined by edges to S, we have $E_1 \subset E_2$. Thus

$$|E_1| \leq |E_2|. \tag{4.3}$$

Moreover by the k-regularity of G we have

$$|E_2| = k|N(S)|. \tag{4.4}$$

Using (4.2), (4.3) and (4.3), we get $k|N(S)| = |E_2| \geq |E_1| = k|S|$ and so

$$k|N(S)| \geq k|S|.$$

Since $k > 0$ this gives $|N(S)| \geq |S|$. Since S was an arbitrary subset of X, it follows from Hall's marriage theorem that G contains a matching M which saturates every vertex in X. Since $|X| = |Y|$ the edges in the matching M also saturate every vertex in Y. Thus M is a perfect matching in G. □

The corollary can be restated as follows:

Corollary 4.5 *If every boy has k girlfriends and every girl has k boyfriends then each boy can marry one of his girlfriends and each girl can marry one of her boyfriends.*

We now present a second proof of Hall's Marriage Theorem. It differs from the one above in that it is a proof by contradiction. Its advantage is that it gives us an algorithm which determines if augmenting paths exist and, if they do, for constructing such paths.

Second Proof of Hall's Marriage Theorem Assume that condition (*) is satisfied, i.e.,

$$|N(S)| \geq |S| \quad \text{for every subset } S \text{ of } X. \qquad (*)$$

We wish to show that G contains a matching which saturates every vertex in X (where G is bipartite with bipartition $V(G) = X \cup Y$). We suppose that G has no such matching and we will obtain a contradiction.

Let M^* be a maximum matching in G. Then by our assumption M^* does not saturate every vertex of X. Let x_0 be a vertex in X which is not saturated by M^*. Since M^* is maximum, by Berge's Theorem (Theorem 4.2) there is no M^*-augmenting path in G. Thus if P is an M^*-alternating path in G which starts at the M^*-unsaturated vertex x_0 it must finish at an M^*-saturated vertex. Thus P is of the form

$$x_0\, y_1\, x_1\, y_2\, x_2 \dots y_n\, x_n \tag{4.5}$$

$$\text{or} \quad x_0\, y_1\, x_1\, y_2\, x_2 \dots x_{n-1}\, y_n \tag{4.6}$$

where the edges $y_1x_1, y_2x_2, \dots$ are in M^*, the other edges, namely $x_0y_1, x_1y_2, \dots$ are not in M^*, and, in (4.6), the vertex y_n is M^*-saturated. Note further that since G is bipartite we have $y_1, \dots, y_n$ all in Y while $x_1, \dots, x_n$ are all in X. Moreover, in (4.5), since $x_{n-1}y_n$ is an edge not in M^* but y_n is M^*-saturated we can extend the path to become one of type (4.6) by adding on the edge y_nx_n where x_n is the vertex matched to y_n under M^*.

Now let S' be the set of all those vertices in X which can be reached from x_0 on an M^*-alternating path. (We exclude x_0 itself from S', i.e., we do not consider the path of length 0.) Thus S' consists of all the possible $x_1, \dots, x_n$ that can occur on paths P as described above. Let T denote the set of all those vertices in Y which can be reached from x_0 on an M^*-alternating path, i.e., T consists of all the possible $y_1, \dots, y_n$ that can occur on paths P as described above. Then, since, as we have seen above, each of these possible $x_1, \dots, x_n$ is matched up with a corresponding y_i, we get

$$|S'| = |T|.$$

We illustrate the proof so far with the bipartite graph G of Figure 4.11 with maximum matching M^* as shown, for the vertex x_0 we have $S' = \{a_1, a_2, a_3, a_4\}$ and $T = \{b_1, b_2, b_3, b_4\}$ (so that $|S'| = 4 = |T|$) because the M^*-alternating paths starting at x_0 are

x_0b_2, x_0b_3	(length 1),
$x_0b_2a_2, x_0b_3a_3$	(length 2),
$x_0b_2a_2b_1, x_0b_2a_2b_3$	(length 3),
$x_0b_2a_2b_1a_1, x_0b_2a_2b_3a_3$	(length 4),
$x_0b_2a_2b_1a_1b_4$	(length 5),
$x_0b_2a_2b_1a_1b_4a_4$	(length 6).

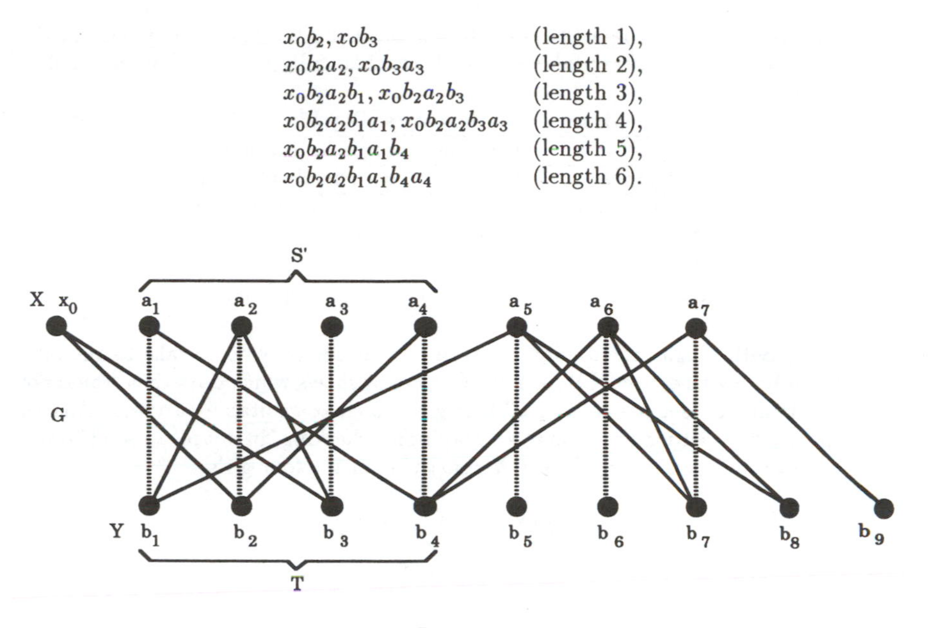

Figure 4.11

Continuing with the proof, we let $S = S' \cup \{x_0\}$. Now if a vertex y is adjacent to a vertex belonging to S we have either

(α) y adjacent to x_0 or

(β) y adjacent to a vertex x_i in X which occurs on an M^*-alternating path starting at x_0.

In case (α), $y \in T$ since y is the terminal vertex of an M^*-alternating path of length 1 (which in fact can be extended to an M^*-alternating path of length 2). In case (β) if P is an M^*-alternating path from x_0 to the vertex x_i then either y is already a vertex on this path or the path can be extended to a longer M^*-alternating path P' by adding on the edge x_iy. Thus again y is in T. This proves that $N(S) \subseteq T$. Moreover, from the definition of T, clearly $T \subseteq N(S)$. Thus $N(S) = T$. However, since $S = S' \cup \{x_0\}$, we have $|S| = |S'| + 1 =$ (from above) $|T| + 1 = |N(S)| + 1$ and so $|N(S)| < |S|$. This contradicts (*). The contradiction shows that G must have a matching which saturates every vertex of X after all. □

Exercises for Section 4.2

4.2.1 Silly Cone Valley High School has vacancies for seven teachers, one for each of the subjects Chemistry, Engish, French, Geography, History, Mathematics and Physics. There are seven applicants for the vacancies and all are qualified to teach more than one subject. The applicants and their subjects are listed in the table below. Draw a bipartite graph to represent this situation and determine the maximum number of (suitably qualified) teachers Silly Cone Valley can employ.

Ms Adventure	Mathematics, Physics
Ms Belief	Chemistry, English, Mathematics
Ms Chance	Chemistry, French, History, Physics
Miss Demeanour	English, French, History, Physics
Ms Entirely	Chemistry, Mathematics
Miss Fire	Mathematics, Physics
Ms Givings	English, Geography, History

4.2.2 Amy, Beth, Cathy and Debby are trying to decide on who to take to the Silly Cone Valley High School Hop. The table below shows which boys they would not mind taking. Is it possible for all four girls to choose suitable partners? Answer this by first setting up a bipartite graph and checking, in detail, to see if Hall's condition is satisfied by the set X consisting of the four girls.

Amy	Matt, Pete
Beth	Nick
Cathy	Pete, Rod, Steve
Debby	Rod, Steve

4.2.3 Prove that a nonempty bipartite graph G has a perfect matching if and only if $|S| \leq |N(S)|$ for all subsets S of $V(G)$. Show, by example, that the bipartite condition can not be omitted.

4.2.4 A **(vertex) covering** of a graph G with vertex set V is a subset K of V such that every edge of G is incident with at least one vertex in K. A covering K is called **minimum** if there is no covering K' of G with $|K'| < |K|$. Figure 4.12 illustrates a covering and a minimum covering in a bipartite graph G.

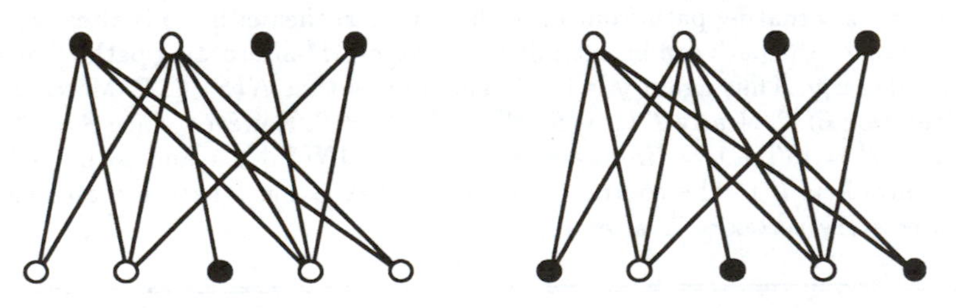

Figure 4.12: A covering and a minimum covering (shown by the white vertices).

Let G be a bipartite graph, M be a matching in G and K be a covering of G.

(a) Prove that $|M| \leq |K|$.

(b) Prove that if $|M| = |K|$ then M is a maximum matching and K is a minimum covering.

(c) Let G have bipartition $V = X \cup Y$ and let M be a maximum matching in G. Let U denote the set of all M-unsaturated vertices in X and let S and T be the sets of those vertices in X and Y respectively joined by an M-alternating path to a vertex in U. By adapting the second proof of Hall's Marriage Theorem, prove that $N(S) = T$.

(d) Keeping the same notation as in (c), prove that $K = (X - S) \cup T$ is a covering of G with $|K| = |M|$.

(e) Prove that the number of vertices in a minimum covering of G equals the number of edges in a maximum matching of G.

4.3 The Personnel Assignment Problem

In a high school, n teachers $x_1, \ldots, x_n$ are available for n classes $y_1, \ldots, y_n$, each teacher being qualified to teach one or more of these classes. Can all the teachers be assigned, one teacher per class, to classes for which they are qualified? This is called the **personnel assignment problem**.

We describe the problem in terms of a bipartite graph G. We define G to have bipartition $V(G) = X \cup Y$ where

$$\begin{aligned} X &= \{x_1, \ldots, x_n\} \quad \text{is the set of teachers,} \\ Y &= \{y_1, \ldots, y_n\} \quad \text{is the set of classes} \end{aligned}$$

and an edge joins teacher x_i to class y_j if and only if x_i is qualified to teach class y_j. The problem is then determining whether G has a matching which saturates each vertex of X.

We now present a method which, for any bipartite graph G with bipartition $V(G) = X \cup Y$, either finds a matching in G which saturates every vertex of X or, if no such matching exists, shows that there is no such matching by finding a subset S of X with $|N(S)| < |S|$. (If there is such a subset S then no such matching exists by Hall's Marriage Theorem.) Note that we do not assume that $|X| = |Y|$ — if $|X| = |Y|$, as in the high school example above, then any matching which saturates X will automatically saturate Y, because of the properties of a matching and a bipartite graph, and so such a matching will be a perfect matching.

The method is as follows:

A matching algorithm for bipartite graphs

Step 1. Begin with any matching M (e.g., any edge will do).

Step 2. Look for an M-unsaturated vertex x_0 in X. If there is none then M is a matching of the type we want.

If there is such a vertex x_0 we look for an M-augmenting path P with origin x_0. If such a path is found then we create a bigger matching M' by a transfer along the path. Since M' is bigger than M it saturates more vertices in X than M does. We then proceed to Step 3 below.

If no such path is found then the argument used in the second proof of Hall's Marriage Theorem produces a subset S of X with $|N(S)| < |S|$ and so G has no matching which saturates X.

Step 3. If all of the vertices of X are saturated by M' we stop since M' is then a matching of the desired type.

Otherwise we repeat Step 2 with M replaced by M'.

Of course, the problem now is how to find the augmenting paths or, if there are none, to show that none exist. The folowing definition will be helpful.

> Let M be a matching in the graph G (still bipartite as above) and let x_0 be an M-unsaturated vertex in X. Then a subgraph H of G is called an **M-alternating tree rooted at x_0** if
> (i) H is a tree,
> (ii) $x_0 \in V(H)$, the vertex set of H, and
> (iii) for every vertex v in the tree H the unique path in H from x_0 to v is an M-alternating path.

An example of a bipartite graph G with matching M and an M-alternating tree is given in Figure 4.13.

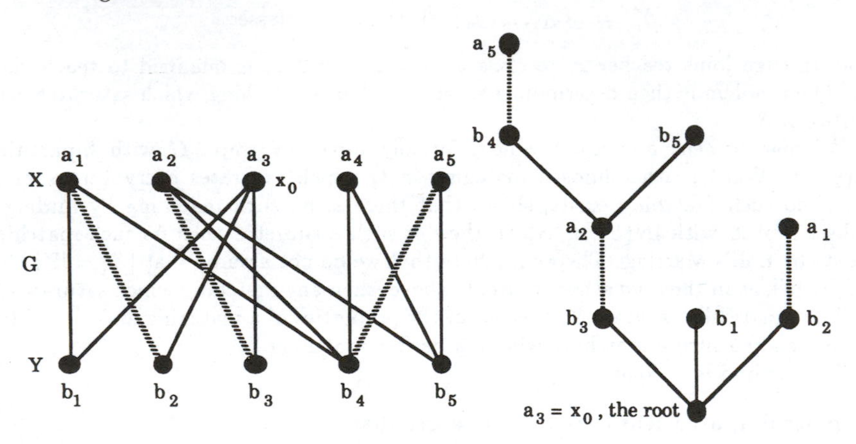

Figure 4.13: A matching M in a bipartite graph and an M-alternating tree.

We now show how to look for an augmenting path with origin x_0 by "growing" an M-alternating tree H rooted at x_0. The method is based on arguments used in a related problem by two Hungarian mathematicians, D. König [37] and J. Egerváry [19]. It is known simply as the **Hungarian method**.

The growth of the M-alternating tree H rooted at x_0 is such that, at any stage, either

(i) all vertices of H except x_0 are M-saturated and matched up in pairs under M or

(ii) H has an M-unsaturated vertex different from x_0.

Figure 4.14 shows examples of the two different cases:

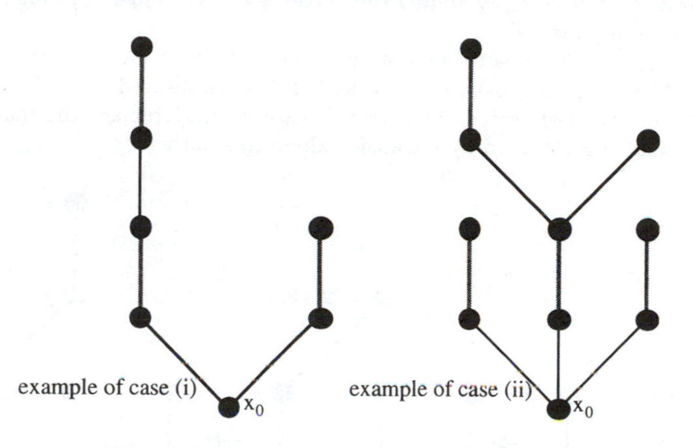

Figure 4.14: The two growth patterns of an alternating tree.

If case (i) holds, as it does at the very first stage in the growth when the tree consists just of the single vertex, the root x_0, then setting S' as those matched vertices in the tree H which are an even distance from x_0 and setting T as those matched vertices in H at an odd distance from x_0 we have

$$|T| = |S'|.$$

More pictorially we can describe T as consisting of those vertices in the tree at the bottoms of the dark (shaded) branches while S' consists of those at the tops of the dark branches. The definitions of S' and T are similar to those used in the second proof of Hall's Marriage Theorem. We now define (again see the second proof)

$$S = S' \cup \{x_0\}.$$

Then $T \subseteq N(S)$.

We now break (i) into the following two subcases:

$$\text{(a) } T = N(S), \quad \text{(b) } T \neq N(S).$$

If (a) holds then, just as in the second proof,

$$|N(S)| = |T| = |S'| = |S| - 1$$

so that

$$|N(S)| < |S|.$$

Thus, in this subcase, G has no matching which saturates X, because of Hall's Marriage Theorem.

If (b) holds then there is a vertex y in G, which is not in T, but which is adjacent to some x in S. Now, since we are dealing with case (i) still, either $x = x_0$ or x is in S' and so already matched up in the tree H under the matching M. In either case the edge xy cannot be in M. Now if y is saturated by M, say the edge yz is in M, then we grow on to the tree H by adding on the edges xy and yz and the new tree will still satisfy case (i). (See, for example, Figures 4.15 and 4.16.) If, on the other hand, y is not M-saturated we grow H by adding the vertex y and the edge xy, thus producing a new tree satisfying case (ii).

If at any stage of the growth we obtain a tree satisfying case (ii) then the tree contains an M-augmenting path starting at the M-unsaturated x_0 and finishing at the other M-unsaturated vertex. Thus in this case we may replace the matching M by a bigger matching obtained by a transfer along this path P.

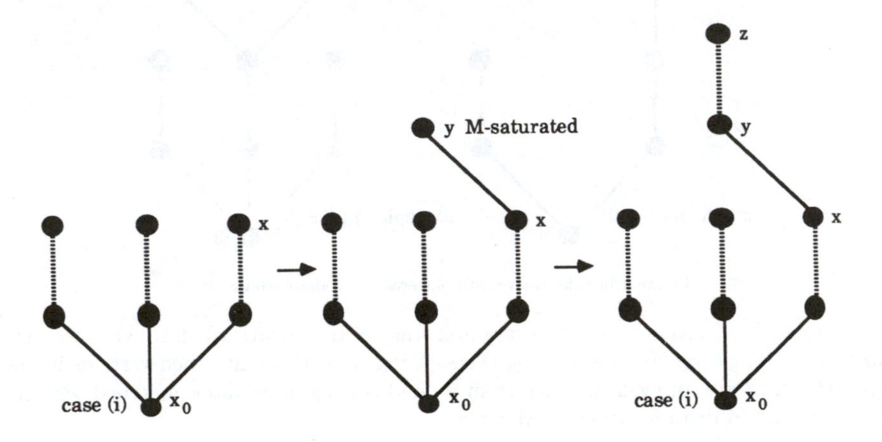

Figure 4.15: A growth of the alternating tree ending in case (i).

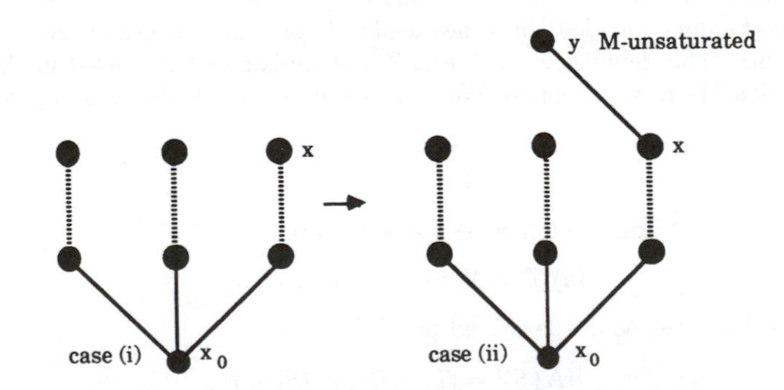

Figure 4.16: A growth of the alternating tree ending in case (ii).

Describing the Hungarian method in algorithmic fashion we now have:

The Hungarian Algorithm

Step 1. Start with an arbitrary matching M.

Step 2. If M saturates every vertex in X, stop. M is then a matching of the desired type.

Otherwise, let x_0 be an M-unsaturated vertex in X and set $S = \{x_0\}$ and $T = \emptyset$.

Step 3. If, in G, $N(S) = T$ then $|N(S)| < |S|$, since $|T| = |S| - 1$.

In this case stop since Hall's Marriage Theorem says that G has no matching which saturates every vertex in X.

Otherwise choose an element y in $N(S)$ which is not in T.

Step 4. If y is M-saturated, let $yz \in M$, i.e., z is the vertex matched to y under M. In this case replace S by $S \cup \{z\}$ and T by $T \cup \{y\}$ and return to Step 3. (Note that the new S and T still satisfy $|T| = |S| - 1$.)

Otherwise, since y is then M-unsaturated, let P be the M-augmenting path from x_0 to y and replace M by the transfer along P of M, denoting this new matching also by M. Then return to Step 2.

We illustrate the algorithm using the graph G of Figure 4.17.

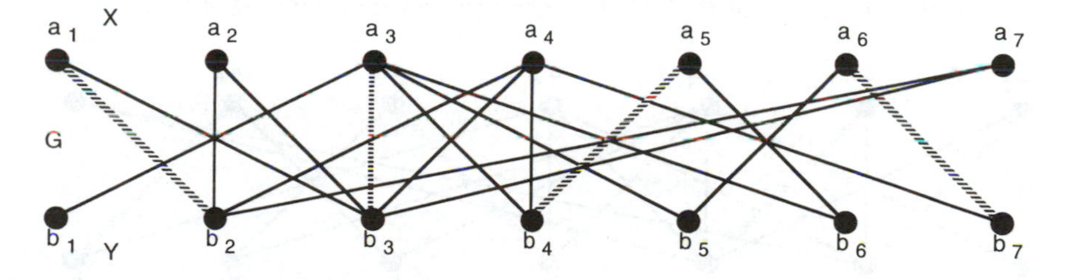

Figure 4.17: The matching M is given by the shaded edges.

Step 1. Start with the matching M given by the shaded edges.

Step 2. The vertex $x_0 = a_2$ is M-unsaturated in X.

Set $S = \{a_2\}$ and $T = \emptyset$.

The tree H at this stage consists simply of the root a_2. (See Figure 4.18 (a).)

Step 3. $N(S) = \{b_2, b_3\}$ so $N(S) \neq T$. Choose b_2 which is in $N(S)$ but not T.

Step 4. $y = b_2$ is M-saturated so the tree H becomes that of Figure 4.18 (b).

S becomes $\{a_1, a_2\}$ and T becomes $\{b_2\}$.

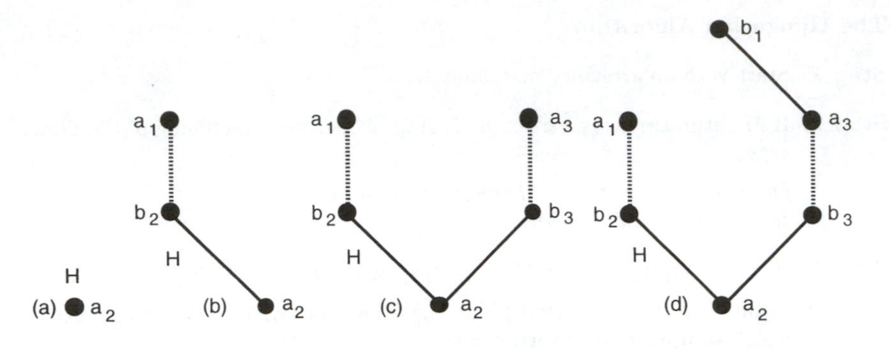

Figure 4.18: The growth of the M-alternating tree H.

Step3. $N(S) = \{b_2, b_3\} \neq T$. Choose $y = b_3$ which is in $N(S)$ but not in T.

Step 4. b_3 is M-saturated so the tree H becomes that of Figure 4.18 (c).

S becomes $\{a_1, a_2, a_3\}$ and T becomes $\{b_2, b_3\}$.

Step 3. $N(S) = \{b_1, b_2, b_3, b_4, b_5, b_6\} \neq T$. Choose $y = b_1$ which is in $N(S)$ but not T.

Step 4. b_1 is M-unsaturated and so the tree H becomes that of Figure 4.18 (d).

We let P be the M-augmenting path from a_2 to b_1, i.e., $a_2b_3a_3b_1$, and replace M by the transfer along P of M. This gives the new matching M of Figure 4.19.

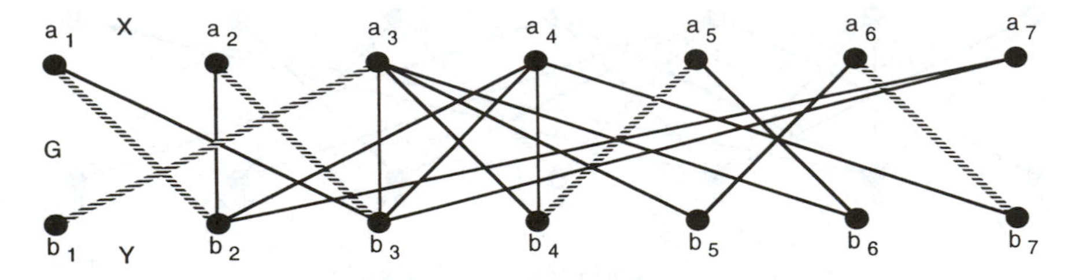

Figure 4.19: The new larger matching M.

We now return to step 2.

Step 2. The vertex $a_4 = x_0$ is M-unsaturated in X.

Set $S = \{a_4\}$ and $T = \emptyset$.

The tree H at this stage consists simply of the root a_4. (See Figure 4.20 (a).)

Step 3. $N(S) = \{b_2, b_3, b_4, b_7\} \neq T$. Choose $y = b_2$ which is in $N(S)$ but not in T.

Step 4. b_2 is M-saturated so the tree H becomes that of Figure 4.20 (b).

S becomes $\{a_1, a_4\}$ and T becomes $\{b_2\}$.

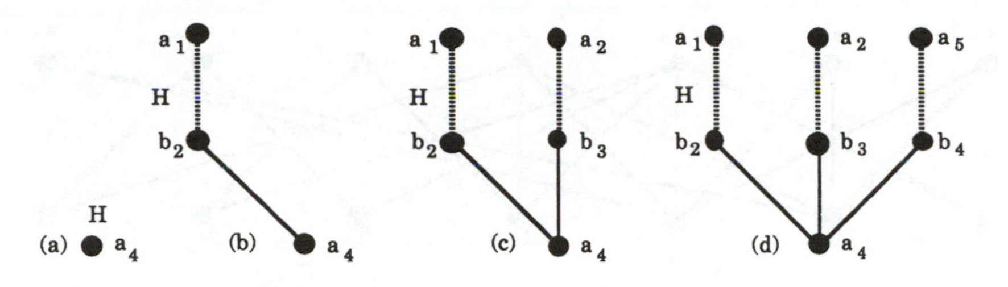

Figure 4.20: The growth of the M-alternating tree H.

Step 3. $N(S) = \{b_2, b_3, b_4, b_7\} \neq T$. Choose $y = b_3$ which is in $N(S)$ but not in T.

Step 4. b_3 is M-saturated so the tree H becomes that of Figure 4.20 (c).
S becomes $\{a_1, a_2, a_4\}$ and T becomes $\{b_2, b_3\}$.

Step 3. $N(S) = \{b_2, b_3, b_4, b_7\} \neq T$. Choose $y = b_4$ which is in $N(S)$ but not in T.

Step 4. b_4 is M-saturated so the tree H becomes that of Figure 4.20 (d).
S becomes $\{a_1, a_2, a_4, a_5\}$ and T becomes $\{b_2, b_3, b_4\}$.

Step 3. $N(S) = \{b_2, b_3, b_4, b_6, b_7\} \neq T$. Choose $y = b_6$ which is in $N(S)$ but not in T.

Step 4. b_6 is M-unsaturated so the tree H becomes that of Figure 4.21.

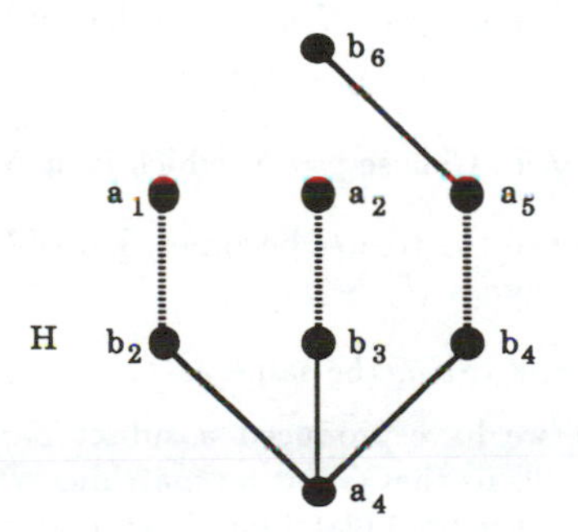

Figure 4.21: The final growth of the M-alternating tree H.

We let P be the M-augmenting path from a_4 to b_6, i.e., $a_4b_4a_5b_6$ and replace M by the transfer along P of M. This gives the new matching M of Figure 4.22.

We now return to step 2.

Step 2. The vertex a_7 is M-unsaturated in X.
Set $S = \{a_7\}$ and $T = \emptyset$.
The tree H at this stage consists simply of the root a_7. (See Figure 4.23 (a).)

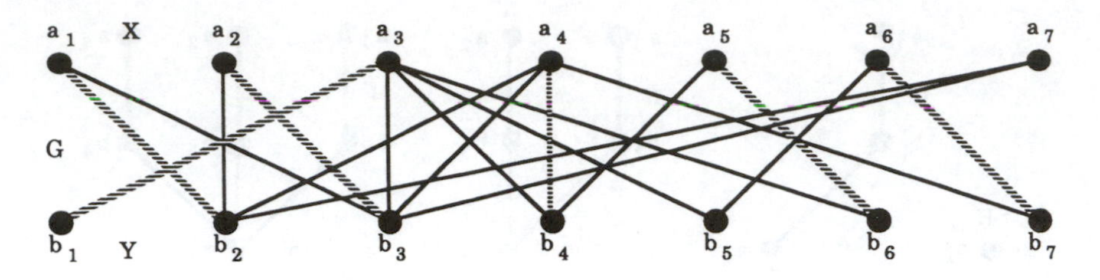

Figure 4.22: The new larger matching M.

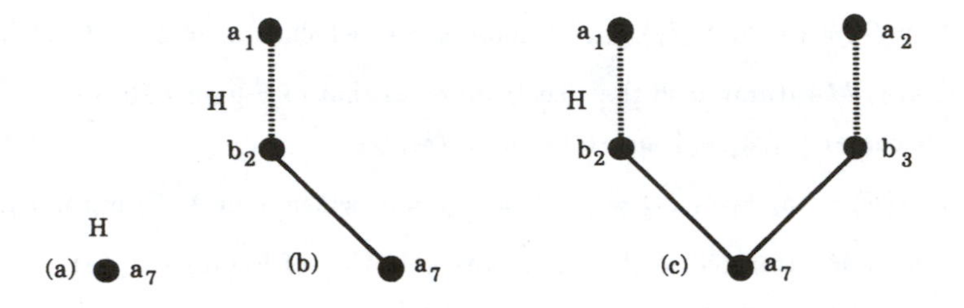

Figure 4.23: The growth of the M-alternating tree H.

Step 3. $N(S) = \{b_2, b_3\} \neq T$. Choose $y = b_2$ which is in $N(S)$ but not in T.

Step 4. b_2 is M-saturated so the tree H becomes that of Figure 4.23 (b). S becomes $\{a_1, a_7\}$ and T becomes $\{b_2\}$.

Step 3. $N(S) = \{b_2, b_3\} \neq T$. Choose $y = b_3$ which is in $N(S)$ but not in T.

Step 4. b_3 is M-saturated so the tree H becomes that of Figure 4.23 (c). S becomes $\{a_1, a_2, a_7\}$ and T becomes $\{b_2, b_3\}$.

Step 3. $N(S) = \{b_2, b_3\}$ and this *is* the same as T.

Thus we *stop* since we have produced a subset S with $|N(S)| < |S|$. Hall's Marriage Theorem tells us that G has no matching which saturates every vertex in X. Moreover, since the final matching M saturates all but one of the vertices of X this matching must be maximum.

Exercises for Section 4.3

4.3.1 Starting with a matching consisting of a *single* edge, perform the Hungarian method on the bipartite graphs of Figure 4.24 to determine whether or not the graphs have matchings which saturate every vertex of the bipartition subset X.

(Use the Hungarian method in a systematic way by considering the vertices in the subsets X and Y in the order given, i.e., $x_1, x_2, \ldots$ and $y_1, y_2, \ldots$, and detail the steps of the algorithm as given the in text.)

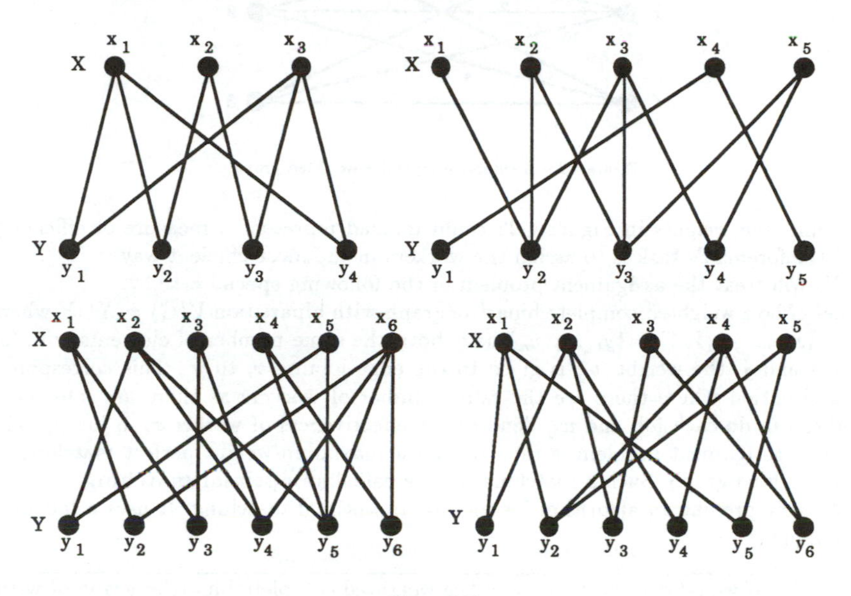

Figure 4.24

4.4 The Optimal Assignment Problem

In production planning a wide variety of assignments are routinely made, e.g., contracts are assigned to subcontractors, individual workers are assigned to tasks, work groups are assigned jobs, machine tools are assigned locations, goods are stored in assigned warehouse locations, etc.

Consider the following simple example involving the assignment of workers A, B, C to three tasks 1, 2, 3 by their foreman. The foreman, on checking his files, finds that each of the three workers has had earlier experience on all three tasks and he knows how long each has taken to complete each task. This gives the following complete bipartite weighted graph of Figure 4.25, the weights denoting the times taken:

Which man should the foreman assign to each task in order to keep the total length of time involved to a minimum? The problem of finding such an assignment is known as the *optimal assignment problem.* As in Bondy and Murty [7], we will actually look at the problem from the viewpoint of maximization, instead of minimization, so, for

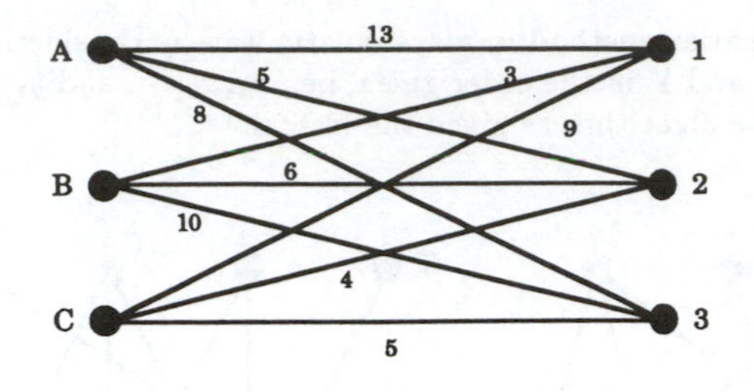

Figure 4.25: A complete bipartite weighted graph.

example, the weights in Figure 4.25 could instead represent a measure of efficiency and the foreman's task is to assign the workers in the most efficient way.

We will treat the assignment problem in the following special case.

Let G be a weighted complete bipartite graph with bipartition $V(G) = X \cup Y$ where $X = \{x_1, \ldots, x_n\}$, $Y = \{y_1, \ldots, y_n\}$ have both the same number of elements and, for each i and j, the weight w_{ij} is given to the edge joining x_i to y_j. This corresponds to a situation where there are the same number of workers as there are jobs, each worker can do each job and w_{ij} denotes the effectiveness of worker x_i in job y_j. The optimal assignment problem is then to find a maximum-weight perfect matching in this weighted graph. Such a matching will be called an **optimal matching**.

We now present an algorithm for finding an optimal matching. It uses a labelling technique.

> If we label each vertex v of the weighted complete bipartite graph G with a real number $\lambda(v)$, then this labelling is called a **feasible vertex labelling** of G if, for all x_i in X and all y_j in Y,
>
> $$\lambda(x_i) + \lambda(y_j) \geq w_{ij}, \text{ the weight of the edge } x_i y_j. \tag{4.7}$$

Thus a feasible vertex labelling is a labelling of the vertices of G such that the sum of the labels of the two ends of an edge is never less than the weight of the edge.

We can always produce a feasible vertex labelling λ for G by defining

$$\left.\begin{array}{rcll} \lambda(x_i) & = & \max\{w_{ij} : 1 \leq j \leq n\} & \text{if } x_i \in X \\ \lambda(y_j) & = & 0 & \text{if } y_j \in Y. \end{array}\right\} \tag{4.8}$$

In other words, we just assign the label 0 to all vertices in Y and, for any $x_i \in X$, $\lambda(x_i)$ is the maximum of all the weights of the edges incident with x_i. This labelling used for the graph of Figure 4.25 is shown in Figure 4.26.

Given a feasible vertex labelling λ for G we let E_λ denote the set of those edges for which *equality* holds in (4.7) above, i.e.,

$$E_\lambda = \{x_i y_j : \lambda(x_i) + \lambda(y_j) = w_{ij}\}$$

(where $x_i y_j$ denotes the unique edge from x_i to y_j).

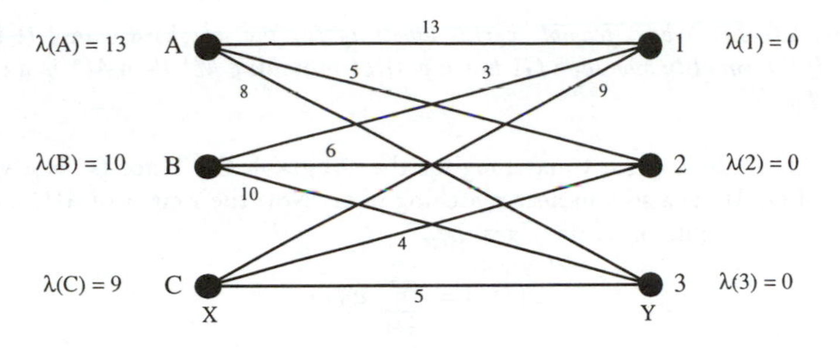

Figure 4.26: A feasible vertex labelling for the graph G of Figure 4.25.

Figure 4.27 shows the equality subgraph G_λ for the graph and feasible vertex labelling λ of Figure 4.26.

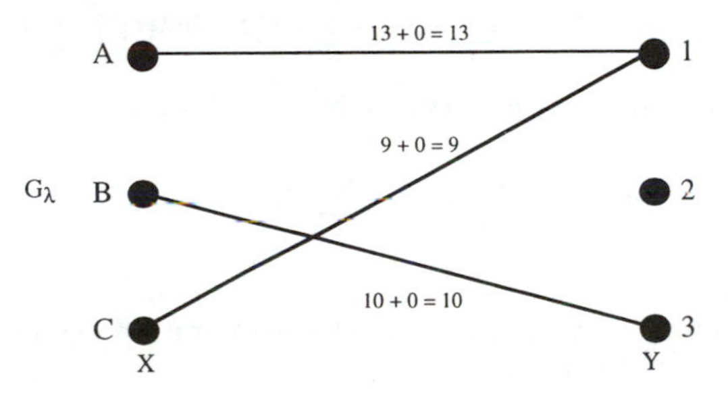

Figure 4.27: G_λ for the graph and feasible vertex labelling λ of Figure 4.26.

As a second example, the labelling μ given by $\mu(A) = 13$, $\mu(1) = 1$, $\mu(B) = 11$, $\mu(2)=0$, $\mu(C)=9$, $\mu(3)=0$ is also a feasible vertex labelling for the graph G of Figure 4.25 and since $\mu(x_i)+\mu(y_j) > w_{ij}$ for every vertex x_i and every vertex y_j, the equality subgraph G_μ corresponding to μ is shown in Figure 4.28.

Figure 4.28: A rather empty equality subgraph G_μ.

The importance of equality subgraphs is given by the following theorem.

Theorem 4.6 *Let λ be a feasible vertex labelling for the weighted complete bipartite graph G. If the equality subgraph G_λ has a perfect matching M^* then M^* is an optimal matching for G.*

Proof Let M^* be a perfect matching for the subgraph G_λ. Since G_λ is a spanning subgraph of G, M^* is also a perfect matching of G. Now the weight of M^*, $w(M^*)$, is the sum of the weights of all its edges, i.e.,

$$w(M^*) = \sum_{e \in M^*} w(e),$$

and since each edge e in M^* is in the equality subgraph G_λ we have, for such an e, $w(e) = \lambda(x) + \lambda(y)$ where x, y are the two end vertices of e. Moreover, since M^* is a perfect matching of G, its set of edges involves *all* the vertices of G (once and only once). Thus

$$w(M^*) = \sum_{e \in M^*} w(e) = \sum_{i=1}^{n} (\lambda(x_i) + \lambda(y_i)) \tag{4.9}$$

On the other hand, if M is *any* perfect matching of G, then

$$w(M) = \sum_{e \in M} w(e) \leq \sum_{i=1}^{n} (\lambda(x_i) + \lambda(y_i)) \tag{4.10}$$

(again since M is perfect it involves all the vertices of G once and only once but, for any edge e in M we can only say $w(e) \leq \lambda(x) + \lambda(y)$, instead of $w(e) = \lambda(x) + \lambda(y)$). Combining (4.9) and (4.10) we get

$$w(M^*) \geq w(M).$$

Since M was an arbitrary perfect matching of G, it follows that M^* is an optimal matching for G, as required. □

We use Theorem 4.6 to give an algorithm for finding an optimal matching in a weighted complete bipartite graph. The algorithm is due to Kuhn (1955) [40] and Munkres (1957) [47].

We start with an arbitrary feasible vertex labelling λ for G (for example, the one defined by (4.8)). We determine the equality subgraph G_λ corresponding to λ. Then, starting with an arbitrary matching M in G_λ, we apply the Hungarian method to G_λ to either find a perfect matching M^* for G_λ or to find a matching M' for G_λ that is not perfect and which cannot be enlarged upon using the Hungarian method.

In the first case, Theorem 4.6 tells us that M^* is an optimal matching for G so we are finished. In the second case, for the matching M' we will have found an M'-alternating tree H which contains no M'-augmenting path and which cannot be grown any more in G_λ. Up to this stage we have just used previously acquired techniques, but now we introduce a new technique to handle this second case. We modify λ to produce a new feasible vertex labelling λ' which has the property that its equality subgraph $G_{\lambda'}$ contains both M' and the M'-alternating tree H but with the extra feature that now H can be grown some more in $G_{\lambda'}$ (although it had achieved maximum growth in G_λ).

The modification uses the fact that the M'-alternating tree H, as described above, is reached in the Hungarian method when there is a subset S of X and a subset T of Y (as produced by the method) with

$$N_{G_\lambda}(S) = T.$$

(Here we have written $N_{G_\lambda}(S)$ rather than $N(S)$ to emphasise that the Hungarian method is performed on G_λ rather than on G.)

With this S and T we now compute

$$d_\lambda = \min\{\lambda(x) + \lambda(y) - w(xy) : x \in S, y \notin T\}. \tag{4.11}$$

We refer to d_λ as the **defect** of λ.

Using the defect we define the new feasible vertex labelling λ' as follows:
For a vertex v in G we let

$$\lambda'(v) = \begin{cases} \lambda(v) - d_\lambda & \text{if } v \in S \\ \lambda(v) + d_\lambda & \text{if } v \in T \\ \lambda(v) & \text{if } v \notin S \cup T. \end{cases} \tag{4.12}$$

Then the equality subgraph $G_{\lambda'}$ does contain M' and the M'-alternating tree H and H can be grown further in $G_{\lambda'}$, as claimed above, but we will not verify this here. (We also leave as an exercise the justification that λ' as defined is indeed a feasible vertex labelling.)

We now go back to the beginning and carry out the process on $G_{\lambda'}$ (instead of G_λ). Repeating the process we make modifications of the labelling whenever necessary until eventually a perfect matching is found in some equality subgraph. When this is found then, by Theorem 4.6, it is an optimal matching of G and so we are finished.

We now summarise the above process in algorithm form and illustrate it with an example.

The Kuhn-Munkres Algorithm

Step 1. Start with an arbitrary vertex labelling λ for G, determine the equality subgraph G_λ and choose an arbitrary matching M in G_λ.

Step 2. If X is M-saturated then M is a perfect matching in G_λ (since $|X| = |Y|$) and so, by Theorem 4.6, M is an optimal matching for G. In this case, stop.

Otherwise, let x_0 be an M-unsaturated vertex in G_λ.

Set $S = \{x_0\}$, $T = \emptyset$.

Step 3. If, in G_λ, $N_{G_\lambda}(S) \neq T$ go to step 4.

Otherwise, $N_{G_\lambda}(S) = T$. Compute the defect d_λ of λ as given by (4.11) and then compute the new feasible vertex labelling λ' according to (4.12) above. Replace λ by λ' and G_λ by $G_{\lambda'}$ and then rename $G_{\lambda'}$ as G_λ.

(At this stage, in the new G_λ we will have $N_{G_\lambda}(S) \neq T$ as at the first line in this step.)

Step 4. Choose a vertex y in $N_{G_\lambda}(S)$ which is not in T. As in the Hungarian method (see Section 4.3) consider whether or not y is M-saturated in G_λ. If y *is* M-saturated, with $yz \in M$, replace S by $S \cup \{z\}$ and T by $T \cup \{y\}$ and return to Step 3.

If y is M-unsaturated, let P be the M-augmenting path from x_0 to y and replace M by the transfer along P of M.

Then return to Step 2.

Example A machine shop possesses seven different drilling machines $y_1, \ldots, y_7$. On a certain day seven jobs $x_1, \ldots, x_7$ arrived that needed drilling. The profit obtained by performing each job x_i on each of the machines y_j is given in the following 7×7 matrix W, where the (i,j)th entry w_{ij} is the profit for the job x_i done on machine y_j.

$$W = \begin{bmatrix} 3 & 5 & 5 & 4 & 1 & 2 & 3 \\ 5 & 7 & 7 & 6 & 5 & 4 & 6 \\ 2 & 0 & 2 & 2 & 2 & 2 & 1 \\ 2 & 4 & 4 & 4 & 3 & 2 & 4 \\ 1 & 2 & 1 & 3 & 1 & 3 & 1 \\ 6 & 8 & 8 & 5 & 8 & 7 & 8 \\ 2 & 4 & 4 & 1 & 0 & 3 & 3 \end{bmatrix}$$

Since each job has to be done at the same time, the shop manager wishes to find the best way to assign each job to a different machine. Having obtained a pre-publication copy of this book, she immediately recognises that the problem is to find an optimal matching for the complete bipartite graph $K_{7,7}$ of Figure 4.29 whose edges are weighted according to the above matrix W.

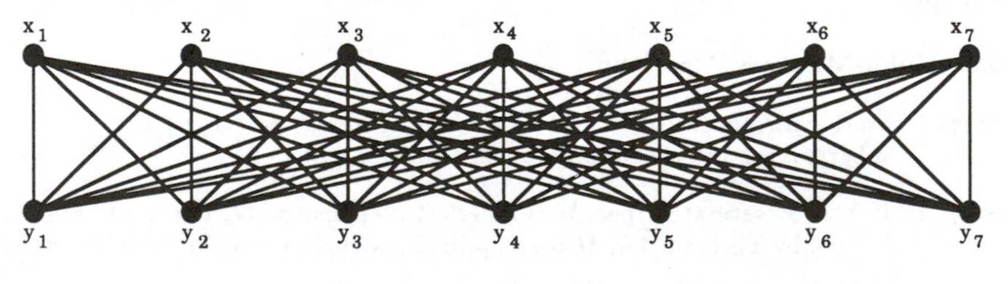

Figure 4.29: $K_{7,7}$.

The graph $G = K_{7,7}$ has 5040 different perfect matchings so she performs the Kuhn-Munkres algorithm on G.

Step 1. We use the feasible vertex labelling λ as defined by (4.8), namely

$$\begin{aligned} \lambda(y_j) &= 0 \text{ for all } j : 1 \leq j \leq 7 \text{ and} \\ \lambda(x_i) &= \max\{w_{ij} : 1 < j < 7\} \text{ for each } i : 1 < i < 7. \end{aligned}$$

In particular, $\lambda(x_i)$ is just the maximum value on the ith row of the matrix W. We write down the matrix W again, with the maximum elements in each row in bold type, recording the labels $\lambda(x_1), \ldots, \lambda(x_7)$ down the right-hand side of W and the labels $\lambda(y_1), \ldots, \lambda(y_7)$ along the bottom of W:

$$\begin{array}{ccc} & & \lambda(x_i) \\ W = & \left[\begin{array}{ccccccc} 3 & \mathbf{5} & \mathbf{5} & 4 & 1 & 2 & 3 \\ 5 & \mathbf{7} & \mathbf{7} & 6 & 5 & 4 & 6 \\ \mathbf{2} & 0 & \mathbf{2} & \mathbf{2} & \mathbf{2} & \mathbf{2} & 1 \\ 2 & \mathbf{4} & \mathbf{4} & \mathbf{4} & 3 & 2 & \mathbf{4} \\ 1 & 2 & 1 & \mathbf{3} & 1 & \mathbf{3} & 1 \\ 6 & \mathbf{8} & \mathbf{8} & 5 & \mathbf{8} & 7 & \mathbf{8} \\ 2 & \mathbf{4} & \mathbf{4} & 1 & 0 & 3 & 3 \end{array}\right] & \begin{array}{c} 5 \\ 7 \\ 2 \\ 4 \\ 3 \\ 8 \\ 4 \end{array} \\ \lambda(y_j) & \begin{array}{ccccccc} 0 & 0 & 0 & 0 & 0 & 0 & 0 \end{array} & \end{array}$$

Then the values i, j corresponding to the bold entries of W give precisely those pairs of vertices x_i, y_j for which

$$\lambda(x_i) + \lambda(y_j) = w_{ij}.$$

Thus the equality subgraph G_λ is the graph of Figure 4.30 which (surprise, surprise!) happens to be the graph of Figure 4.22 (but with $x_1, x_2, \ldots, y_1, y_2, \ldots$ instead of $a_1, a_2, \ldots, b_1, b_2, \ldots$):

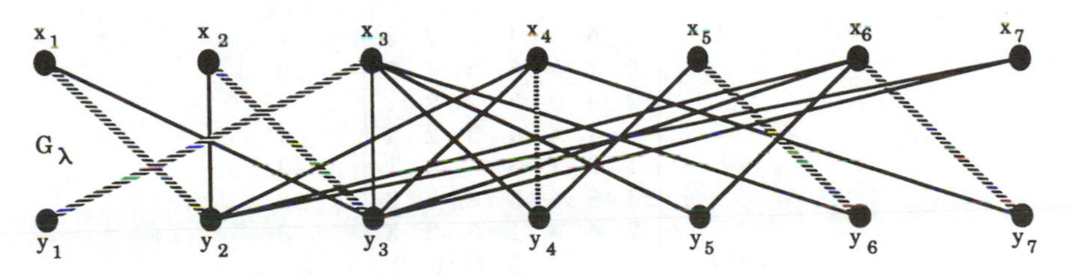

Figure 4.30: The equality subgraph G_λ with matching M.

We choose the matching M for G_λ as given by the shaded edges. Again, this corresponds to the matching shown in Figure 4.22.

Steps 2, 3, 4, 3, 4, 3: X is not M-saturated. Performing the Hungarian method on G_λ, we find the subset $S = \{x_1, x_2, x_7\}$ of X and the subset $T = \{y_2, y_3\}$ of Y such that

$$N_{G_\lambda}(S) = T.$$

Hence we now compute the defect d_λ of λ using (4.11):

$$
\begin{aligned}
d_\lambda &= \min\{\ \lambda(x)+\lambda(y)-w(xy) : x \in S, y \notin T\} \\
&= \min\{\ \lambda(x_1)+\lambda(y_1)-w(x_1y_1),\ \lambda(x_1)+\lambda(y_4)-w(x_1y_4), \\
&\qquad\quad \lambda(x_1)+\lambda(y_5)-w(x_1y_5),\ \lambda(x_1)+\lambda(y_6)-w(x_1y_6), \\
&\qquad\quad \lambda(x_1)+\lambda(y_7)-w(x_1y_7),\ \lambda(x_2)+\lambda(y_1)-w(x_2y_1), \\
&\qquad\quad \vdots \\
&\qquad\quad \lambda(x_2)+\lambda(y_7)-w(x_2y_7),\ \lambda(x_7)+\lambda(y_1)-w(x_7y_1), \\
&\qquad\quad \vdots \\
&\qquad\quad \lambda(x_7)+\lambda(y-7)-w(x_7y_7)\}, \\
&= \min\{\ 5+0-3, \underline{5+0-4}, \ldots, 4+0-3\} \\
&= \ 1.
\end{aligned}
$$

(Note: Using the definition of the defect d_λ, one can prove that d_λ is always strictly positive, i.e., $d_\lambda > 0$. In this example, d_λ is a difference of integer numbers and the smallest possible such difference which is strictly positive is, of course, 1. Thus, as soon as we reach 1 in the above calculation, we know that this is d_λ.)

We now compute the new feasible vertex labelling λ' using (4.12):

$$
\lambda'(v) = \begin{cases} \lambda(v) - d_\lambda & \text{if } v \in S = \{x_1, x_2, x_7\} \\ \lambda(v) + d_\lambda & \text{if } v \in T = \{y_2, y_3\} \\ \lambda(v) & \text{if } v \notin S \cup T. \end{cases}
$$

We write down the matrix W again, recording the new labels $\lambda'(x_1), \ldots, \lambda'(x_7)$ down the right-hand side of W and the new labels $\lambda'(y_1), \ldots, \lambda'(y_7)$ along the bottom of W. The bold entries of W correspond to the edges x_iy_j where $\lambda'(x_i) + \lambda'(y_j) = w(x_iy_j)$:

$$
\begin{array}{ccc}
 & & \lambda'(x_i) \\
W \ = & \begin{bmatrix}
3 & \mathbf{5} & \mathbf{5} & \mathbf{4} & 1 & 2 & 3 \\
5 & \mathbf{7} & \mathbf{7} & \mathbf{6} & 5 & 4 & \mathbf{6} \\
\mathbf{2} & 0 & 2 & \mathbf{2} & \mathbf{2} & \mathbf{2} & 1 \\
2 & 4 & 4 & \mathbf{4} & 3 & 2 & \mathbf{4} \\
1 & 2 & 1 & \mathbf{3} & 1 & \mathbf{3} & 1 \\
6 & 8 & 8 & 5 & \mathbf{8} & 7 & \mathbf{8} \\
2 & \mathbf{4} & \mathbf{4} & 1 & 0 & \mathbf{3} & \mathbf{3}
\end{bmatrix} & \begin{matrix} 4 \\ 6 \\ 2 \\ 4 \\ 3 \\ 8 \\ 3 \end{matrix} \\
\lambda'(y_j) & \begin{matrix} 0 & 1 & 1 & 0 & 0 & 0 & 0 \end{matrix} &
\end{array}
$$

Renaming λ' by λ and computing the new equality subgraph G_λ we get the graph of Figure 4.31 (where the edges correspond to the bold entries of W as shown above).

Note that, as claimed earlier, the new equality subgraph G_λ contains the matching M of the old equality subgraph and also the M-alternating tree, corresponding to the subset $S = \{x_1, x_2, x_7\}$ and $T = \{y_2, y_3\}$, shown in Figure 4.32 (a). However, in our new G_λ, $N_{G_\lambda}(S) = \{y_2, y_3, y_4, y_6, y_7\} \neq T$.

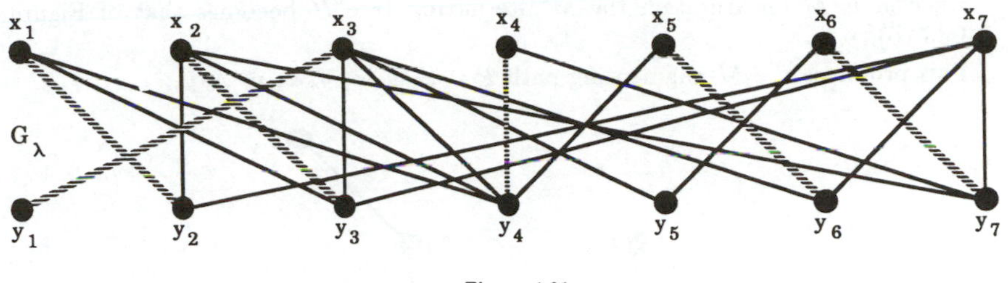

Figure 4.31

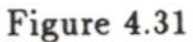

Step 4. Since $N_{G_\lambda}(S) \neq T$, choose $y = y_4$ which is in $N_{G_\lambda}(S)$ but not in T.
Since y_4 is M-saturated the alternating tree H becomes that of Figure 4.32 (b). S becomes $\{x_1, x_2, x_4, x_7\}$ and T becomes $\{y_2, y_3, y_4\}$.

Step 3. $N_{G_\lambda}(S) = \{y_2, y_3, y_4, y_6, y_7\} \neq T$.

Step 4. Choose $y = y_6$ which is in $N_{G_\lambda}(S)$ but not in T.
Since y_6 is M-saturated the alternating tree H becomes that of Figure 4.32 (c). S becomes $\{x_1, x_2, x_4, x_5, x_7\}$ and T becomes $\{y_2, y_3, y_4, y_6\}$.

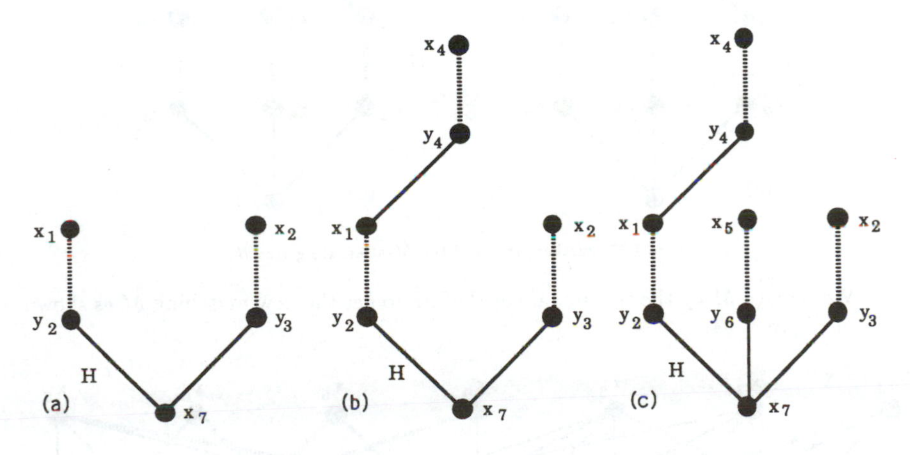

Figure 4.32: The growth of the M-alternating tree H.

Step 3. $N_{G_\lambda}(S) = \{y_2, y_3, y_4, y_6, y_7\} \neq T$.

Step 4. Choose $y = y_7$ which is in $N_{G_\lambda}(S)$ but not in T.
Since y_7 is M-saturated the alternating tree H becomes that of Figure 4.33 (a). S becomes $\{x_1, x_2, x_4, x_5, x_6, x_7\}$ and T becomes $\{y_2, y_3, y_4, y_6, y_7\}$.

Step 3. $N_{G_\lambda}(S) = \{y_2, y_3, y_4, y_5, y_6, y_7\} \neq T$.

Step 4. Choose $y = y_5$ which is in $N_{G_\lambda}(S)$ but not in T.

Since y_5 is M-unsaturated, the M-alternating tree H becomes that of Figure 4.33 (b).
This produces the M-augmenting path $P = x_7\ y_2\ x_1\ y_4\ x_4\ y_7\ x_6\ y_5$.

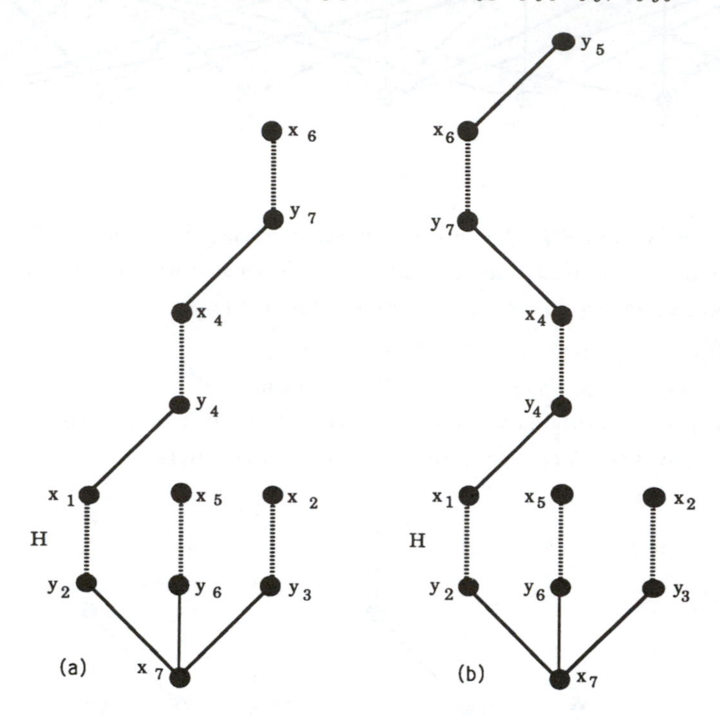

Figure 4.33: Further growth of the M-alternating tree H.

We replace M by the transfer along P of M to get the new matching M as shown in Figure 4.34.

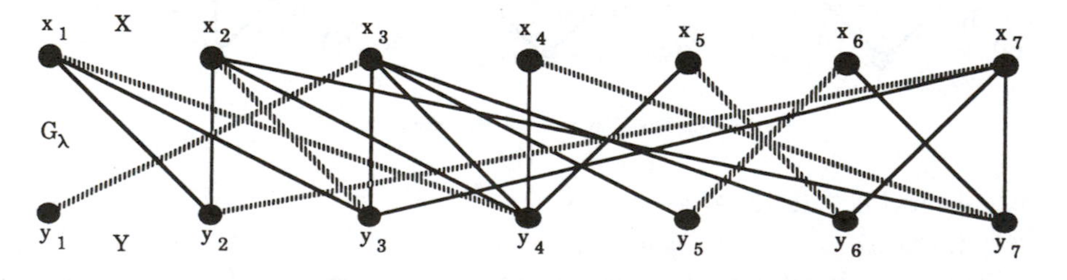

Figure 4.34: The new improved matching M.

Step 2. X is M-saturated by this new matching M. Thus M is a perfect matching of G_λ and so M is an optimal matching for G.

Hence the maximum profit is obtained by assigning the jobs to the machines as shown in the following table:

job	machine	profit
x_1	y_4	4
x_2	y_3	7
x_3	y_1	2
x_4	y_7	4
x_5	y_6	3
x_6	y_5	8
x_7	y_2	4
	Total profit = 32	

We now briefly consider what would happen if we had chosen a different initial vertex labelling in step 1. Let us suppose we had chosen the feasible vertex labelling λ given by $\lambda(y_j) = 0$ for all j, $1 \leq j \leq 7$, and $\lambda(x_i) = 100$ for all i, $1 \leq i \leq 7$. Then for all i, j,

$$\lambda(x_i) + \lambda(y_j) = 100 + 0 = 100 > w_{ij}.$$

Thus for this λ the equality subgraph is the spanning subgraph of G with every edge deleted, as shown in Figure 4.35.

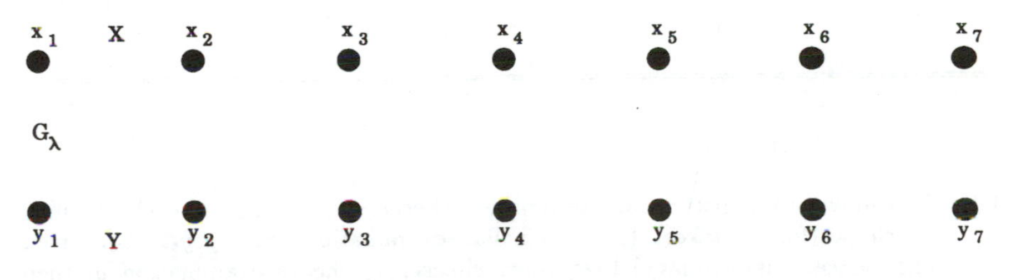

Figure 4.35: A rather boring equality subgraph.

The next step in the algorithm is to choose a matching M in G_λ. Since our G_λ has no edges our choice for M here must be $M = \emptyset$.
We next choose an M-unsaturated vertex x_0 in X. Here x_1 will do. Then we set $S = \{x_1\}$ and $T = \emptyset$ (just as in the Hungarian method). However $N_{G_\lambda}(S) = N_{G_\lambda}(\{x_1\}) = \emptyset = T$ and so, according to Step 3, we now compute the defect d_λ of λ, as given by

$$d_\lambda = \min\{\lambda(x) + \lambda(y) - w(xy) : x \in S, y \notin T\}.$$

This gives

$$\begin{aligned} d_\lambda &= \min\{\lambda(x_1) + \lambda(y_j) - w(x_1y_j) : 1 \leq j \leq 7\} \\ &= \min\{100 + 0 - w(x_1y_j) : 1 \leq j \leq 7\} \\ &= \min\{100-3, 100-5, 100-5, 100-4, 100-1, 100-2, 100-3\} \\ &= 95. \end{aligned}$$

Using (4.12), we now define the new feasible vertex labelling λ' by:

$$\begin{aligned}\lambda'(x_1) &= 100 - 95 = 5, \\ \lambda'(v) &= \lambda(v) \quad \text{for all other vertices } v.\end{aligned}$$

This new labelling does give a better-looking equality subgraph $G_{\lambda'}$, shown in Figure 4.36, since now there are edges present.

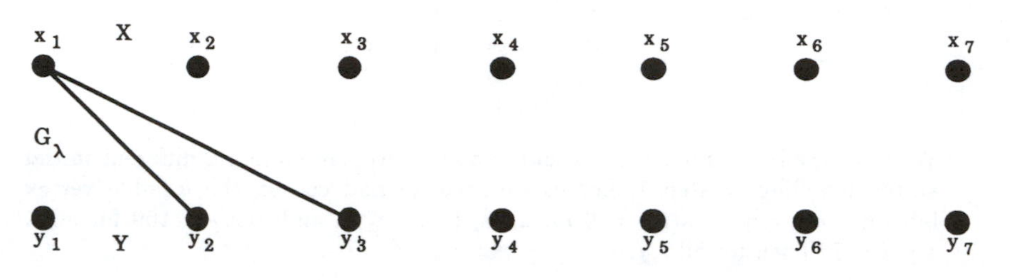

Figure 4.36: A better-looking equality subgraph.

Further repeated modification of the labelling introduces more edges and we eventually obtain a solution just as with our first choice of labelling given earlier.

Exercises for Section 4.4

4.4.1 In a university registry four examination checkers $x_1, \ldots, x_4$ are each qualified to check (for mistakes) four particular examinations $y_1, \ldots, y_4$. From past experience it is estimated that when checker x_i checks examination y_j then w_{ij} mistakes will be found, producing the following matrix

$$W = (w_{ij}) = \begin{bmatrix} 9 & 11 & 12 & 11 \\ 6 & 3 & 8 & 5 \\ 7 & 6 & 13 & 11 \\ 9 & 10 & 10 & 7 \end{bmatrix}.$$

Mr Greene, the Chief Supervisor of Examinations, wishes to allocate one examination per checker so as to find the greatest possible number of mistakes over the four examinations.

Starting with the feasible vertex labelling defined by the equations (4.8), use the Kuhn-Munkres algorithm to find such an allocation.

4.4.2 Repeat Exercise 4.4.1, now assuming firstly that there are only 3 checkers and 3 markers and then secondly that there are 5 checkers and 5 markers, with the corresponding matrices as follows.

$$W_1 = \begin{bmatrix} 7 & 6 & 11 \\ 9 & 4 & 8 \\ 6 & 7 & 5 \end{bmatrix}, \quad W_2 = \begin{bmatrix} 9 & 7 & 11 & 9 & 7 \\ 10 & 10 & 6 & 5 & 10 \\ 9 & 7 & 8 & 8 & 6 \\ 10 & 8 & 7 & 10 & 8 \\ 4 & 10 & 13 & 12 & 9 \end{bmatrix}.$$

4.4.3 Let λ' be the labelling calculated from a feasible vertex labelling λ using its defect $d_{\lambda'}$, as given by equation (4.12).

(a) Prove that λ' is a feasible vertex labelling.

(b) Prove that the equality subgraph $G_{\lambda'}$, contains the tree H used in calculating the defect and that H can be grown further in $G_{\lambda'}$.

4.5 A Chinese Postman Problem Postscript

In Section 3.2 we showed how to solve the Chinese postman problem when the corresponding graph was Euler or it had precisely two odd vertices. For a more general situation, where we have more than two odd vertices, we can compute the shortest distances between every pair of odd vertices (for example, by using Dijkstra's algorithm). This then gives a complete weighted graph where the vertices are the original odd vertices and the weights represent the shortest distances.

If we now find a perfect matching of minimal weight for this graph (note that there is an even number of vertices in this weighted graph so there do exist perfect matchings), then each edge in this perfect matching can be used to build an Euler supergraph of the original graph. An application of Fleury's algorithm on this supergraph then produces a Chinese postman tour for the graph just as in the case for two odd vertices.

Exercise for Section 4.5

4.5.1 Solve the Chinese Postman Problem for the graph of Figure 3.18 by constructing a complete weighted graph on the four odd vertices, with weights representing the shortest distances, and then finding a perfect matching for this weighted graph. What is the resulting postman's tour?

Chapter 5

Planar Graphs

5.1 Plane and Planar Graphs

> A **plane graph** is a graph drawn in the plane (of the paper, blackboard, etc.) in such a way that any pair of edges meet only at their end vertices (if they meet at all).
>
> A **planar graph** is a graph which is isomorphic to a plane graph, i.e., it can be (re)drawn as a plane graph.

For example, in Figure 5.1 all five graphs are planar but G_1 and G_4 are not plane graphs. (In fact G_2 and G_3 are redrawings of G_1 while G_5 is a redrawing of G_4.)

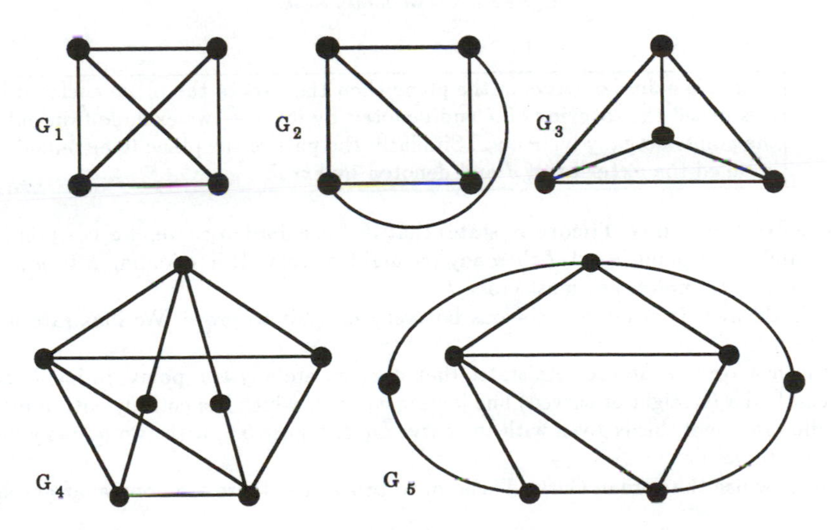

Figure 5.1: Five planar graphs.

Not all graphs are planar. To see this we need to speak first about a major theorem in mathematics.

A **Jordan curve** in the plane is a continuous non-self-intersecting curve whose origin and terminus coincide.

For example, in Figure 5.2 the curve C_1 is not a Jordan curve because it intersects itself, C_2 is not a Jordan curve since its origin and terminus do not coincide, i.e., its two end points do not meet up, but C_3 is a Jordan curve. Another Jordan curve is shown in Figure 5.3.

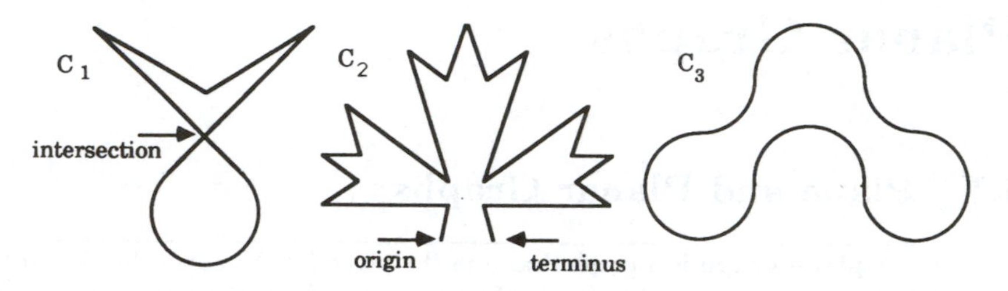

Figure 5.2: C_1 and C_2 are not Jordan curves but C_3 is.

Figure 5.3: A brave Jordan curve.

If J is a Jordan curve in the plane then the part of the plane enclosed by J is called the **interior** of J and denoted by int J — we exclude from int J the points actually lying on J. Similarly the part of the plane lying outside J is called the **exterior** of J and denoted by ext J.

The **Jordan Curve Theorem** states that if J is a Jordan curve, if x is a point in int J and y is a point in ext J then any (straight or curved) line joining x to y must meet J at some point, i.e., must cross J.

This theorem is intuitively obvious but very difficult to prove. We illustrate it in Figure 5.4.

Another form of the theorem states that if x_1, x_2 are any two points in int J then we can find a (straight or curved) line joining x_1 to x_2 which lies entirely within int J. An illustration of this is given with the curve C_4 of Figure 5.3, with two points joined by an internal line.

We now use the Jordan Curve Theorem to prove that there are nonplanar graphs.

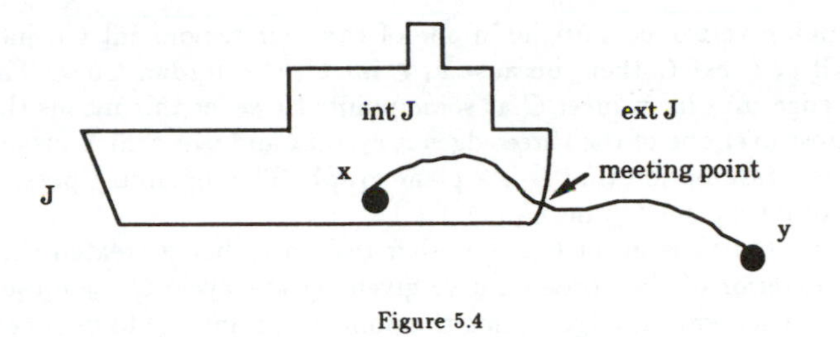

Figure 5.4

Theorem 5.1 *K_5, the complete graph on five vertices, is nonplanar.*

Proof Recall that one of the usual ways of drawing K_5 is as shown in Figure 5.5.

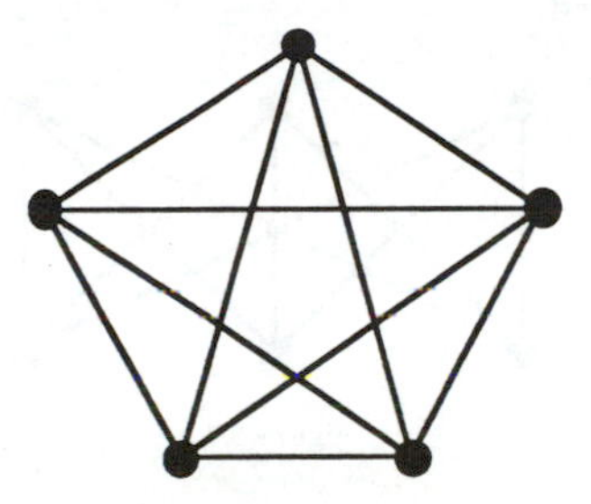

Figure 5.5: K_5.

We assume that K_5 is planar and derive a contradiction from this assumption. Let G be a plane graph corresponding to K_5 and denote the vertices of G by v_1, v_2, v_3, v_4, v_5. Since G is complete, any pair of distinct vertices are joined by an edge. Let C be the cycle $v_1v_2v_3v_1$ in G. Then C forms a Jordan curve in the plane. Since v_4 does not lie on C it must either lie in int C or in ext C. Let us suppose that v_4 is in int C. (The other possibility, that v_4 is in ext C, has a similar argument.) Then the edges v_4v_1, v_4v_2 and v_4v_3 divide int C into three regions int C_1, int C_2 and int C_3 where C_1, C_2, and C_3 are the cycles $v_1v_2v_4v_1$, $v_2v_3v_4v_2$, and $v_1v_3v_4v_1$ respectively. (See Figure 5.6.)

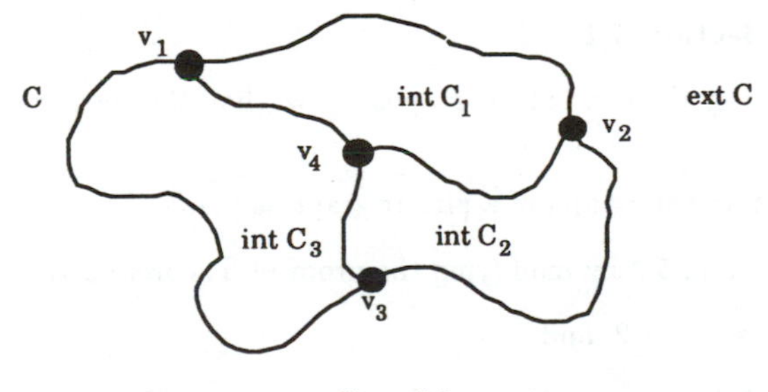

Figure 5.6

The remaining vertex v_5 must lie in one of the four regions int C_1, int C_2, int C_3 and ext C. If $v_5 \in$ ext C then, because $v_4 \in$ int C, the Jordan Curve Theorem tells us that the edge v_4v_5 must meet C at some point. However this means that the edge v_4v_5 must cross over one of the three edges v_1v_2, v_2v_3 and v_3v_1 which make up C. This contradicts our assumption that G is a plane graph. The remaining possibility is that v_5 is in one of int C_1, int C_2, int C_3.

We suppose that v_5 is in int C_1, the other two cases being treated similarly. Now v_3 is in the exterior of the Jordan Curve given by the cycle $C_1 = v_1v_2v_4v_1$. By the Jordan Curve Theorem the edge joining the point v_5 (in int C_1) to v_3 (in ext C_1) must cross the curve C_1 and so must cross one of the three edges v_1v_2, v_2v_4, v_4v_1, again contradicting the assumption that G is plane. This final contradiction shows that our initial assumption must be false. Hence K_5 is not planar. □

Recall that the usual way of drawing $K_{3,3}$ is as shown in Figure 5.7. It too is nonplanar as we now record.

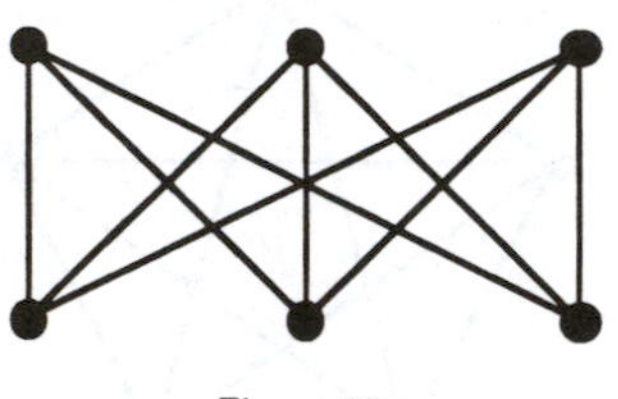

Figure 5.7

Theorem 5.2 *The complete bipartite graph $K_{3,3}$ is nonplanar.*

Proof The proof is similar to that of the previous theorem and is left as Exercise 5.1.3. □

We will give different proofs for these last two theorems in the next section.

Note that, since (clearly) any supergraph of a nonplanar graph is also nonplanar, these two theorems guarantee a plentiful supply of nonplanar graphs.

Exercises for Section 5.1

5.1.1 By redrawing them as plane graphs show that the graphs of Figure 5.8 are planar.

5.1.2 Prove that any subgraph of a planar graph is planar.

5.1.3 Prove Theorem 5.2 by modifying the proof of Theorem 5.1.

5.1.4 Using Exercise 5.1.2, find

(a) all $n \geq 2$ for which the complete graph K_n is planar,

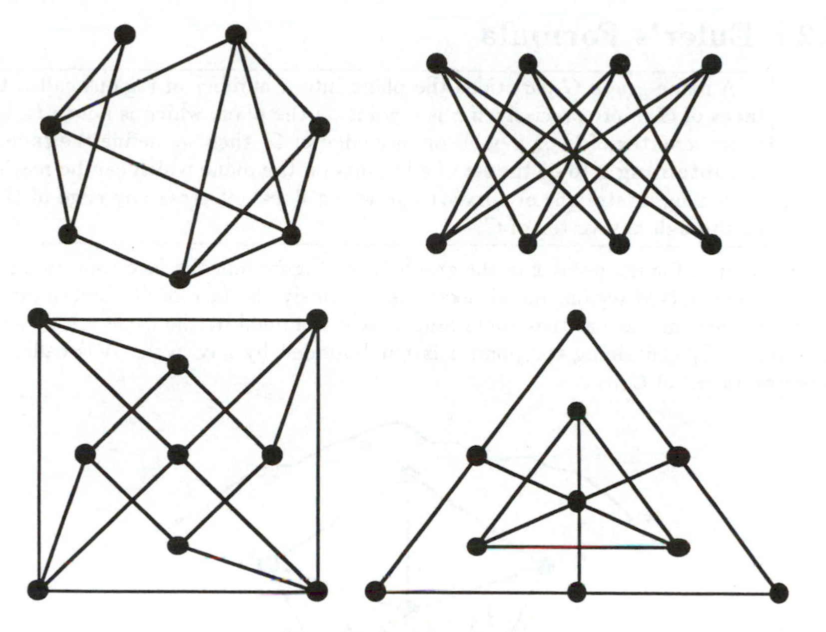

Figure 5.8: Four planar graphs.

(b) all $m, n \geq 1$ for which the complete bipartite graph $K_{m,n}$ is planar.

5.1.5 (a) Show that if e is any edge of K_5 then $K_5 - e$ is planar.

(b) Show that if e is any edge of $K_{3,3}$ then $K_{3,3} - e$ is planar.

5.1.6 A nonplanar graph G is called **critical planar** if $G - v$ is planar for every vertex v of G.

(a) Which complete graphs K_n are critical planar?

(b) Which complete bipartite graphs $K_{m,n}$ are critical planar?

(c) Prove that any critical planar graph must be connected and have no cut vertex.

5.2 Euler's Formula

A plane graph G partitions the plane into a number of regions called the **faces** of G. More precisely, if x is a point on the plane which is not in G, i.e., is not a vertex of G or a point on any edge of G, then we define the **face of** G **containing** x to be the set of all points on the plane which can be reached from x by a (straight or curved) line which does not cross any edge of G or go through any vertex of G.

For example, for the point x in the graph G_1 of Figure 5.9, the face containing x is shown as the dotted region. In this example obviously the face of G_1 containing the point y is the same face as that containing x. It is bounded by the cycle $v_2v_4v_3v_6v_5v_4$. The face of G_1 containing the point z is not bounded by any cycle. It is called the **exterior face**. of G_1.

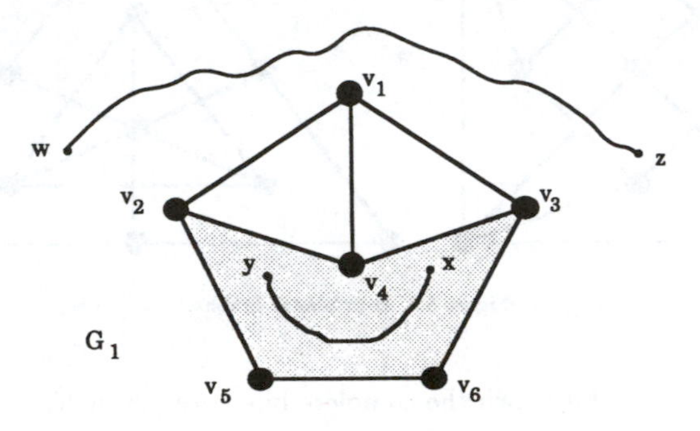

Figure 5.9: A plane graph with four faces.

Any plane graph has exactly one exterior face. Any other face is bounded by a closed walk in the graph and is called an **interior face**.

As another example, in Figure 5.10 we have a graph G_2 with nine faces $f_1, \ldots, f_9$. Here f_6 is the exterior face.

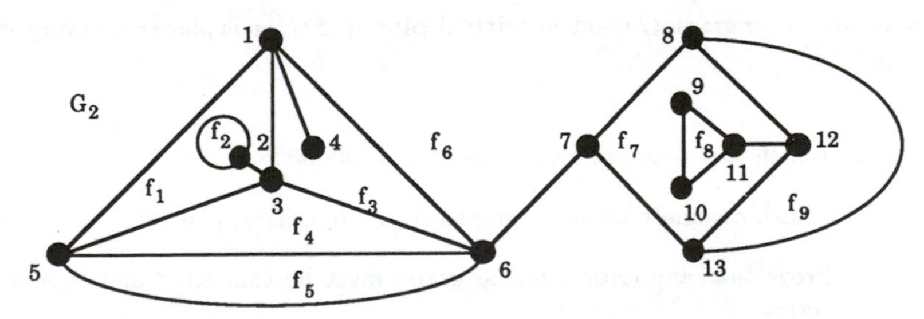

Figure 5.10: A plane graph with nine faces.

The number of faces of a plane graph G is denoted by $f(G)$ or just simply by f. Thus, for the above, $f(G_1) = 4, f(G_2) = 9$.

In our next result we give a simple formula showing the relationship between the number of vertices, edges and faces in a connected plane graph. It is one of the best-known formulae in mathematics and was proved by Euler [22] in 1752.

Theorem 5.3 (Euler's Formula) *Let G be a connected plane graph, and let n, e, and f denote the number of vertices, edges and faces of G, respectively. Then*

$$n - e + f = 2.$$

Proof We will give two proofs.

(First proof.) In this proof we use induction on f, the number of faces of G.

If $f = 1$ then G has only one face, the exterior face. If G contains any cycle C then, in the region of the plane bounded by C, there is at least one bounded face of G, impossible since G has only the exterior face, which is unbounded. Thus G has no cycles. Hence, since G is connected, it is a tree. Then, by Theorem 2.4, the number of edges e of G is $n - 1$. Hence

$$n - e + f = n - (n - 1) + 1 = 2$$

and this proves the theorem in the case when $f = 1$.

Now suppose that $f > 1$ and that the theorem is true for all connected plane graphs with less than f faces. Since $f > 1$, G is not a tree and so, by Theorem 2.8, G has an edge k which is not a bridge. Then the subgraph $G - k$ is still connected and, since any subgraph of a plane graph is clearly a plane graph, $G - k$ is also a plane graph. Moreover, since the edge k must be part of a cycle (see Theorem 2.7), it separates two faces of G from each other and so in $G - k$ these two faces combine to form one face of $G - k$. We illustrate this in Figure 5.11.

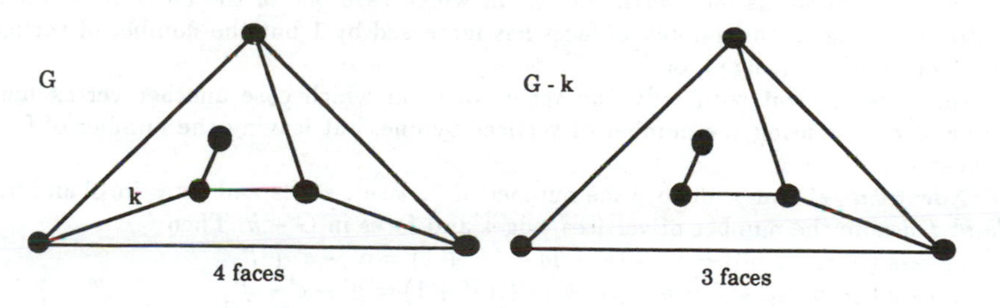

Figure 5.11: Two faces combine when a cycle edge is deleted.

Thus, letting $n(G - k)$, $e(G - k)$ and $f(G - k)$ denote the number of vertices, edges and faces respectively of $G - k$, we have $n(G - k) = n$, $e(G - k) = e - 1$ and $f(G - k) = f - 1$. Moreover, by our induction assumption, since $G - k$ has less than f faces we have

$$n(G - k) - e(G - k) + f(G - k) = 2$$

and so $n - (e - 1) + (f - 1) = 2$ which gives $n - e + f = 2$, as required. Hence, by induction, the result is true for all connected plane graphs.

(Second proof.) This time we use induction on e, the number of edges of G.

If $e = 0$ then G must have just one vertex, i.e., $n = 1$, and one face, the exterior face, i.e., $f = 1$. Thus

$$n - e + f = 1 - 0 + 1 = 2$$

and so the result is true for $e = 0$.

Although it is not necessary to do this, let us now look at the case when $e = 1$. Then the number of vertices of G is either 1 or 2, the first possibility occurring when the edge is a loop. These two possibilities give rise to two faces and one face respectively, as shown in Figure 5.12.

Figure 5.12: The connected plane graphs with one edge.

Thus

$$n - e + f = \left\{ \begin{array}{ll} 1 - 1 + 2, & \text{in the loop case} \\ 2 - 1 + 1, & \text{in the non-loop case} \end{array} \right\} = 2, \text{ as required.}$$

Now suppose that the result is true for any connected plane graph G with $e-1$ edges (for a fixed $e \geq 1$). Let us add one new edge k to G to form a connected supergraph of G which we denote by $G + k$. There are three ways of doing this:

(i) k is a loop, in which case we have created a new face (bounded by the loop), but the number of vertices remains unchanged, or

(ii) k joins two distinct vertices of G, in which case one of the faces of G is split into two, so again the number of faces has increased by 1 but the number of vertices has remained unchanged, or

(iii) k is incident with only one vertex of G in which case another vertex must be added, increasing the number of vertices by one, but leaving the number of faces unchanged.

Now let n', e' and f' denote the number of vertices, edges and faces in G and n, e and f denote the number of vertices, edges and faces in $G + k$. Then

in case (i), $n - e + f = n' - (e' + 1) + (f' + 1) = n' - e' + f'$,

in case (ii), $n - e + f = n' - (e' + 1) + (f' + 1) = n' - e' + f'$,

in case (iii), $n - e + f = (n' + 1) - (e' + 1) + f' = n' - e' + f'$

and by our induction assumption, $n' - e' + f' = 2$. Thus, in each case, $n - e + f = 2$. Now any plane connected graph with e edges is of the form $G + k$, for some plane connected graph G with $e - 1$ edges and a new edge k. Thus it follows by induction that the formula is true for all plane graphs. □

Corollary 5.4 *Let G be a plane graph with n vertices, e edges, f faces and k connected components. Then*

$$n - e + f = k + 1.$$

Proof We leave this to the reader as Exercise 5.2.8. □

Corollary 5.5 *Let G_1 and G_2 be two plane graphs which are both redrawings of the same planar graph G. Then $f(G_1) = f(G_2)$, i.e., G_1 and G_2 have the same number of faces.*

Proof Let $n(G_1), n(G_2)$ denote the number of vertices and $e(G_1), e(G_2)$ the number of edges in G_1, G_2 respectively. Then, since G_1 and G_2 are both isomorphic to G we have $n(G_1) = n(G_2)$ and $e(G_1) = e(G_2)$. Using Euler's Formula we get

$$f(G_1) = e(G_1) - n(G_1) + 2 = e(G_2) - n(G_2) + 2 = f(G_2),$$

as required. □

The next theorem tells us that a simple planar graph can not have "too many" edges. In the proof we use the following definition.

> Let φ be a face of a plane graph G. We define the **degree** of φ, denoted by $d(\varphi)$, to be the number of edges on the boundary of φ.

Note that $d(\varphi) \geq 3$ for any interior face φ of a *simple* plane graph.

Theorem 5.6 *Let G be a simple planar graph with n vertices and e edges, where $n \geq 3$. Then*

$$e \leq 3n - 6.$$

Proof By redrawing G if necessary, we can assume that G is a plane graph (as distinct from planar). We first suppose that G is connected.

If $n = 3$, i.e., we have three vertices, then, since G is simple, G can have at most three edges, i.e., $e \leq 3$. Thus

$$e \leq (3 \times 3) - 6 = 3n - 6,$$

and so the result is true in this case.

Thus we can now assume that $n \geq 4$. If G is a tree then $e = n - 1$ and so, since $n \geq 4$, we get $e \leq 3n - 6$, as required. If G is not a tree then, since it is connected, it must contain a cycle. It follows that there is a cycle in G all of whose edges lie on the boundary of the exterior face of G. Then, since G is simple, we have $d(\varphi) \geq 3$ for each face φ of G. Let

$$b = \sum_{\varphi \in \Phi} d(\varphi),$$

where Φ denotes the set of all faces of G. Then, since each face has at least three edges on its boundary, we have

$$b \geq 3f$$

(where f is the number of faces of G). However, when we were summing up to get b, each edge of G was counted either once or twice (twice when it occurred as a boundary edge for two faces) and so

$$b \leq 2e.$$

Thus

$$3f \leq b \leq 2e.$$

In particular $3f \leq 2e$ and so $-f \geq -2e/3$. Now, by Euler's theorem, $n = e - f + 2$ and so

$$n \geq e - \frac{2e}{3} + 2 = \frac{e}{3} + 2.$$

Thus $3n \geq e + 6$, i.e., $e \leq 3n - 6$, as required.

Now suppose that G is not connected. Let $G_1, \ldots, G_t$ be its connected components and for each $i, 1 \leq i \leq t$, let n_i and e_i denote the number of vertices and edges respectively in G_i. Then, since each G_i is a planar simple graph, we have, from the above argument, that $e_i \leq 3n_i - 6$ for each $i, 1 \leq i \leq t$. Moreover

$$n = \sum_{i=1}^{t} n_i \text{ and } e = \sum_{i=1}^{t} e_i \text{ and so}$$

$$e \leq \sum_{i=1}^{t}(3n_i - 6) = 3\sum_{i=1}^{t} n_i - 6t \leq 3n - 6,$$

as required. □

Corollary 5.7 *If G is a simple planar graph then G has a vertex v of degree less than 6, i.e., there is a v in $V(G)$ with $d(v) \leq 5$.*

Proof If G has only one vertex this vertex must have degree 0. If G has only two vertices then both must have degree at most 1. Thus we can suppose that $n \geq 3$, i.e., that G has at least three vertices. Now if the degree of every vertex of G is at least six we have

$$\sum_{v \in V(G)} d(v) \geq 6n.$$

However, by Theorem 1.1, $\sum_{v \in V(G)} d(v) = 2e$. Thus $2e \geq 6n$ and so $e \geq 3n$. This is impossible since, by the above theorem, $e \leq 3n - 6$. This contradiction shows that G must have at least one vertex of degree less than 6, as required. □

We now give the promised alternative proofs of Theorems 5.1 and 5.2.

Corollary 5.8 *K_5 is nonplanar.*

Proof Here $n = 5$ and $e = (5 \times 4)/2 = 10$ so that $3n - 6 = 9$. Thus $e > 3n - 6$ and so, by the theorem, $G = K_5$ can not be planar. □

Corollary 5.9 *$K_{3,3}$ is nonplanar.*

Proof Since $K_{3,3}$ is bipartite it contains no odd cycles (by Theorem 1.3) and so in particular no cycle of length three. It follows that every face of a plane drawing of $K_{3,3}$,if such exists, must have at least four boundary edges. Thus, using the argument of the proof of Theorem 5.6, we get $b \geq 4f$ and then $4f \leq 2e$, i.e., $2f \leq e = 9$. This gives $f \leq 9/2$. However, by Euler's Formula, $f = 2 - n + e = 2 - 6 + 9 = 5$, a contradiction. □

Exercises for Section 5.2

5.2.1 Verify Euler's Formula for the plane graphs of Figures 5.13 and 5.14 by counting the vertices, edges and faces.

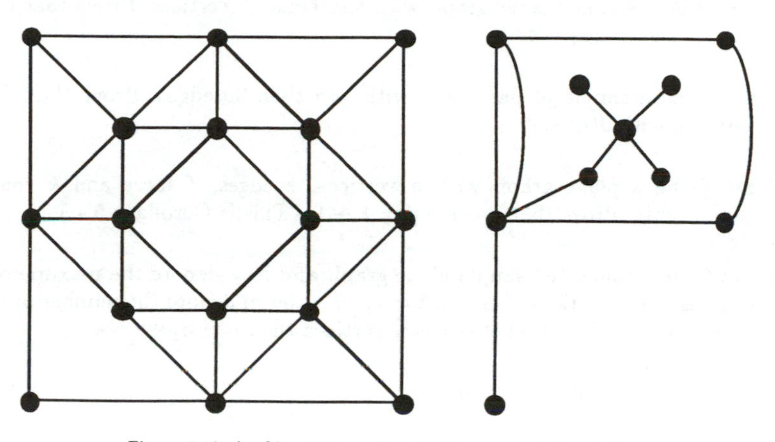

Figure 5.13: At this stage our imagination was starting to flag.

5.2.2 Verify Euler's Formula for plane drawings of the wheel W_n, and the complete bipartite graph $K_{2,n}$ where $n \geq 3$.

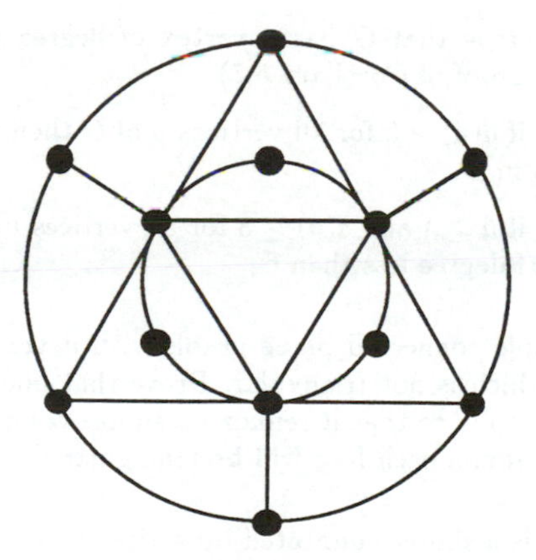

Figure 5.14

5.2.3 Draw an example of a simple plane graph in which the degree of every vertex is at least 5.

5.2.4 Let G be a plane 4-regular graph with 10 faces. Determine how many vertices G has and give a drawing of such a graph.

5.2.5 Let G be a simple graph with at least 11 vertices. Prove that either G or its complement $\overline{G}$ must be nonplanar.

5.2.6 Let G be a simple planar graph with less than 12 vertices. Prove that G has a vertex v with $d(v) \leq 4$.

5.2.7 Let G be a simple planar graph with less than 30 edges. Prove that G has a vertex v with $d(v) \leq 4$.

5.2.8 Let G be a plane graph with n vertices, e edges, f faces and k connected components. Prove that $n - e + f = k + 1$. (This is Corollary 5.4.)

5.2.9 Let G be a connected simple plane graph and let t denote the maximum of the degrees of the vertices of G. For $i = 1, \ldots, t$, let n_i denote the number of vertices of degree i in G, so that if G has n vertices then $n = n_1 + \cdots + n_t$.

(a) Prove that if $n \geq 3$ then

$$5n_1 + 4n_2 + 3n_3 + 2n_4 + n_5 \geq n_7 + 2n_8 + \cdots + (t-6)n_t + 12.$$

(b) Prove that the inequality in (a) becomes an equality if and only if the degree of each face of G is 3.

(c) Using (a) prove that G has a vertex of degree at most 5 (so giving an alternative proof of Corollary 5.7).

(d) Prove that if $d(v) \geq 5$ for all vertices v of G then there at least 12 vertices of degree 5 in G.

(e) Prove that if $n \geq 4$ and $d(v) \geq 3$ for all vertices v of G then G has at least 4 vertices of degree less than 6.

5.2.10 Let G be a simple connected plane graph with n vertices and e edges and at least one face which is not triangular. Prove that one can add a maximum of $3n - e - 6$ edges to G so that it remains a simple connected plane graph. (Note that at this maximum each face will be triangular.)

5.2.11 Prove that if G is a simple connected bipartite plane graph with e edges, where $e \geq 3$, and n vertices then $e \leq 2n - 4$.

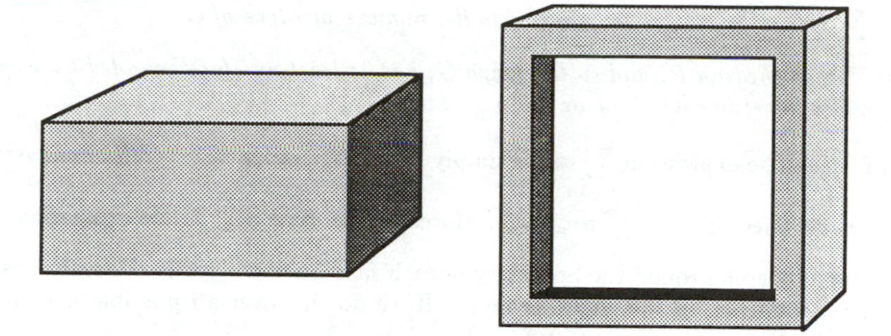

Figure 5.15: A brick and a window frame.

5.3 The Platonic Bodies

In three-dimensional space a **polyhedron** is a solid bounded by surfaces, simply called **faces**, each of which is a plane. Such a solid is called **convex** if any two of its interior points can be joined by a straight line lying entirely within the interior. Thus Figure 5.15 shows two polyhedra and, of these, the brick is convex while the window frame is not.

The vertices and edges of a polyhedron, which form a "skeleton" of the solid, give a simple graph in three-dimensional space. It can be shown that if the polyhedron is convex then this graph is planar. To see this informally, imagine the faces of the polyhedron to be made of rubber, with one face, say at the base, missing. Then, taking hold of the edges at the missing face, we are able to stretch out the rubber to form one plane sheet. The vertices and edges in this transformation now form a plane graph. Moreover each face of this plane graph corresponds to a face of the solid, with the exterior face of the graph corresponding to the missing face we used in the stretching process.

Clearly the plane graph is also connected, the degree of every vertex is at least 3 and the degree of every face is also at least 3 (and so, in particular, the graph is also simple). This encourages the following definition.

A simple connected plane graph G is called **polyhedral** if $d(v) \geq 3$ for every vertex v of G and $d(\varphi) \geq 3$ for every face φ of G.

Of course, Euler's Formula of Theorem 5.3 applies to any polyhedral graph and so in turn applies to any convex polyhedron. In this "new" rôle it is more commonly known as the **Euler Polyhedron Formula**.

We now prove two simple but useful properties of polyhedral graphs (and convex polyhedra.)

Theorem 5.10 *Let P be a convex polyhedron and G be its corresponding polyhedral graph. For each $n \geq 3$ let v_n denote the number of vertices of G of degree n and let f_n denote the number of faces of G of degree n.*

(a) $\sum_{n\geq 3} nv_n = \sum_{n\geq 3} nf_n = 2e$, *where e is the number of edges of G.*

(b) The polyhedron P, and so the graph G, has at least one face bounded by a cycle of length n for either n = 3, 4 or 5.

Proof (*a*) The expression $\sum_{n\geq 3} nv_n$ is simply $\sum_{v\in V(G)} d(v)$ since $d(v) \geq 3$ for each vertex v. Thus, by Theorem 1.1, $\sum_{n\geq 3} nv_n = 2e$. Moreover for each $n \geq 3$, the expression nf_n is obtained by going round the boundary of each face having a cycle of length n as its boundary, counting up the edges as we go. If we do this over all possible n then we count each edge twice and so $\sum_{n\geq 3} nf_n = 2e$.

(*b*) Assume to the contrary that P has no faces bounded by cycles of length 3, 4 or 5. Then $f_3 = f_4 = f_5 = 0$ and so, by (*a*),

$$2e = \sum_{n\geq 6} nf_n \geq \sum_{n\geq 6} 6f_n = 6\sum_{n\geq 6} f_n = 6f,$$

where f is the total number of faces of P. Hence $f \leq \frac{1}{3}e$. Moreover, again by (a),

$$2e = \sum_{n\geq 3} nv_n \geq \sum_{n\geq 3} 3v_n = 3\sum_{n\geq 3} v_n = 3v,$$

where v denotes the number of vertices of P. Hence $v \leq \frac{2}{3}e$. However, by Euler's Formula, $e = v + f - 2$. Hence, using the inequalities above,

$$e \leq \frac{2}{3}e + \frac{1}{3}e - 2 = e - 2.$$

Since this is impossible, P must have a face of the desired type. □

The ancient Greek geometers discovered five **regular polyhedra**. Here a polyhedron is called **regular** if it is convex, and its faces are congruent regular polygons (so that the polyhedral angles are all equal). The five regular polyhedra known to the Greeks are shown on the right hand side of Figure 5.16, alongside their corresponding polyhedral graphs. They are called the **Platonic bodies** or **Platonic solids**. For quite some time the Greek geometers tried to find other regular polyhedra until Theaetetus (414–368 B.C.) proved that there are in fact no others. We now prove this using the Euler Polyhedron Formula.

Theorem 5.11 *The only regular polyhedra are the tetrahedron, the cube, the octahedron, the dodecahedron and the icosahedron.*

Proof Let P be a regular polyhedron and let G be its corresponding polyhedral graph. Let v, e and f denote the number of vertices, edges and faces of G, respectively. Since the faces of G are congruent to each other, each one is bounded by the same number of edges. Thus for some $r \geq 3$, $d(\varphi) = r$ for every face φ of G. Similarly, since the polyhedral angles are all equal to each other, the graph G is regular, say

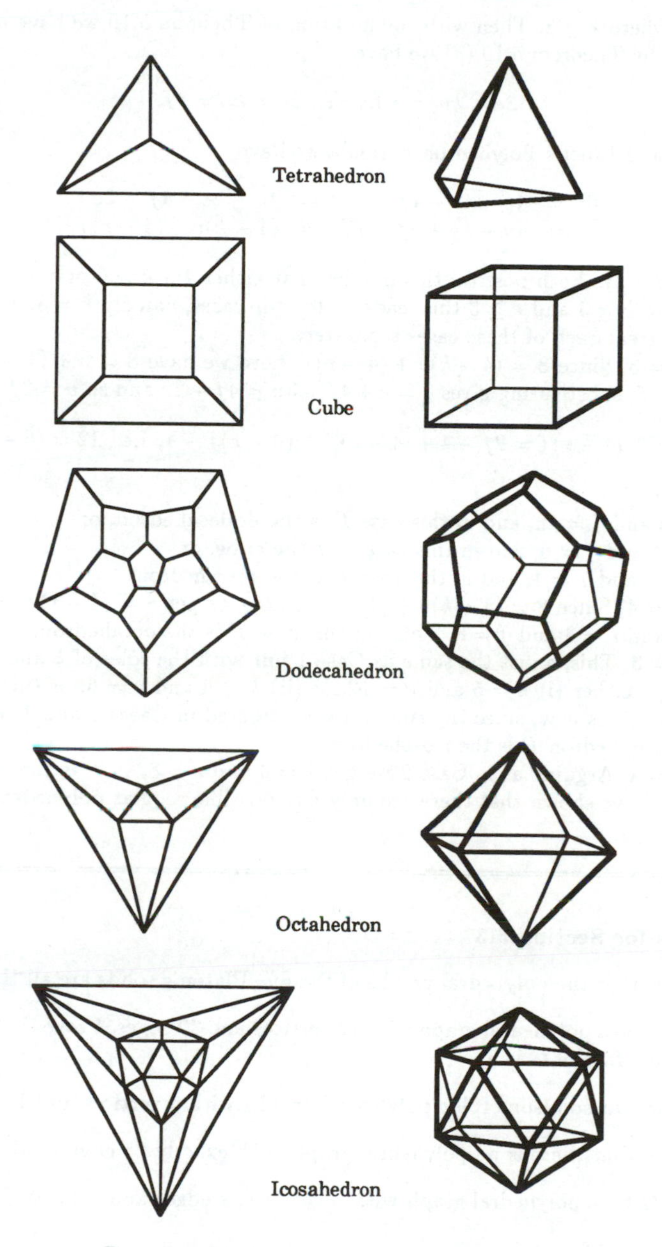

Figure 5.16: The five Platonic bodies and their graphs.

k-regular, where $k \geq 3$. Then with the notation of Theorem 5.10 we have $v = v_k$ and $f = f_r$ and by Theorem 5.10 (a) we have

$$2e = kv_k = rf_r, \text{ i.e., } 2e = kv = rf.$$

Using this and Euler's Polyhedron Formula we have

$$\begin{aligned} 8 &= 4v - 4e + 4f &&= 4v - 2e + 4f - 2e \\ &= 4v - kv + 4f - rf &&= (4-k)v + (4-r)f. \end{aligned}$$

Since v and f are both positive this implies that either $4 - k \geq 0$ or $4 - r \geq 0$. Since we also have $k \geq 3$ and $r \geq 3$ this leads us to four cases, namely $k = 3$, 4 and $r = 3$, 4. We now treat each of these cases separately.
Case 1: $k = 3$. Since $8 = (4-k)v + (4-r)f$, here we have $8 = v + (4-r)f$. Since $2e = 3v = rf$, substituting gives $8 = v + 4f - 3v = 4f - 2v$ and so $v = 2f - 4$. Hence

$$8 = v + (4-r)f = 2f - 4 + (4-r)f = (6-r)f - 4, \text{ i.e., } 12 = (6-r)f.$$

Thus either
(i) $r = 5$ and $f = 12$, and in this case P is the dodecahedron, or
(ii) $r = 4$ and $f = 6$, and in this case P is the cube, or
(iii) $r = 3$ and $f = 4$, and in this case P is the tetrahedron.
Case 2: $k = 4$. Since $8 = (4-k)v + (4-r)f$, here we get $8 = (4-r)f$. Since $r \geq 3$ this results in $r = 3$ and $f = 8$. Thus, in this case P is the octahedron.
Case 3: $r = 3$. This is just the same as Case 1 but with the rôles of k and r reversed. Thus we get either (i) $k = 5$ and $v = 12$, or (ii) $k = 4$ and $v = 6$, or (iii) $k = 3$ and $v = 4$. Only (i) is new, since (ii) and (iii) were treated in Cases 2 and 1 respectively. In (i) the polyhedron P is the icosahedron.
Case 4: $r = 4$. Arguing as in Case 2 we get $k = 3$ and $v = 8$, so P is the cube again.
Thus we have shown that there are only five possible regular polyhedra. □

Exercises for Section 5.3

5.3.1 Prove that the polyhedral graphs of the five Platonic solids are all Hamiltonian.

5.3.2 Let G be a polyhedral graph with 12 vertices and 30 edges. Prove that the degree of each face of G is 3.

5.3.3 Draw two non-isomorphic polyhedral graphs with 6 vertices and 10 edges.

5.3.4 Prove that there is no polyhedral graph with exactly 25 edges and 10 faces.

5.3.5 Let G be a polyhedral graph with v vertices, e edges and f faces. Prove that

(a) $3f \leq 2e$,

(b) $2v \geq 4 + f$,

(c) $2f \geq 4 + v$,

(d) $3f - 6 \geq e$,

(e) $v \geq 4$, $e \geq 6$ and $f \geq 4$.

5.3.6 Let P be a convex polyhedron whose faces are either pentagons or hexagons. Use the Euler Polyhedron Formula to prove that P must have no less than 12 pentagonal faces.

5.4 Kuratowski's Theorem

Let G be any graph. A **subdivision** of G is a graph obtained from G by inserting vertices (of degree 2) into some of the edges of G.

For example, in Figure 5.17 the graph H is a subdivision of the graph G. On the other hand, the graph L of Figure 5.18 cannot be a subdivision of this G. To see why, note first that any subdivision of a graph will either create longer cycles or leave cycles unaffected. Then, since L has a cycle of length 3, whereas G has only a cycle of length 4, L cannot be a subdivision of G.

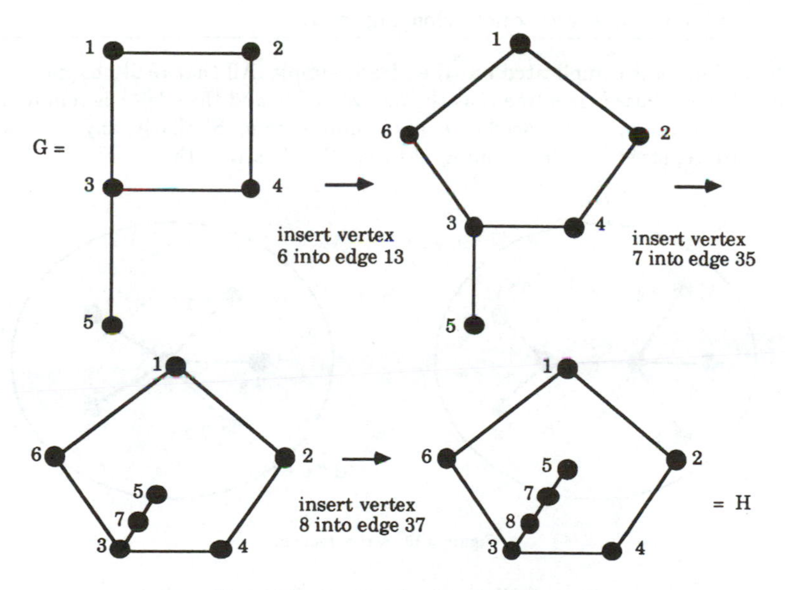

Figure 5.17: An example of a subdivision.

Clearly any subdivision of a plane (or planar) graph is again plane (or planar). Moreover any subdivision of a nonplanar graph is again nonplanar.

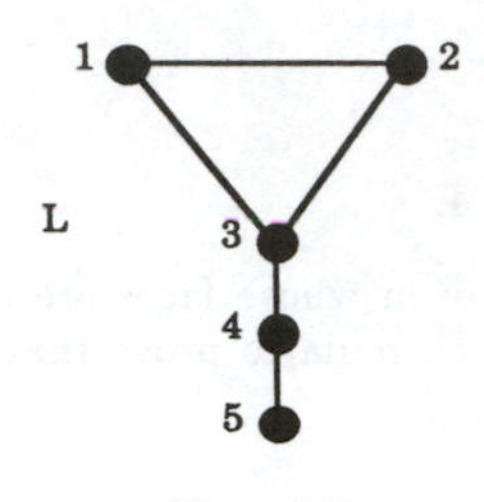

Figure 5.18

The concept of subdivision gives an important characterisation of planarity due to Kuratowski [41]. We will sketch a proof of this here.

First we recall a graph operation which we introduced briefly in Exercise 2.3.6.

> Let $e = uv$ be an edge of the simple connected graph G, let $d(u) = k$ and $d(v) = l$ and let $N(u) = \{v, u_1, \ldots, u_{k-1}\}$ and $N(v) = \{u, v_1, \ldots, v_{l-1}\}$. A **contraction** on the edge e changes G to a new graph $G * e$ where
>
> $$V(G * e) = (V(G) - \{u, v\}) \cup \{w\},$$
>
> $$E(G * e) = E(G - \{u, v\}) \cup \{wu_1, \ldots, wu_{k-1}, wv_1, \ldots, wv_{l-1}\}$$
>
> and w is a new vertex not belonging to G.

The definition is complicated but the idea is simple. All that really happens is that u and v become fused (identified) as the one vertex w and the edge e is removed. Any edges joining u to u_i are replaced by edges joining w to u_i. Similarly, any edges joining v to v_j are replaced by edges joining w to v_j. We illustrate this in Figures 5.19 and 5.20.

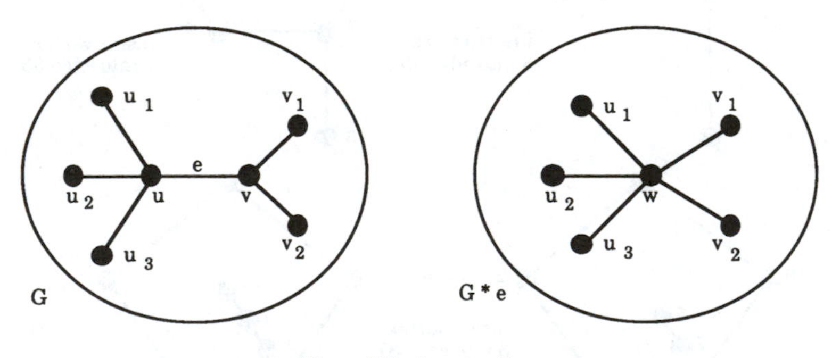

Figure 5.19: A contraction.

We make one simplifying adjustment to the definition. If $N(u) \cap N(v) \neq \vee$ then the contraction on the edge $e = uv$ will create parallel edges incident with w. In this case we delete all but one of the edges in a collection of parallel edges. This amounts to taking the underlying simple graph.

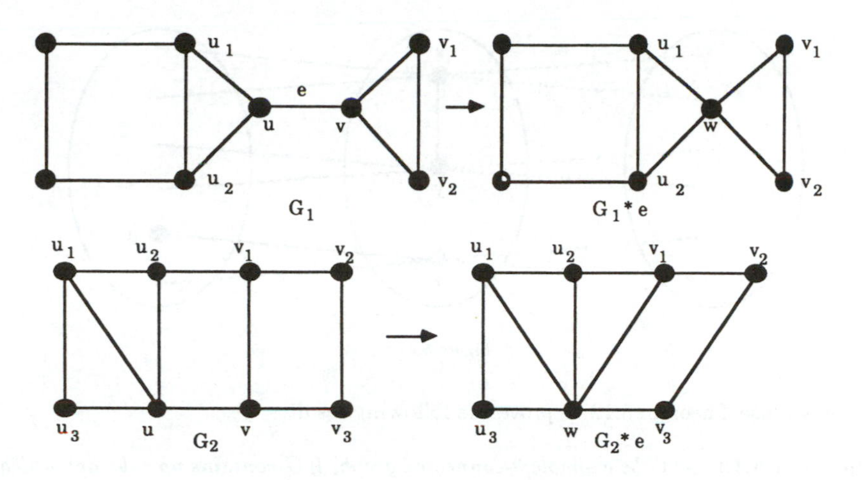

Figure 5.20: Two more contractions.

As can be seen from Exercise 5.5.3, it is not always true that G and $G * e$ have the same vertex connectivity. When this is so it is straightforward to see that $\kappa(G * e) < \kappa(G)$. We will be interested in when these two graphs do have the same connectivity.

> If G and $G * e$ have the same connectivity we say that e is a **contractible** edge.

Our next result is some guarantee that contractible edges exist.

Theorem 5.12 *Let G be a simple 3-connected graph with at least five vertices. Then G has a contractible edge.*

Proof Let $e = xy$ be any edge of G. If e is not contractible then $G * e$ is not 3-connected and so there must be a vertex z in G such that $G - \{x, y, z\}$ is disconnected.

Choose such an edge $e = xy$ in G with such a corresponding subset $S = \{x, y, z\}$ so that $G - S$ has a component U with as many vertices as possible. Since G is 3-connected, the vertex z is adjacent to some vertex u not in U, because otherwise the removal of just x and y would disconnect G. We illustrate this in Figure 5.21.

Let H be the subgraph of G induced by U, x and y. If H is 1-connected then it has a cut vertex, c, say. But then $G - \{c, z\}$ will be disconnected, which is a contradiction since G is 3-connected. Hence H is 2-connected.

Now consider the edge $f = uz$. If this is not contractible then, as before, there is a vertex v in G such that $G - \{u, v, z\}$ is disconnected. Now if v lies in H then, since u and z do not lie in H, v will be a cut vertex of H, a contradiction since H is 2-connected. Hence $v \notin H$ and so $H \subseteq G - \{u, v, z\}$. Clearly H is then a connected subgraph of the disconnected $G - \{x, y, z\}$. This is a contradiction to our choice of the edge $e = xy$. Thus G must have a contractible edge, as required. □

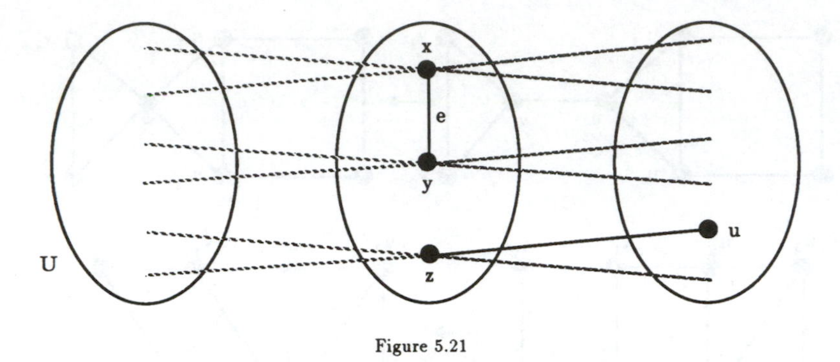

Figure 5.21

We will use Theorem 5.12 to prove the following result.

Theorem 5.13 *Let G be a simple 3-connected graph. If G contains no subgraph which is a subdivision of K_5 or $K_{3,3}$ then G is planar.*

Proof We use induction on $|V(G)|$. Clearly the result is true for $|V(G)| = 4$. So we assume that every 3-connected graph with fewer than $|V(G)|$ vertices, which contains no subgraph which is a subdivision of K_5 or $K_{3,3}$, is planar.

By Theorem 5.12 we know that G contains an edge e such that $G * e$ is 3-connected. Let w be the vertex corresponding to e in $G * e$.

It is straightforward to see that if $G * e$ contains a subgraph which is either a subdivision of K_5 or of $K_{3,3}$ then so does G, in contradiction to our hypothesis. Thus $G * e$ has no such subgraph and so, since it has fewer vertices than G, we can assume, by induction, that it is planar and, without loss of generality, in fact plane.

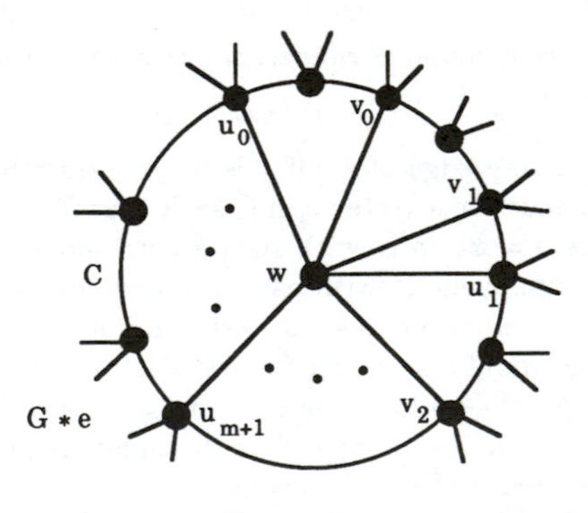

Figure 5.22

Consider the plane drawing of $G*e$ in Figure 5.22. Here C is the cycle surrounding w which surrounds a face in $(G*e)-w$. Let $u_i \in N(U) \cap V(C)$ and let $v_j \in N(V) \cap V(C)$. Further let $P_{(i)}$ be the subpath of C which joins u_{i-1} to u_i for $i = 1, 2, \ldots, n$, where $u_n = u_0$. Let v_j be any vertex of $(V(G) \cap V(C)) - \{u\}$ which is adjacent to v.

There are four cases which may arise when we return w to uv (and hence from $G*e$ to G). We look at these in turn.

Case 1. Suppose that every v_j lies on precisely one $P_{(i)}$. Then we can insert u, v and their adjacent vertices in the plane drawing of $G*e$ to give a plane drawing of G. (See Figure 5.23 (a).)

Case 2. Suppose that v is adjacent to v_j, an internal vertex of P_i, and v is adjacent to some vertex not in $P_{(i)}$. (See Figure 5.23 (b).) This implies that G contains a subgraph with a subdivision of $K_{3,3}$. Since this contradicts the hypothesis on G, this case does not occur.

Case 3. Suppose that v is only adjacent to neighbours of u (otherwise see Case 2) and that it is adjacent to at least three such neighbours. From Figure 5.23 (c) we see that then G contains a subdivision of K_5. This is a contradiction.

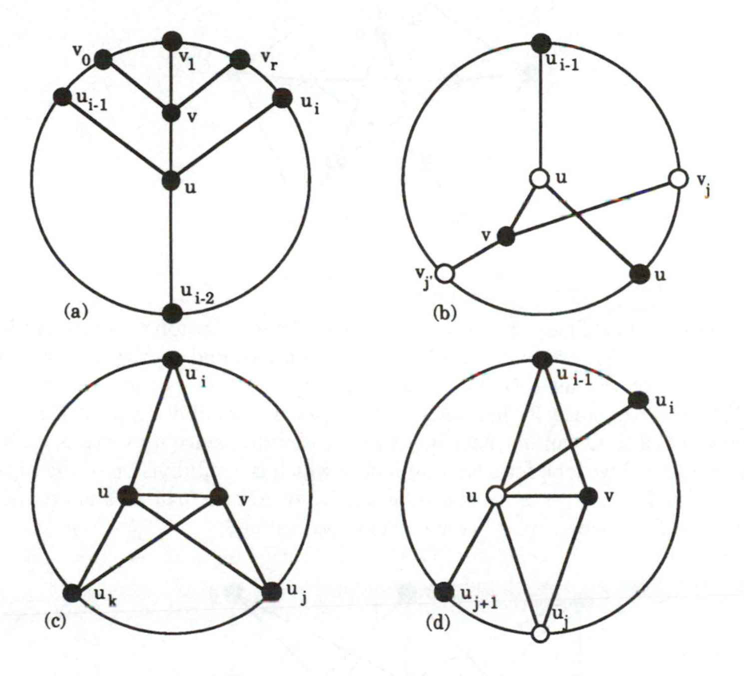

Figure 5.23

Case 4. Suppose v is adjacent to two neighbours of u and that these two neighbours are not end vertices of a $P_{(i)}$. From Figure 5.23 (d) we see that this implies that G contains a subgraph with a subdivision of $K_{3,3}$. □

The proofs of Theorems 5.12 and 5.13 were given by Thomassen [59] in 1981. We now present the main result of this section.

Theorem 5.14 (Kuratowski, 1930) *A graph G is planar if and only if it has no subgraph isomorphic to a subdivision of K_5 or $K_{3,3}$.*

Proof As noted earlier, clearly any subdivision of a nonplanar graph is nonplanar and so if G does have a subgraph which is a subdivision of either K_5 or $K_{3,3}$ then G is not planar.

Conversely, suppose that G has no such subgraph. Then, if G is 3-connected, it follows from Theorem 5.13 that G is planar. We leave the rest of the proof, i.e., when G is not 3-connected, to the reader in the form of Exercise 5.4.7. □

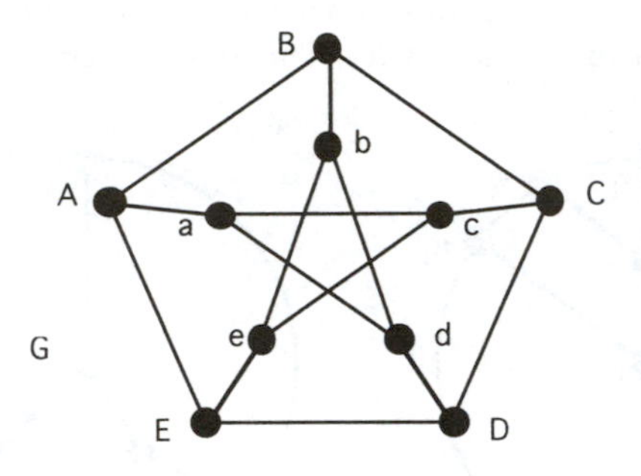

Figure 5.24: The Petersen graph.

Using Kuratowski's Theorem we prove that the Petersen graph G, shown in Figure 5.24, is nonplanar. Perhaps one's first thought is to try to find a subgraph of G which is a subdivision of K_5 since G has some similarity with K_5 because of the 5-cycle $ABCDEA$. However since K_5 has vertices of degree 4 any subdivision of K_5 will also have such vertices so G can not have any such subdivision since its vertices each have degree 3. So instead we look for a subgraph of G which is a subdivision of the bipartite graph $K_{3,3}$. Trial and error gives the subgraph H of Figure 5.25 as a subdivision of $K_{3,3}$. Thus the Petersen graph G is not planar, as claimed.

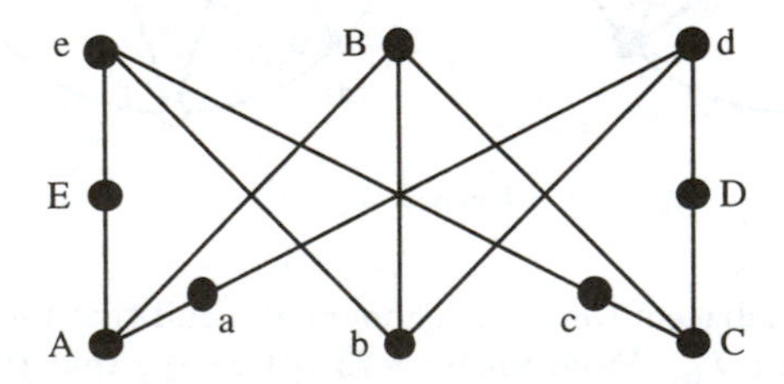

Figure 5.25: A subgraph of the Petersen graph which is a subdivision of $K_{3,3}$.

Finally in this section we draw the reader's attention to a recent article by Thomassen [60] illustrating a link between the Jordan Curve Theorem of Section 5.1 and Kuratowski's Theorem.

Exercises for Section 5.4

5.4.1 Perform contractions on the edges e in the graphs of Figure 5.26.

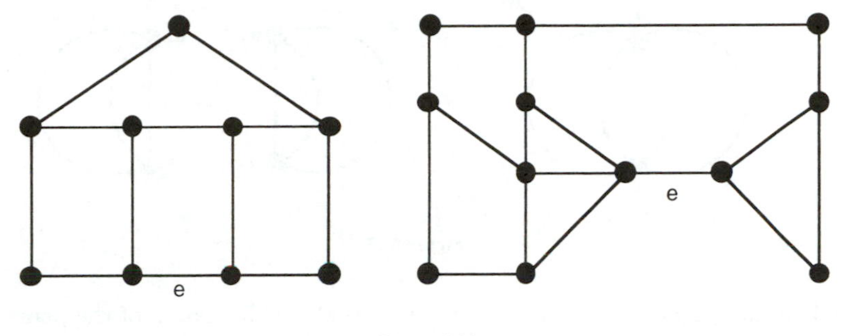

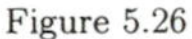

Figure 5.26

5.4.2 Show that $P_n * e = P_{n-1}$ for each $n \geq 2$. Is $C_n * e = C_{n-1}$ for each $n \geq 3$?

5.4.3 Show by example that the connectivity of G may be different from that of a contraction $G * e$.

5.4.4 (a) Is it true that if G is 1-connected, then there exists $e \in E(G)$ such that $G * e$ is 1-connected?

(b) Is it true that if G is 2-connected, then there exists $e \in E(G)$ such that $G * e$ is 2-connected?

5.4.5 Determine which complete tripartite graphs are planar.

5.4.6 Determine the values of n for which the n-cube Q_n is planar.

5.4.7 The **crossing number** of a graph G, denoted by $cr(G)$, is the smallest possible number of pairs of crossing edges that occur in a drawing of G in the plane. Here we assume that in such drawings adjacent edges never cross, no edge crosses itself, we do not have three or more edges crossing at any point, and two nonadjacent edges cross no more than once. From this definition we have that $cr(G) = 0$ if and only if G is planar.

(a) Show that $cr(K_5) = cr(K_{3,3}) = 1$.

(b) Show that $cr(K_6) = 3$, by drawing K_6 with three crossings, introducing new vertices at the crossing points, and using Theorem 5.6.

5.4.8 Use the proof of Theorem 5.13 to show that every 3-connected planar graph can be drawn with edges as straight lines and with every internal face convex. (Here a face is called **convex** if any two points inside it can be joined by a straight line which does not cross any of its boundary edges.)

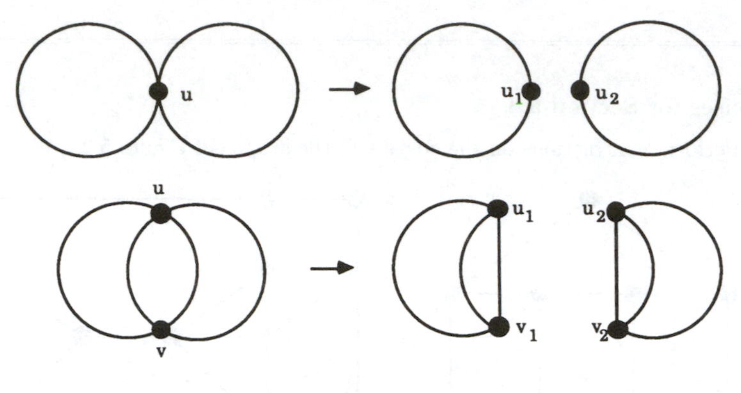

Figure 5.27

5.4.9 Use the diagrams of Figure 5.27 to show that the discussion of the planarity of all graphs can be reduced to the planarity of 3-connected graphs.

5.4.10 Use the previous exercise to complete the proof of Kuratowski's Theorem, i.e., show that if G is connected and contains no subgraph which is a subdivision of K_5 or $K_{3,3}$, then G is planar.

5.4.11 Use Kuratowski's Theorem to show that the graphs of Figures 5.28 and 5.29 are nonplanar.

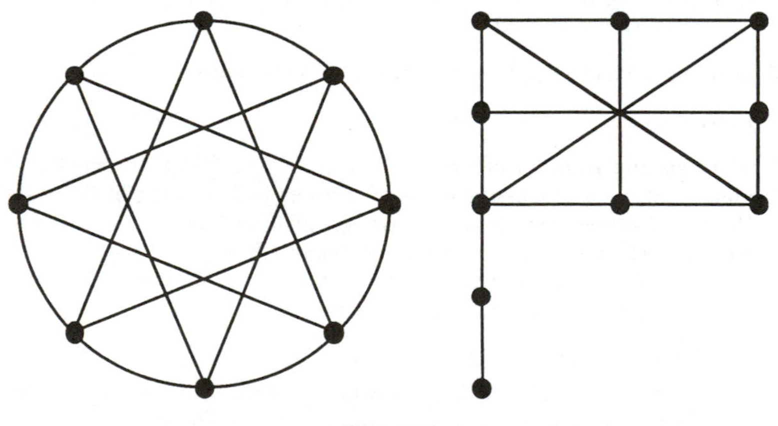
Figure 5.28

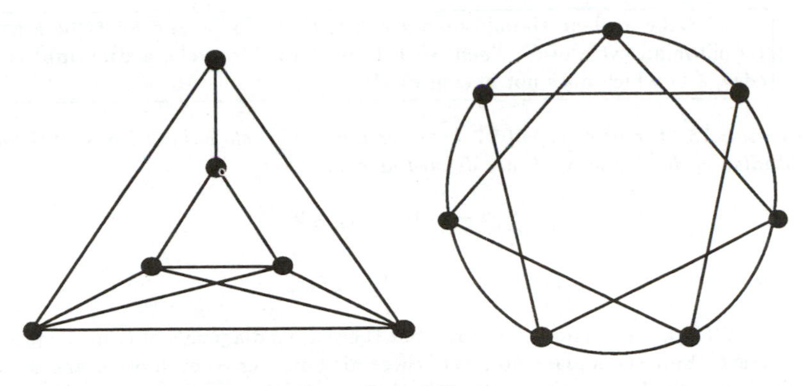

Figure 5.29

5.5 Non-Hamiltonian Plane Graphs

Given a plane graph G which has a Hamiltonian cycle C we let α_i denote the number of faces of degree i lying *inside* the cycle C and β_i denote the number of faces of degree i lying *outside* the cycle C

For example, in the plane graph G of Figure 5.30 the cycle C shown in shaded lines is Hamiltonian and the numbers are the degrees of the corresponding faces. (Note that we must also consider the exterior face.)

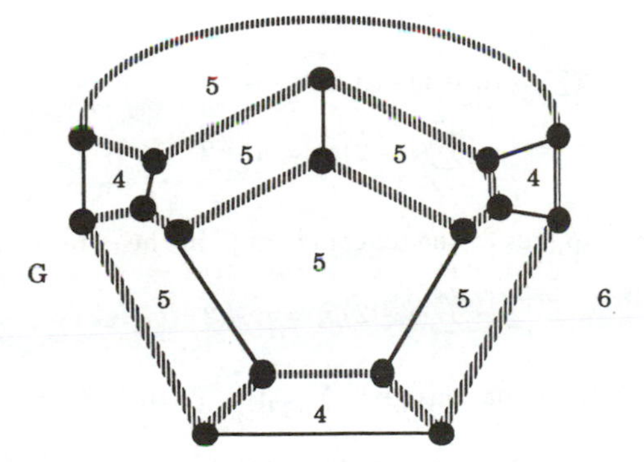

Figure 5.30: A Hamiltonian plane graph to illustrate Grinberg's Theorem.

Here the possible values for i are 4, 5 and 6 and we have

$$\alpha_4 = 1, \beta_4 = 2, \ \alpha_5 = 4, \beta_5 = 2, \ \alpha_6 = 0, \beta_6 = 1.$$

There is a useful formula, due to Grinberg [31], which involves the numbers α_i and β_i. Before we state it, in Theorem 5.15, we introduce a term used in its proof.

> Let G be a plane Hamiltonian graph without loops and let C be a fixed Hamiltonian cycle of G. Then, with respect to this cycle, a **diagonal** is an edge of G which does not belong to C.

Theorem 5.15 (Grinberg, 1968) *Let G be a plane graph without loops. If G has a Hamiltonian cycle C and α_i, β_i are defined as above then*

$$\sum_i (i-2)(\alpha_i - \beta_i) = 0$$

(where the summation is over all possible values of i).

Proof Let G have n vertices. Suppose that there are s diagonals of G in the interior of the cycle C. Since G is plane no two of these diagonals cross each other and so each diagonal is an edge for exactly two faces in the interior of C. If we consider drawing in these diagonals one at time, after each one is drawn we have increased the number of faces in the interior of C by one. Thus, with all s diagonals drawn, we see that there are $s+1$ faces of G within the interior of C. Hence

$$\sum_{i\geq 2} \alpha_i = s + 1.$$

Now let d denote the sum of the degrees of the faces within the interior of C. Then $d = \sum_{i\geq 2} i\alpha_i$. However in obtaining this total for d each diagonal is counted twice, since it is an edge for exactly two of the faces, and each of the n edges on the cycle C is counted once. Thus we also have

$$d = 2s + n.$$

Hence $\sum_{i\geq 2} i\alpha_i = d = 2(\sum_{i\geq 2} \alpha_i - 1) + n$ and so

$$\sum_{i\geq 2} (i-2)\alpha_i = n - 2.$$

A similar argument applies to the diagonals of G in the exterior of the cycle C and this produces

$$\sum_{i\geq 2} (i-2)\beta_i = n - 2.$$

Combining the last two equations gives $\sum_{i\geq 2}(i-2)(\alpha_i - \beta_i) = 0$, as required. □

To illustrate Grinberg's Theorem, for the graph G of Figure 5.30 we have

$$\begin{aligned}\sum_i (i-2)(\alpha_i - \beta_i) &= (4-2)(\alpha_4 - \beta_4) + (5-2)(\alpha_5 - \beta_5) + (6-2)(\alpha_6 - \beta_6) \\ &= 2(1-2) + 3(4-2) + 4(0-1) = -2 + 6 - 4 = 0,\end{aligned}$$

as expected from the Theorem.

The main use of Grinberg's Theorem is to establish that certain planar graphs are non-Hamiltonian. We illustrate this using the graph of Figure 5.31 (known, not

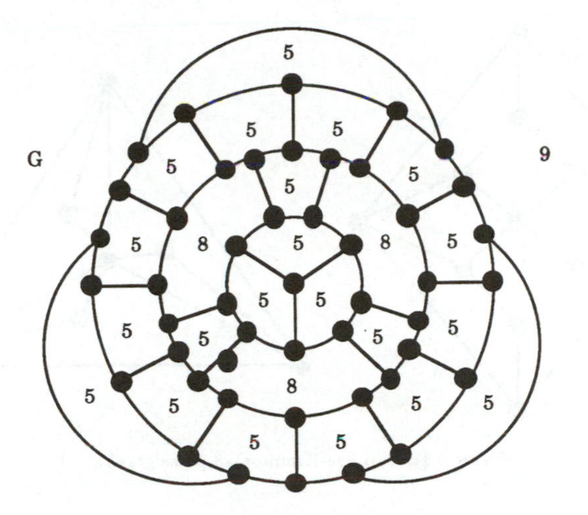

Figure 5.31: The Grinberg graph.

coincidentally, as the **Grinberg graph**.) As with the graph of Figure 5.30, the numbers shown here are the degrees of each face. We show that the graph is not Hamiltonian by assuming that it is and then deducing from this assumption that we have a contradiction to Grinberg's Theorem.

Thus let us assume that G has a Hamiltonian cycle C. Then, by the Theorem,

$$\sum_i (i-2)(\alpha_i - \beta_i) = 0.$$

Here the values for i are 5, 8 and 9 so the equation becomes

$$3(\alpha_5 - \beta_5) + 6(\alpha_8 - \beta_8) + 7(\alpha_9 - \beta_9) = 0$$

and so

$$3\{(\alpha_5 - \beta_5) + 2(\alpha_8 - \beta_8)\} = 7(\beta_9 - \alpha_9).$$

This shows that 3 must divide $7(\beta_9 - \alpha_9)$. However the exterior face is the only face with 9 edges on its boundary and so either

$$\beta_9 - \alpha_9 = 1 - 0 \text{ or } \beta_9 - \alpha_9 = 0 - 1, \text{ i.e., } 7(\beta_9 - \alpha_9) = \pm 7.$$

This means that, if our assumption is true, then 3 divides ± 7, codswallop. Our assumption must therefore be false. Hence G is not Hamiltonian.

Exercises for Section 5.5

5.5.1 Use Grinberg's Theorem to show the plane graphs of Figures 5.32 and 5.33 are not Hamiltonian.

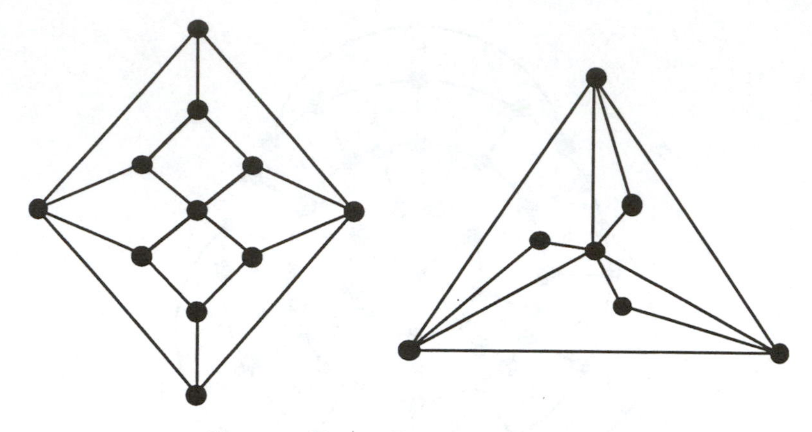

Figure 5.32: Two non-Hamiltonian plane graphs.

5.5.2 Use Grinberg's Theorem to prove that there are no plane Hamiltonian graphs with

(a) faces of degree 5, 8 and 7, with just one face of degree 7,

(b) faces of degree 4, 6 and 9 with just one face of degree 9.

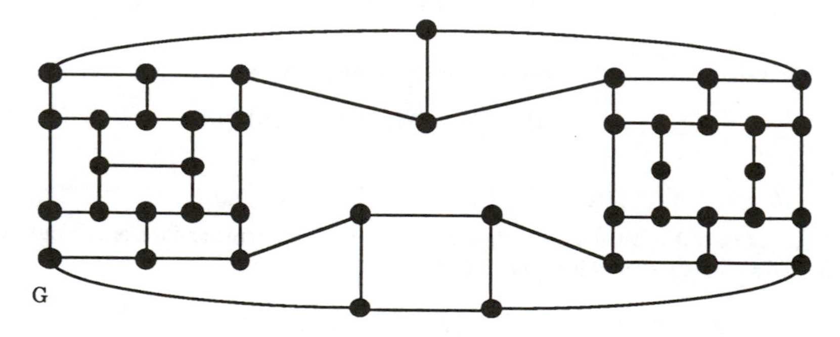

Figure 5.33: Another non-Hamiltonian plane graph.

5.5.3 Use Grinberg's Theorem to show that the plane Hamiltonian graph of Figure 5.34 has no Hamiltonian cycle containing both the edges e and f.

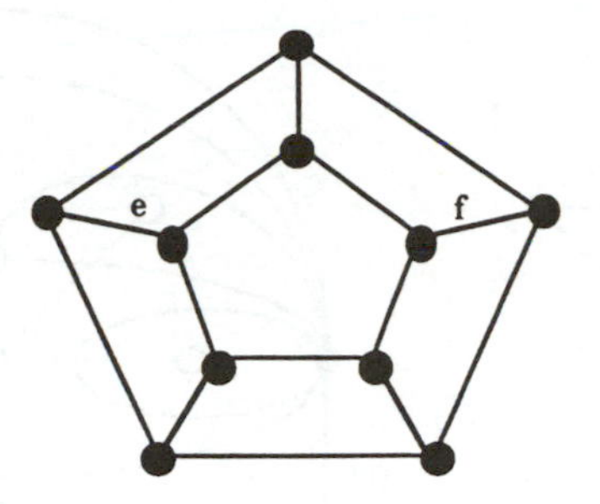

Figure 5.34

5.6 The Dual of a Plane Graph

In this section we introduce a concept which will prove useful when we study the colouring of maps in Section 6.7.

> Let G be a plane graph. We define the **dual** of G to be the graph G^* constructed as follows. To each face f of G there is a corresponding vertex f^* of G^* and to each edge e of G there is a corresponding edge e^* in G^* such that if the edge e occurs on the boundary of the two faces f and g then the edge e^* joins the corresponding vertices f^* and g^* in G^*. (If the edge e is a bridge then we treat it as though it occurs twice on the boundary of the face f in which it lies and then the corresponding edge e^* is a loop incident with the vertex f^* in G^*.)

Figure 5.35 shows a plane graph and its dual.

It turns out that the dual G^* of the plane graph G is also planar. We indicate why this is so by describing how we can draw G^* as a plane graph. Given our plane drawing of G, place the vertex f^* of G^* inside its corresponding face f. If the edge e lies on the boundary of two faces f and g of G, join the two vertices f^* and g^* by the edge e^* drawing so that it crosses the edge e exactly once and crosses no other edge of G. (This procedure is still possible if the edge e is a bridge.) We used this drawing procedure in Figure 5.35.

If the edge e is a loop in G then it is the only edge on the common boundary of two faces, one of which, say f, lies within the area of the plane surrounded by e with the other, say g, lying outside this area. The face f may not be the only face enclosed by e but, clearly from the definition of G^*, any path from a vertex h^*, corresponding to such a face h, to the vertex g^* must use the edge e^*. Thus e^* is a bridge in G^*.

Conversely, if the edge e^* is a bridge in G^*, joining vertices f^* and g^*, then e^* is the only path in G^* from f^* to g^*. This implies, again from the definition of G^*, that the edge e in G must completely enclose one of the faces f and g and so e must be a loop.

To summarise, the edge e is a loop in G if and only if e^* is a bridge in G^*.

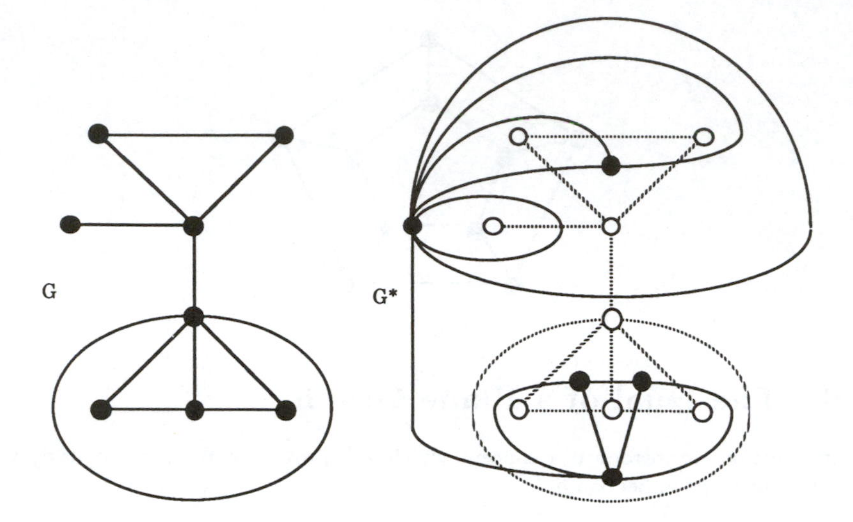

Figure 5.35: A plane graph and its dual.

The occurrence of parallel edges in G^* is easily described. A moment's thought should convince the reader that, given two faces f and g of G, then there are k parallel edges between f^* and g^* in G^* if and only if f and g have k edges on their common boundary.

The reader may have noticed that we have defined the dual of a *plane* graph instead of a *planar* graph. The reason for this is that different plane drawings G_1 and G_2 of the same planar graph G may result in non-isomorphic duals G_1^* and G_2^*. Exercise 5.6.2 illustrates this.

We now record some simple consequences of the dual construction.

Theorem 5.16 *Let G be a connected plane graph with n vertices, e edges and f faces. Let n^*, e^* and f^* denote the number of vertices, edges and faces respectively of G^*. Then $n^* = f$, $e^* = e$ and $f^* = n$.*

Proof The first two equalities follow immediately from the definition of G^*. The third then follows from Euler's Formula since both G and G^* are connected plane graphs. □

Now suppose that the face φ of the plane graph G^*, corresponding to the vertex v of G, has $e_1^*, \ldots, e_n^*$ as its boundary edges. Then, by our construction of G^*, each of these edges e_i^* crosses the corresponding edge e_i of G and, as illustrated in Figure 5.35, these edges are all incident with the vertex v. It follows that φ contains the vertex v.

Since G^* is a plane graph, we may also construct the dual of G^*, called the **double dual** of G and denoted by G^{**}. From the discussion of the previous paragraph, the following result may come as no surprise.

Theorem 5.17 *Let G be a plane connected graph. Then G is isomorphic to its double dual G^{**}.*

Proof As seen above, any face φ of the dual G^* contains at least one vertex of G, namely its corresponding vertex v. In fact this is the only vertex of G that φ contains since, by Theorem 5.16, the number of faces of G^* is the same as the number of vertices of G. Hence, in the construction of the double dual G^{**}, we may choose the vertex v to be the vertex in G^{**} corresponding to the face φ of G^*. This choice gives us the required isomorphism. □

It is quite easy to see that Theorems 5.16 and 5.17 are not true for disconnected plane graphs and we leave this to the reader in the form of Exercises 5.6.7 and 5.6.8.

Exercises for Section 5.6

5.6.1 Draw the duals of the plane graphs of Figure 5.36.

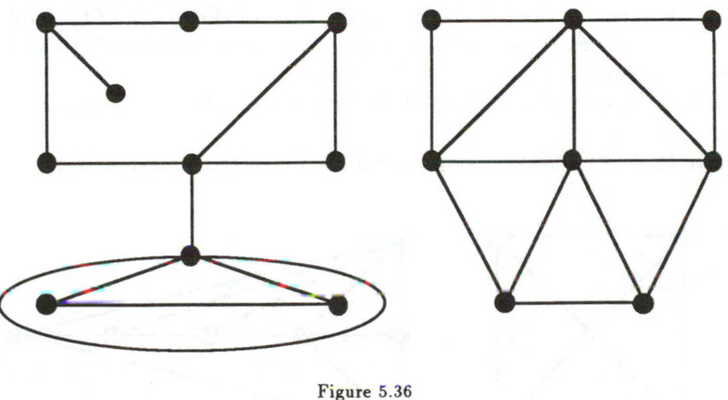

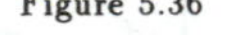

Figure 5.36

5.6.2 Show that the plane graphs G_1 and G_2 of Figure 5.37 are isomorphic but have non-isomorphic duals.

5.6.3 A connected plane graph G is called **self-dual** if it is isomorphic to its dual G^*.

(a) Show that the graphs of Figure 5.38 are self-dual.

(b) Prove that if G is self-dual with n vertices and e edges then $2n = e + 2$.

(c) For each $n \geq 4$, give an example of a self-dual graph with n vertices. (Try generalising one of the examples in (a).)

(d) Prove that the converse to (b) is false in general by giving an example of a connected plane graph with five vertices and eight edges which is not self-dual.

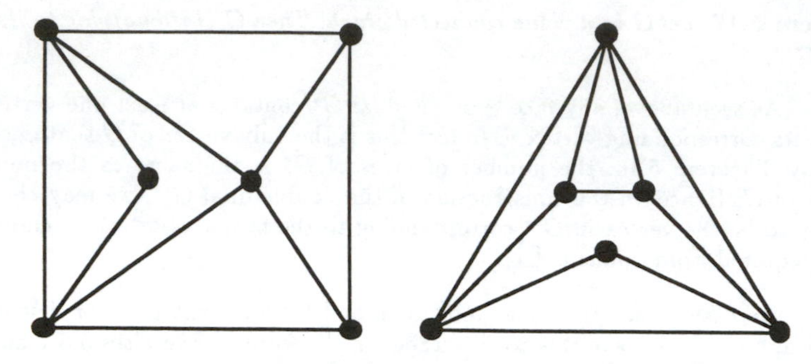

Figure 5.37: Two isomorphic plane graphs with non-isomorphic duals.

5.6.4 Let G be a plane connected graph. Prove that G is bipartite if and only if G^* is an Euler graph.

5.6.5 Prove that the graph of the octahedron is the dual of the graph of the cube and that the graph of the icosahedron is the dual of the graph of the dodecahedron. (You do not need to draw the duals in each case. Instead consider degrees and use the argument of the proof of Theorem 5.11.)
What is the dual of the graph of the tetrahedron, the remaining Platonic body?

5.6.6 Prove that G^* is always connected, even when G is disconnected.

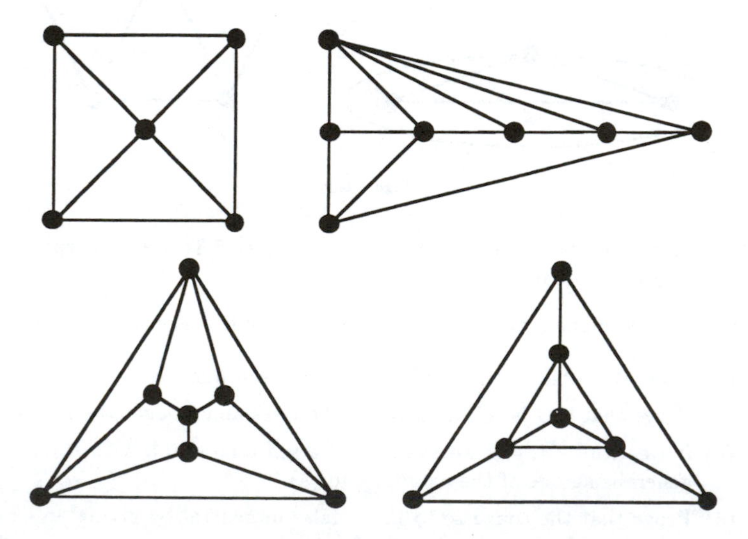

Figure 5.38: Four self-dual graphs.

5.6.7 Let G be an arbitrary plane graph. Using the notation of Theorem 5.16, prove that $f^* = n$ only if G is connected. Express f^* in terms of n and the number of connected components of G.

5.6.8 Prove that the plane graph G is isomorphic to its double dual if and only if G is connected.

Chapter 6

Colouring

6.1 Vertex Colouring

Lame Duck Airlines of Pocatello, Idaho, is considering operating the following seven weekly flights, all starting from Pocatello.

Flight	Route
A	Burley $\longrightarrow$ Twin Falls $\longrightarrow$ Boise $\longrightarrow$ Lewiston
B	Billings $\longrightarrow$ Great Falls $\longrightarrow$ Missoula $\longrightarrow$ Lewiston
C	Idaho Falls $\longrightarrow$ Boise $\longrightarrow$ Lewiston $\longrightarrow$ Pullman
D	Idaho Falls $\longrightarrow$ Yellowstone $\longrightarrow$ Great Falls
E	Idaho Falls $\longrightarrow$ Sun Valley $\longrightarrow$ Boise
F	Butte $\longrightarrow$ Missoula $\longrightarrow$ Lewiston
G	Butte $\longrightarrow$ Helena $\longrightarrow$ Great Falls

To allow time for plane repairs, Frank Drake, owner of Lame Duck, wants the flights to take place only on Mondays, Wednesdays and Fridays. He also wants no more than one flight per day visiting any of the towns. To see if this is possible he constructs the graph of Figure 6.1, which has seven vertices, one representing each of the proposed flights, and where an edge joins two vertices if the corresponding flights have a town in common.

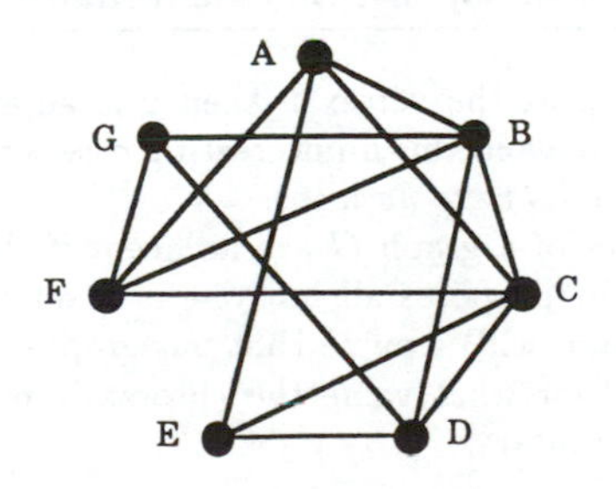

Figure 6.1: The Lame Duck Airlines network.

If there is a solution to his problem then he should be able to assign the labels Monday, Wednesday and Friday to the vertices, one per vertex, so that no two adjacent vertices have the same label.

Replacing the three days by colours red, white and blue, the problem becomes one of colouring the seven vertices so that adjacent vertices have different colours. This is a typical example of how colours can be used in graphs to model problems where one wishes to avoid some form of "interference" or ensure some "independence" — here two flights interfere with each other if they have a town in common. Further examples are given in the exercises at the end of this section.

We now formalise the above ideas.

> Let G be a graph. A **(vertex) colouring** of G assigns colours, usually denoted by 1, 2, 3, ..., to the vertices of G, one colour per vertex, so that adjacent vertices are assigned different colours.
>
> A **k-colouring** of G is a colouring which consists of k different colours and in this case G is said to be **k-colourable**.

For example, in the Lame Duck Airlines problem, the question being asked amounts to whether or not the graph of Figure 6.1 has a 3-colouring. Figure 6.2 shows that it has a 4-colouring.

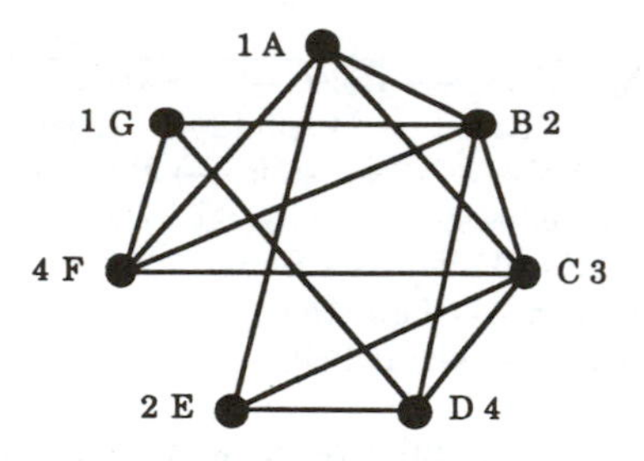

Figure 6.2: Lame Duck flies again.

> The minimum number n for which there is an n-colouring of the graph G is called the **chromatic index** (or **chromatic number**) of G and is denoted by $\chi(G)$. If $\chi(G) = k$ we say that G is **k-chromatic**.

If the graph G has a loop at the vertex v then v is adjacent to itself and so no colouring of G is possible. To avoid this uninteresting case we will assume that in any vertex colouring context graphs have *no loops*.

Also, two distinct vertices of a graph G are adjacent if there is at least one edge between them and so for our purposes all but one of a set of parallel edges may be ignored. In short, we can (and will) assume that our graphs are simple.

We now try to get a feel for what value the chromatic number of a graph takes. Our first result is quite elementary.

Theorem 6.1 *(a) If the graph G has n vertices, $\chi(G) \leq n$.*
(b) If H is a subgraph of the graph G then $\chi(H) \leq \chi(G)$.
(c) $\chi(K_n) = n$ for all $n \geq 1$.
(d) If the graph G contains K_n as a subgraph, $\chi(G) \geq n$.
(e) If the graph G has $G_1, \ldots, G_n$ as its connected components then

$$\chi(G) = \max_{1 \leq i \leq n} \chi(G_i).$$

Proof This is Exercise 6.1.1. □

From Theorem 6.1(d) we can now solve the Lame Duck Airlines problem. The problem's graph G in Figure 6.1 has K_4 as a subgraph (induced by the vertices A, B, C and F) and so $\chi(G) \geq 4$. Since we have displayed a 4-colouring of G in Figure 6.2 this means that $\chi(G) = 4$. Thus the seven flights, can be scheduled on four days but not three, subject to the stated restrictions.

Let us now look at some simple examples. Firstly if the graph G has no edges then each vertex can be given the same colour, i.e., $\chi(G) = 1$. Clearly the converse also holds. Thus $\chi(G) = 1$ if and only if G is an empty graph.

Now let $G = C_n$, the cycle of length n, with vertices $v_1, \ldots, v_n$ appearing in order round the cycle. If we assign colour 1 to v_1, v_2 must be coloured differently, say by colour 2. But then we may colour v_3 by 1 again. Continue in this fashion round the cycle and we see that if n is odd then v_n needs a different colour, 3. Thus $\chi(C_n) = 2$ if n is even, 3 if n is odd.

We can in fact completely characterise graphs with chromatic index 2. Our next result shows that they are old friends.

Theorem 6.2 *Let G be a nonempty graph. Then $\chi(G) = 2$ if and only if G is bipartite.*

Proof Let G be bipartite with bipartition $V = X \cup Y$. Assigning colour 1 to all vertices in X and colour 2 to all vertices in Y gives a 2-colouring for G and so, since G is nonempty, $\chi(G) = 2$.

Conversely, suppose that $\chi(G) = 2$. Then G has a 2-colouring. Denote by X the set of all those vertices coloured 1 and by Y the set of all those vertices coloured 2. Then no two vertices in X are adjacent and similarly for Y. Thus any edge in G must join a vertex in X and a vertex in Y. Hence G is bipartite with bipartition $V = X \cup Y$. □

Corollary 6.3 *Let G be a graph. Then $\chi(G) \geq 3$ if and only if G has an odd cycle.*

Proof This follows from Theorem 1.3. □

Unlike the $n = 2$ case there is no easy characterisation of graphs with chromatic index 3, or, for that matter, higher index. However there are various results which give upper bounds for the chromatic index of an arbitrary graph G, provided we know the degrees of all their vertices of G. For the first of these we need some notation.

For a graph G we let

$$\Delta(G) = \max\{d(v) : v \text{ is a vertex of } G\}.$$

Thus $\Delta(G)$ is the **maximum vertex degree** of G.

Theorem 6.4 *For any graph G, $\chi(G) \leq \Delta(G) + 1$.*

Proof We use induction on n, the number of vertices in G. The theorem is clearly true for $n = 1$ since here $G = K_1$, $\chi(G) = 1$ and $\Delta(G) = 0$.

Now suppose that the result is true for all graphs with $n - 1$ vertices, where n is a fixed integer greater than 1, and let G be some graph with n vertices. Choose a vertex v of G. Then the subgraph $G - v$ has $n - 1$ vertices and so, by the induction assumption, $\chi(G - v) \leq \Delta(G - v) + 1$. This allows us to choose a vertex colouring of $G - v$ involving $\Delta(G - v) + 1$ colours. Now our vertex v has at most $\Delta(G)$ neighbours in G and so these neighbours use up at most $\Delta(G)$ colours in the colouring of $G - v$. Thus if $\Delta(G) = \Delta(G - v)$ there is at least one colour not used by v's neighbours and we can use such a colour for v, giving a $(\Delta(G) + 1)$-colouring for G. On the other hand, if $\Delta(G) \neq \Delta(G - v)$ then $\Delta(G - v) < \Delta(G)$ and simply colouring v with a brand new colour gives a $(\Delta(G - v) + 2)$-colouring of G which is good enough since $\Delta(G - v) + 2 \leq \Delta(G) + 1$. Hence, in both cases, we have $\chi(G) \leq \Delta(G) + 1$. □

Notice that if $G = K_n$ or a cycle of odd length we actually have $\chi(G) = \Delta(G) + 1$. However we can often improve upon Theorem 6.4. For this purpose we now describe a technique which allows us, in certain circumstances, to improve upon a colouring.

Let G be a graph with a colouring involving at least two different colours, denoted by i and j. Let $H(i, j)$ denote the subgraph of G induced by all the vertices of G coloured either i or j. Let K be a connected component of this subgraph. Then, as the reader can easily check, if we interchange the colours i and j on the vertices of K but leave the colours of all other vertices of G unchanged then we get a new colouring of G, involving the same colours we started with. Such a subgraph K is called a **Kempe chain** and the recolouring technique is called the **Kempe chain argument**. Figure 6.3 illustrates this for a graph with vertex colours 1, 2, ..., 5.

We will use the Kempe chain argument in the proof of the improvement to Theorem 6.4. Although this improvement, due to Brooks [9], seems a modest one, the proof is quite intricate.

Theorem 6.5 (Brooks, 1941) *Let G be a connected graph with $\Delta(G) \geq 3$. If G is not complete then $\chi(G) \leq \Delta(G)$.*

Proof We use induction on n, the number of vertices of G. Since $\Delta(G) \geq 3$ our induction starts at $n = 4$ and so, since G is not complete, in this case G is one of the graphs shown in Figure 6.4. It is straightforward to see that the chromatic index is at most 3 in each of these cases.

Now assume that $n \geq 5$ and that the result is true for $n - 1$. By looking back at the proof of Theorem 6.4 we see that if G has a vertex v of degree less that $\Delta(G)$ then

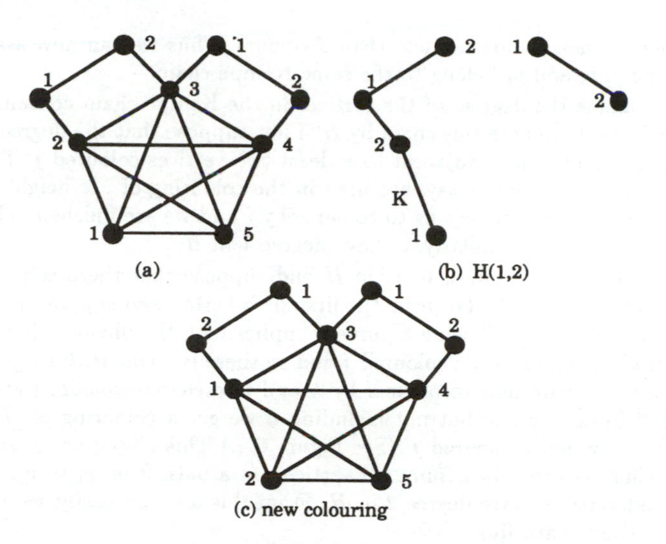

Figure 6.3: The Kempe chain argument.

we can colour G by $\Delta(G)$ colours, since the neighbours of v use up at most $\Delta(G) - 1$ colours. Thus in this case the result is proved. This means that we can now assume that every vertex of G has degree $\Delta(G)$, i.e., setting $d = \Delta(G)$, that G is d-regular. We will finish the proof if we show that G has a k-colouring.

Let v be some fixed vertex of G. Then, by our induction hypothesis, the subgraph $G - v$ has a d-colouring. If the neighbours of v in G do not use up all the d colours in this d-colouring of $G - v$ then any unused colour may be applied to v giving a d-colouring of G, as required. Thus we are left to deal with the case where the d neighbours of v use up all the d colours in $G - v$'s colouring. Let these neighbours be $v_1, \dots, v_d$, coloured by the colours $1, \dots, d$ respectively.

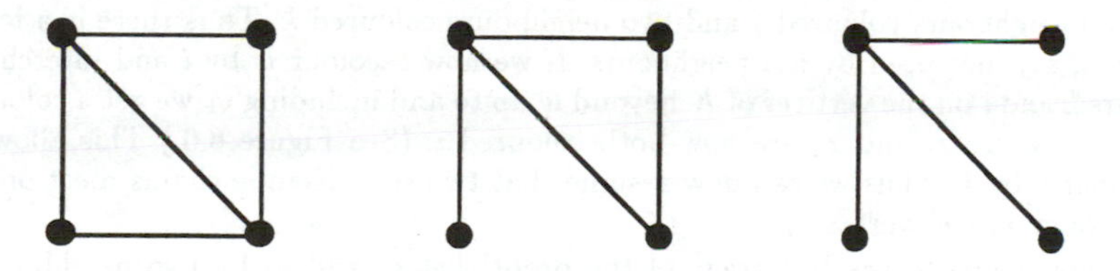

Figure 6.4: The incomplete simple graphs on four vertices.

Now suppose that there are two neighbours v_i and v_j such that the corresponding Kempe chains $H_{v_i}(i,j)$ and $H_{v_j}(i,j)$ containing v_i and v_j are different, i.e., v_i and v_j belong to different components of the subgraph $H(i,j)$ induced by the colours i and j. Then, by the Kempe chain argument, we can interchange the colours in $H_{v_i}(i,j)$ to give a d-colouring of $G - v$ where v_i now has colour j, the same as v_j's colour. However this brings us back to the situation we dealt with in the previous paragraph, namely when v's neighbours use less than d colours. Thus we can now assume that, for each i and j, v_i and v_j belong to the same Kempe chain.

We now examine the degrees of the vertices in the Kempe chain containing v_i and v_j. For simplicity we denote this chain by H. First suppose that the degree of v_i in H is greater than 1. Then v_i is adjacent to at least two vertices coloured j. This implies that there is a third colour, k say, not used in the colouring of v_i's neighbours. If we now recolour v_i by k this allows us to colour v by i and we are finished. Thus we can now assume that v_i and, similarly, v_j have degree 1 in H.

Now let P be a path from v_i to v_j in H and suppose that there is a vertex in P with degree at least 3 in H. Let u be the first such vertex and suppose it is coloured i. (If it is coloured j, the following argument applies with the obvious changes.) Then at least 3 of u's neighbours are coloured j and so there is a colour, k say, not used by these neighbours. If we now recolour u by k and interchange colours i and j on the vertices of P from x_i up to but not including u we get a colouring of $G - v$ where x_i and x_j are now both coloured j. (See Figure 6.5.) This allows us to colour v by i as before. Thus we can now assume all vertices on a path from v_i to v_j, apart from these two end vertices, have degree 2 in H. From this one can easily see that H just consists of a single path from v_i to v_j.

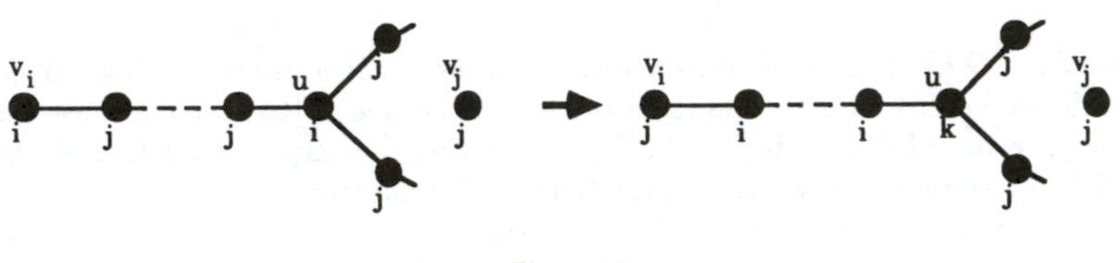

Figure 6.5

We are now in the situation where all the Kempe chains are paths. Let H and K be those chains corresponding to v_i, v_j and v_i, v_k respectively, where $j \neq k$. Suppose that w is a vertex different from v_i but a member of both chains. Then w is coloured i, has two neighbours coloured j and two neighbours coloured k. Thus there is a fourth colour, l say, not used by w's neighbours. If we now recolour w by l and interchange colours k and i on the vertices of K beyond w up to and including v_k we get a colouring of $G - v$ where x_i and x_k are now both coloured i. (See Figure 6.6.) This allows us to colour v by k. Thus we can now assume that two such Kempe chains meet only at their common end vertex v_i.

We now come to the last stage of the proof! Let v_i and v_j be two neighbours of v which are not adjacent and let x be the vertex coloured j adjacent to v_i on the Kempe chain H from v_i to v_j. With $k \neq j$, let K denote the Kempe chain from v_i to v_k. Then, by the Kempe chain argument, we can interchange the colours in K without interfering with the colours of the other vertices. This results in v_i coloured k and v_k coloured i. Since w is adjacent to v_i it must be in the Kempe chain for the colours k and j. However it is also the the Kempe chain for the colours i and j. This contradicts our assumption that Kempe chains have at most one vertex in common, the end vertex.

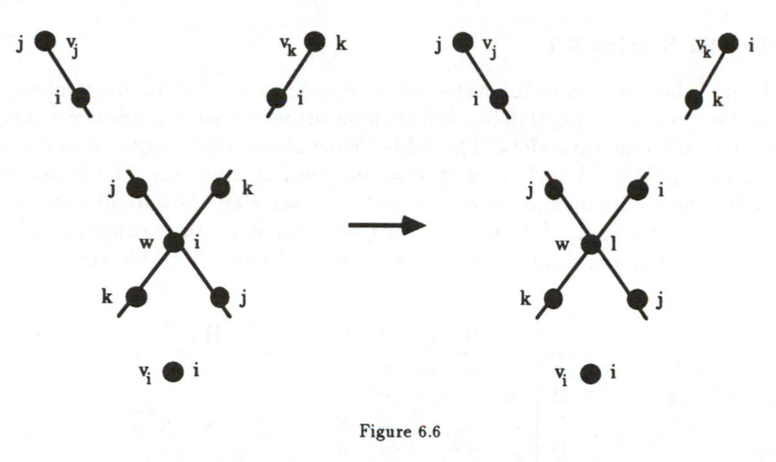

Figure 6.6

This contradiction implies that any two v_i and v_j must be adjacent. In other words all neighbours of v are neighbours of each other. This implies that G is the complete graph K_d, a contradiction to the hypothesis on G. Thus our proof is complete. □

Combined with Theorem 6.1(d), Brooks' Theorem provides estimates for the chromatic index. For example, for the graph G_1 of Figure 6.7, $\Delta(G_1) = 8$ while G_1 has K_4 as a subgraph. Thus $4 \leq \chi(G_1) \leq 8$ (and it is not difficult to see that $\chi(G_1) = 4$). Similarly for G_2 of Figure 6.7, known as the **Birkhoff diamond**, $\Delta(G_2) = 5$ while G_2 has K_3 as a subgraph. Thus $3 \leq \chi(G_2) = 5$. We leave it to the reader to show that $\chi(G_2) = 4$.

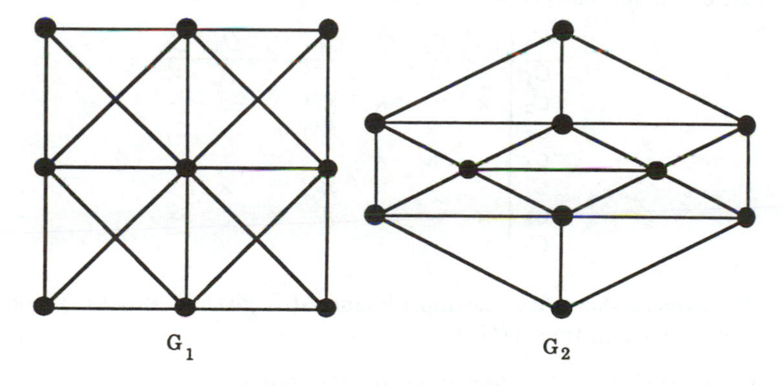

Figure 6.7: Two 4-chromatic graphs.

On the other hand, as Exercise 6.1.3 shows, the upper bound of $\Delta(G)$ may be quite different from $\chi(G)$.

Exercises for Section 6.1

6.1.1 A high school wishes to timetable for examinations in nine different subjects. Of course if there is a pupil doing two of these subjects their examinations must be held in different time slots. The table below shows (by crosses) which pairs of subjects, labelled A to I, have at least one pupil in common. The school wishes to find the minimum number of time slots necessary and how to allocate subjects to times accordingly. Interpretting this problem as a graph colouring problem, find the minimum number of time slots needed and a suitable time allocation of the subjects.

	A	B	C	D	E	F	G	H	I
A		×	×	×					
B	×			×					
C	×				×			×	×
D	×	×			×	×			
E			×	×		×	×		
F				×	×		×		
G					×	×		×	
H			×				×		×
I			×					×	

6.1.2 An industrial company wishes to store seven different chemicals $C_1, \ldots, C_7$, but since some of these can not be stored together safely different locations are needed. The table below indicates (by crosses) which pairs of chemicals cannot be stored together. Use graph colouring to find the minimum number of locations needed and how one can assign the chemicals to such an optimal solution.

	C_1	C_2	C_3	C_4	C_5	C_6	C_7
C_1		×				×	×
C_2	×		×	×			
C_3		×		×	×		
C_4		×	×		×	×	
C_5			×	×		×	×
C_6	×			×	×		×
C_7	×				×	×	

6.1.3 (This exercise shows that the upper bound $\Delta(G)$ given by Brooks' Theorem can be quite different from $\chi(G)$.)

(a) Show that for the wheel graph W_n the chromatic index is 3 if n is odd and 4 if n is even. What is $\Delta(W_n)$?

(b) Find $\chi(G)$ and $\Delta(G)$ when G is the k-cube. (See Exercise 1.4.13.)

6.1.4 (a) A graph G is called **k-partite** if the vertex set V can be partitioned into k nonempty sets $V_1, \ldots, V_k$ such that every edge of G joins vertices from different subsets. Thus $k = 2$ gives us the bipartite graphs. Generalise part of Theorem 6.2 by showing that if G is k-partite then $\chi(G) \leq k$.

(b) The k-partite graph G is called **complete k-partite** if, for each $i \neq j$, every vertex of the subset v_i is adjacent to every vertex of the subset v_j. Show that any k-chromatic graph is a subgraph of a complete k-partite graph.

6.1.5 Prove that $\chi(G) = \Delta(G) + 1$ if and only if G is either a complete graph or a cycle of odd length.

6.1.6 Show that if G contains exactly one odd cycle then $\chi(G) = 3$.

6.1.7 Let G be a graph in which any pair of odd cycles have a common vertex. Prove that $\chi(G) \leq 5$. (Assume that $\chi(G) \geq 6$ and consider the subgraphs H_1 and H_2 where H_1 is induced by the vertices coloured 1, 2 or 3 and H_2 is induced by the other vertices.)

6.2 Vertex Colouring Algorithms

There is no good algorithm for colouring the vertices of a graph using the minimum number of colours and consequently there is no good algorithm for determining the chromatic index of a graph. However there are a number of colouring algorithms which give approximations to minimal colourings and we present some of these in this section.

The first of these is called the **simple sequential algorithm**. This starts with any ordering of the vertices of the graph G, say $v_1, \ldots, v_n$. Now assign colour 1 to v_1. Moving to vertex v_2 colour it 1 if it is not adjacent to v_1; otherwise, colour it 2. Proceeding to v_3, colour it 1 if it is not adjacent to v_1; if it is adjacent to v_1, colour it 2 if it is not adjacent to v_2; otherwise colour it 3. Proceed in this manner, colouring each vertex with the first available colour that has not been used by any of its neighbours.

We describe this in stepwise fashion.

The Simple Sequential Colouring Algorithm

Step 1. List the vertices of G as $x_1, \ldots, x_n$. List the colours available as $1, 2, \ldots, n$.

Step 2. For each $i = 1, \ldots, n$, let $C_i = \{1, 2, \ldots, i\}$, the list of colours that could colour vertex x_i.

Step 3. Set $i = 1$.

Step 4. Let c_i be the first colour in C_i and assign this to vertex x_i.

Step 5. For each j with $i < j$ and x_i adjacent to x_j in G set $C_j = C_j - \{c_i\}$. (This means that x_j will not be given the same colour as x_i.) Change i to $i+1$ and if $i+1 \leq n$ return to Step 4.

Step 6. Record each vertex and its colour.

We illustrate the algorithm using the graph of Figure 6.8.

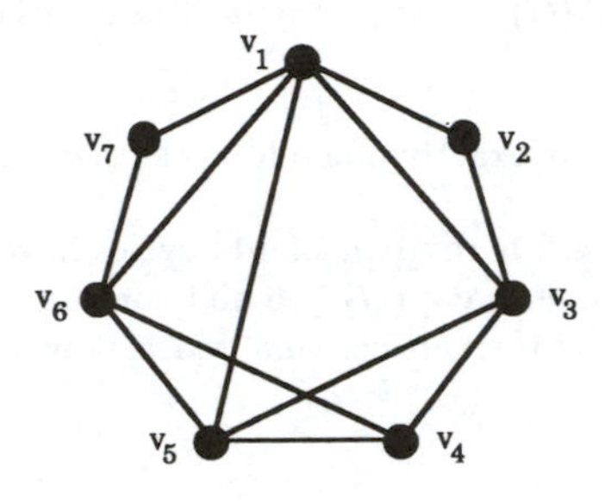

Figure 6.8

Step 1. List the vertices as shown: $v_1, \ldots, v_7$. The colours available are 1, 2, ...,7.

Step 2. $C_1 = \{1\}, C_2 = \{1,2\}, \ldots, C_7 = \{1,2,\ldots,7\}$.

Step 3. $i = 1$.

Step 4. 1 is the first colour in C_1 so assign it to vertex v_1.

Step 5. v_2, v_3, v_5, v_6 and v_7 are adjacent to v_1 so we get
$C_2 = \{1,2\} - \{1\} = \{2\}$, $C_3 = \{2,3\}$, $C_4 = \{2,3,4\}$,
$C_5 = \{2,3,4,5\}$, $C_6 = \{2,3,4,5,6\}$, $C_7 = \{2,3,4,5,6,7\}$.

i becomes 2 and we return to

Step 4. 2 is the first colour in C_2 so we assign this to v_2.

Step 5. v_3 is adjacent to v_2 so C_3 becomes $\{2,3\} - \{2\} = \{3\}$.

Change i to 3 and return to

Step 4. 3 is the first colour in C_3 so we assign this to v_3.

Step 5. v_4 and v_5 are adjacent to v_3 so C_4 becomes $\{1,2,3,4\} - \{3\} = \{1,2,4\}$ and C_5 becomes $\{2,4,5\}$.

i becomes 4 and we return to

Step 4. 1 is the first colour in C_4 so we assign this to v_4.

Step 5. v_5 and v_6 are adjacent to v_4 so C_5 stays as $\{2,4,5\}$ and C_6 stays as $\{2,3,4,5,6\}$.

i becomes 5 and we return to

Step 4. 2 is the first colour in C_5 so we assign this to v_5.

Step 5 v_6 is adjacent to v_5 so C_6 becomes $\{3,4,5,6\}$.

i becomes 6 and we return to

Step 4. 3 is the first colour in C_6 so we assign this to v_6.

Step 5. v_7 is adjacent to v_6 so C_7 becomes $\{2,4,5,6,7\}$.

i becomes 7 and we return to

Step 4. 2 is the first colour in C_7 so we assign this to v_7.

i becomes 8.

Step 6. v_1 and v_4 are coloured 1.
v_2,v_5 and v_7 are coloured 2.
v_3 and v_6 are coloured 4.

Although in this example the algorithm has clearly given a minimal colouring this need not always be the case even for relatively small graphs. For example, for the graph G of Figure 6.9, with the vertex ordering $x_1, x_2, \ldots, x_6$ as shown, the algorithm gives a 3-colouring (also shown), but $\chi(G) = 2$ since G is bipartite.

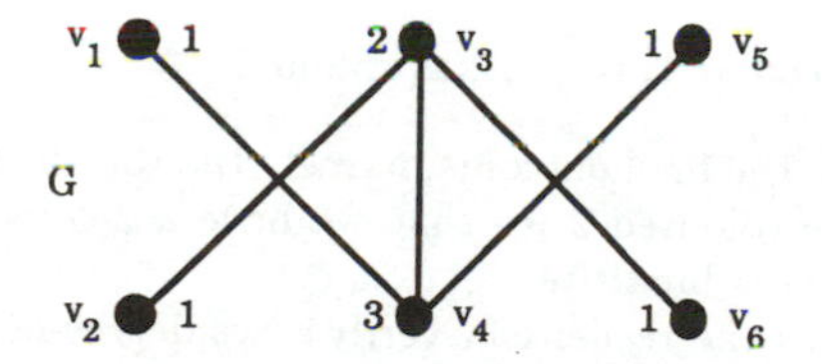

Figure 6.9: The simple sequential colouring algorithm does not give a minimal colouring here.

Not surprisingly the actual listing of the vertices used can affect the outcome of the algorithm. We now describe a listing which may encourage a smaller number of colours used in the final colouring.

Roughly speaking, if a vertex has large degree and many of its neighbours have already been coloured by the algorithm process, then it is more likely that we have to colour it with a previously unused colour. For this reason we initially list the vertices of the graph in decreasing order of their degrees. This modification of the simple sequential colouring algorithm is called the **largest-first sequential algorithm** or the **Welsh and Powell algorithm** (after its proposers [63]). As we said, the only difference between it and the simple version is in its initial step. We now formally record this new Step 1.

The Largest-First Sequential Algorithm (Welsh and Powell)

Step 1. List the vertices of G as $x_1, \ldots, x_n$ so that $d(x_1) \geq d(x_2) \geq \cdots \geq d(x_n)$.

List the colours available as $1, 2, \ldots, n$.

Now proceed with Steps 2 through to 6 of the simple sequential colouring algorithm.

We note that there may be some choice in the listing in Step 1 — in particular if the graph is regular! We illustrate the Welsh and Powell algorithm using the graph of Figure 6.9.

Step 1. Take $x_1 = v_3$, $x_2 = v_4$, $x_3 = v_1$, $x_4 = v_5$, $x_5 = v_2$, $x_6 = v_6$. This gives $d(x_1) \geq d(x_2) \geq \cdots \geq d(x_6)$, as required.

The colours available are 1, 2, ...,6.

Step 2. $C_1 = \{1\}$, ..., $C_6 = \{1,2,3,4,5,6\}$.

Step 3. $i = 1$.

Step 4. x_1 is assigned colour 1, i.e., v_3 has colour 1.

Step 5. $C_2 = \{2\}$, $C_5 = \{2,3,4,5\}$, $C_6 = \{2,3,4,5,6\}$.

i becomes 2.

Step 4. x_2 is assigned colour 2, i.e., v_4 has colour 2.

Step 5. $C_3 = \{1,3\}$, $C_4 = \{1,3,4\}$. i becomes 3.

Step 5. x_3 is assigned colour 1, i.e., v_1 has colour 1

and so on until we reach the final outcome, namely the top three vertices are coloured 1 and the bottom three coloured 2, so that we have a 2-colouring. This is the best possible outcome since G is bipartite.

In fact it turns out (as the reader can verify) *any* degree-descending listing of the vertices of this graph G will give a 2-colouring of G so this is indeed an improvement on the arbitrary simple sequential colouring algorithm.

The Welsh and Powell algorithm does not always give a minimal colouring, even for bipartite graphs, as can be seen in Exercise 6.2.1.

However it can be shown that the algorithm always uses at most $\Delta(G) + 1$ colours. (Recall Theorem 6.4.)

We now describe a third algorithm, called the **smallest-last sequential algorithm**, due to Matula, Marble and Isaacson [43], which often performs better than the Welsh and Powell algorithm. It is also a modification of the simple sequential colouring algorithm. This time we create a list by first choosing x_n, the last vertex in the list, as a vertex of minimum degree in G. Then we choose x_{n-1}, the second last vertex in the list, to be a vertex of minimum degree in the vertex-deleted subgraph $G - x_n$. Continue in this way creating vertex-deleted subgraphs and choosing vertices of minimum degree. This gives us the following initial step.

The Smallest-Last Sequential Algorithm

Step 1. (a)Choose x_n to be a vertex of minimum degree in G.

(b)For $i = n-1, n-2, \ldots, 1$, choose x_i to be a vertex of minimum degree in the vertex-deleted subgraph $G - \{x_n, x_{n-1}, \ldots, x_{i+1}\}$.

(c)List the vertices as $x_1, \ldots, x_n$.

(d)List the colours available as $1, \ldots, n$.

Now proceed with Steps 2 through to 6 of the simple sequential colouring algorithm.

We forego a detailed worked example on the smallest-last sequential algorithm. Instead we refer the reader to the graph of Figure 6.10. There the given listing $x_1, x_2, \ldots, x_7$ has been produced by Step 1, parts (a) and (b), of the algorithm (although other listings are also possible).The algorithm then continues to produce a 3-colouring for the graph with vertices x_1, x_3 and x_7 coloured 1, x_2 and x_5 coloured 2 and x_4 and x_6 coloured 3. (The reader should verify this!)

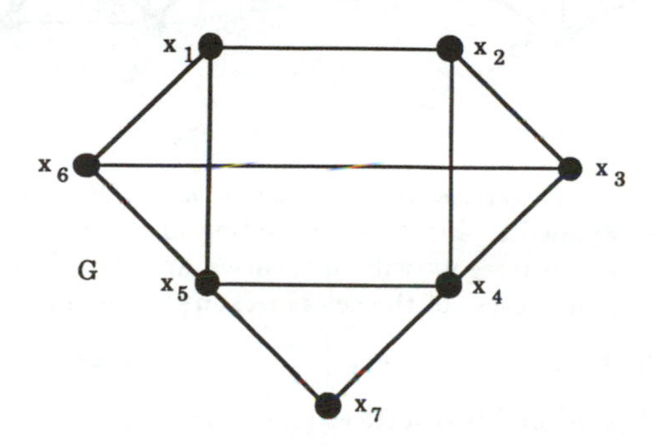

Figure 6.10: The smallest-last sequential algorithm at work.

Exercises for Section 6.2

6.2.1 Find a colouring of the regular bipartite graph G in Figure 6.11 using the simple sequential colouring algorithm on the list $x_1, x_2, \ldots, x_8$ as indicated. Find a colouring of G by also using the smallest-last sequential algorithm.

6.2.2 Prove that for any $k \geq 3$ there is a bipartite graph G with vertices listed as $x_1, \ldots, x_n$ such that the simple sequential algorithm produces a k-colouring of G using this list.

6.2.3 Carry out each of the three algorithms on the graphs in Figure 6.12, using the listing (if given) for the simple sequential colouring algorithm.

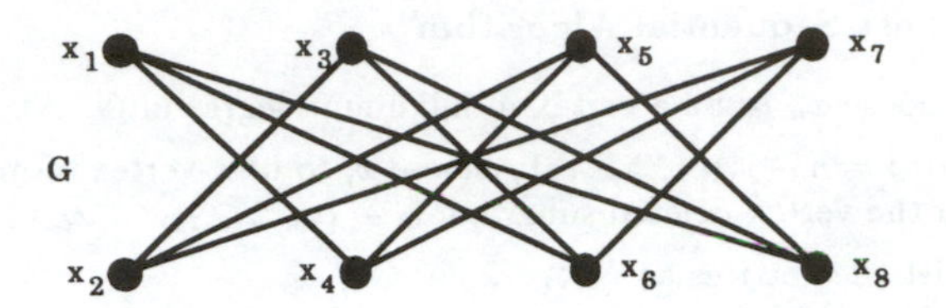

Figure 6.11

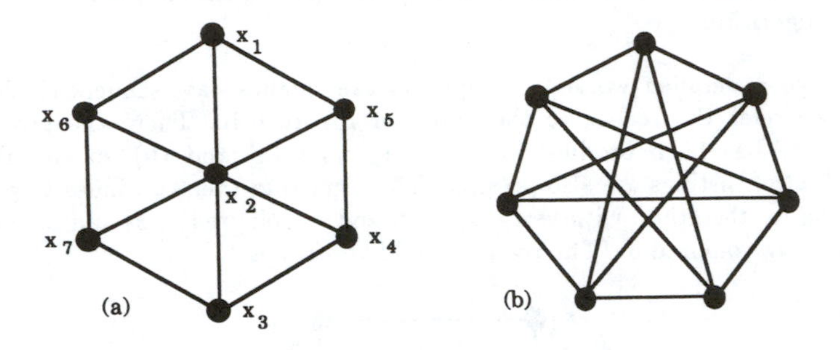

Figure 6.12

6.2.4 (The graphs in this exercise are featured in a recent article by Butcher [10] where they are applied to a method of bidding in the game of contract bridge.) Let m and n be two integers with n positive and $0 \leq k \leq n$. Let $\mathcal{P}(k,n)$ be the collection of all subsets of the set $\{1,\ldots,n\}$ consisting of k elements. Then $|\mathcal{P}(k,n)|$ is the binomial coefficient $\binom{n}{k}$. We now define a graph $\Gamma_{k,n}$ whose vertex set is $\mathcal{P}(k,n)$ and two vertices (subsets) are joined by an edge if one can be obtained from the other by replacing just one integer.

For example, for $k = 2$ and $n = 4$ the 6 vertices are $v_1 = \{1,2\}$, $v_2 = \{1,3\}$, $v_3 = \{2,3\}$, $v_4 = \{3,4\}$, $v_5 = \{1,4\}$, $v_6 = \{2,4\}$ and $\Gamma_{2,4}$ is shown in Figure 6.13.

Also, for $k = 2$ and $n = 6$ the 10 vertices are $v_1 = \{1,2\}$, $v_2 = \{1,3\}$, $v_3 = \{2,3\}$, $v_4 = \{2,4\}$, $v_5 = \{3,4\}$, $v_6 = \{3,5\}$, $v_7 = \{4,5\}$, $v_8 = \{1,4\}$, $v_9 = \{2,5\}$, $v_{10} = \{2,5\}$ and $\Gamma_{2,5}$ is also given in Figure 6.13.

(a) Prove that $\Gamma_{k,n}$ is a regular graph. What is the degree of each of its vertices?

(b) $\Gamma_{0,n}$ and $\Gamma_{1,n}$ are well-known graphs in disguise. Reveal their identity!

(c) Show that $\Gamma_{k,n}$ and $\Gamma_{n-k,n}$ are isomorphic.

(d) Using the listing $v_1,\ldots,v_n$ as shown in Figure 6.13, find a colouring of $\Gamma_{2,4}$ and $\Gamma_{2,5}$ using the simple sequential algorithm.

(e) Find a colouring of $\Gamma_{2,5}$ using the simple sequential algorithm and the listing v_1, v_{10}, v_2, v_9, v_3, v_8, v_4, v_7, v_5, v_6.

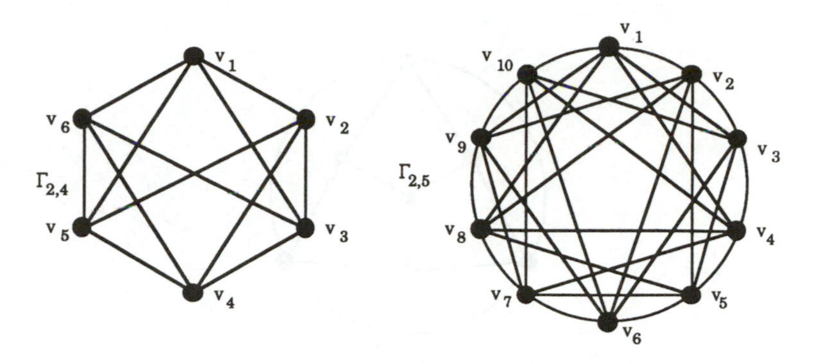

Figure 6.13: $\Gamma_{2,4}$ and $\Gamma_{2,5}$.

6.3 Critical Graphs

In this section we look at graphs with chromatic index k but which are very close to having smaller index. These graphs have proved useful in attempts to find suitable upper bounds for the chromatic index.

> A graph G is called **k-critical** if $\chi(G) = k$ and $\chi(G - v) < k$ for each vertex v of G.

Thus a graph is k-critical if it needs k colours but each of its vertex deleted subgraphs can be coloured with less than k colours.

To get a feel for the definition let us look at k-critical graphs for small k values. A moment's thought shows that K_1 is the only 1-critical graph. A 2-critical graph G will be bipartite (by Theorem 6.2) and such that, for every vertex v, $G - v$ is 1-colourable, i.e., $G - v$ is an empty graph. Such a graph can not have more than one edge and from this we see that K_2 is the only 2-critical graph.

It turns out that the 3-critical graphs are precisely the odd cycle graphs C_{2n+1}. We leave this as Exercise 6.3.1, but note here that part (b) of the following theorem should be useful in this.

Since, as we have remarked earlier, there is no easy characterisation of k-chromatic graphs when $k \geq 3$, it is not surprising that there is similarly no easy description of k-critical graphs when $k \geq 4$.

Figure 6.14 shows a 4-critical graph — the verification of this is left to the reader in Exercise 6.3.3 — while another example is given later in Figure 6.15.

The next result, due to Dirac [17], describes some of the important features of a k-critical graph.

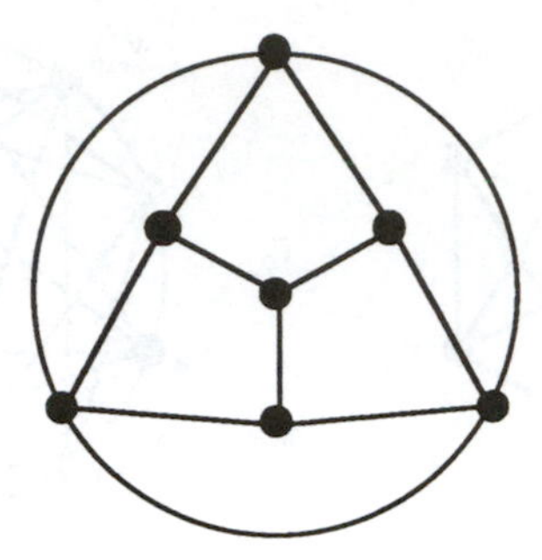

Figure 6.14: A 4-critical graph.

Theorem 6.6 (Dirac, 1952) *Let G be a k-critical graph. Then*

(a) G is connected,

(b) the degree of every vertex of G is at least $k-1$,

(c) G has no pair of subgraphs G_1 and G_2 for which $G = G_1 \cup G_2$ and $G_1 \cap G_2$ is a complete graph,

(d) $G - v$ is connected for every vertex v of G (provided $k > 1$).

Proof (a) Suppose that G is not connected. Since $\chi(G) = k$, by Theorem 6.1(e), there is a connected component C of G with $\chi(C) = k$. Then, if v is any vertex of G not in C, C will be a component of the subgraph $G - v$ and so, again by Theorem 6.1(e), $\chi(G - v) = \chi(C) = k$. This gives a contradiction since G is k-critical. Hence G must be connected.
(b) Suppose that v is a vertex of G with $d(v) < k-1$. Since G is k-critical, the subgraph $G-v$ has a $(k-1)$-colouring. Since v has at most $k-2$ neighbours these neighbours do not use up all the colours in this $(k-1)$-colouring of $G-v$. By colouring v with one of these unused colours we extend the colouring of $G-v$ to a $(k-1)$-colouring of G. This is a contradiction since $\chi(G) = k$. Hence every vertex v has degree at least $k-1$, as required.
(c) Suppose that $G = G_1 \cup G_2$ where G_1 and G_2 are subgraphs with $G_1 \cap G_2 = K_t$. Since G is k-critical, G_1 and G_2 both have chromatic index at most $k-1$. Choose a $(k-1)$-colouring for G_1 and one for G_2. In the overlap $G_1 \cap G_2$, since this is complete, every vertex has a different colour (in each of the $(k-1)$-colourings). This enables us to rearrange the colours in the $(k-1)$-colouring of G_2 so that it gives the same colour to each vertex in $G_1 \cap G_2$ as the colouring of G_1 gives. Combining the two colourings then produces a $(k-1)$-colouring of all of G. This is impossible since $\chi(G) = k$. Thus no such subgraphs G_1, G_2 exist.
(d) Suppose that $G - v$ is disconnected for some vertex v of G. Then $G - v$ has subgraphs H_1 and H_2 with $H_1 \cup H_2 = G - v$ and $H_1 \cap H_2 = \vee$. Set G_1 and G_2 as the subgraphs of G induced by H_1 together with v and H_2 together with v respectively. Then $G = G_1 \cup G_2$ and $G_1 \cap G_2 = K_1$ (with v_k as the single vertex). This contradicts (c) and so $G - v$ cannot be disconnected. □

We will use Theorem 6.6(b) to prove a result about k-chromatic graphs. First we show that any k-chromatic graph G contains a k-critical subgraph. To see this, note that if G is not k-critical then $\chi(G - v) = k$ for some vertex v of G. If $G - v$ happens to be k-critical then we have got our subgraph. If $G - v$ is not k-critical then $G - \{v, w\} = (G - v) - w$ has chromatic index k, for some vertex w in $G - v$. If this new subgraph is k-critical then, as before, we are finished. If not, we continue this vertex deletion procedure and we will eventually reach a k-critical subgraph.

Theorem 6.7 *Let G be a graph with $\chi(G) = k$. Then G has at least k vertices v such that $d(v) \geq k - 1$.*

Proof Let H be a k-critical subgraph of G. Then every vertex of H has degree at least $k - 1$ in H, by Theorem 6.6(b), and so has at least this degree in G also. Since H is k-chromatic it has at least k vertices and so this completes the proof. □

We note that Theorem 6.4 is an easy consequence of Theorem 6.7.

We finish this section with a result proved by Welsh and Powell [63] and related to their algorithm (presented in Section 6.2).

Theorem 6.8 *Let G be a graph and let $v_1, v_2, \ldots, v_n$ be a listing of the vertices of G such that*

$$d(v_1) \geq d(v_2) \geq \cdots \geq d(v_n).$$

Then

$$\chi(G) \leq \max_{1 \leq i \leq n} \{\min\{i, d(v_i) + 1\}\}.$$

Proof If G is an empty graph, $\chi(G) = 1$ and $d(v_i) = 0$ for each i. Thus

$$\max_{1 \leq i \leq n} \{\min\{i, d(v_i) + 1\}\} = \max_{1 \leq i \leq n} \{\min\{i, 1\}\} = 1$$

and the result follows in this case.

Now let G be a nonempty graph with $\chi(G) = k$. Then G has a k-critical subgraph. By Theorem 6.7, H has at least k vertices of degree at least k and consequently so has G. Thus in our given listing of G's vertices we have $d(v_i) \geq k - 1$ for $i = 1, \ldots, k$. In particular, $\min\{k, d(v_k) + 1\} = k$. Thus

$$\max_{1 \leq i \leq n} \{\min\{i, d(v_i) + 1\}\} \geq \chi(G),$$

as required. □

Exercises for Section 6.3

6.3.1 Prove that the 3-critical graphs are just the odd cycles C_{2n+1}. (Hint: use Theorems 6.6(b) and 3.1.)

6.3.2 Prove that the wheel W_{2n} is a 4-critical graph for each $n \geq 2$.

6.3.3 Show that the graph of Figure 6.14 is a 4-critical graph.

6.3.4 Let $G = G_1 + G_2$, the join of the two graphs G_1 and G_2.

(a) Prove that $\chi(G) = \chi(G_1) + \chi(G_2)$.

(b) Prove that G is a k-critical graph if and only if G_1 is k_1-critical and G_2 is k_2-critical, where $k = k_1 + k_2$.

6.3.5 Let G be a k-critical graph, where $k \geq 3$, and let u and v be two distinct vertices of G. Show that u has a neighbour which is not a neighbour of v. Hence show that G has at least $k + 2$ vertices.

6.3.6 Show that the graph of Figure 6.15 is 4-critical.

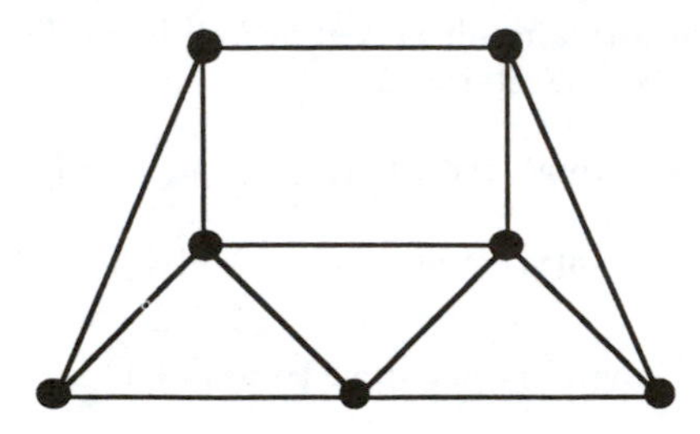

Figure 6.15: A 4-critical graph.

6.3.7 Use Theorem 6.8 to show that for $\overline{G}$, the complement of G,

$$\chi(\overline{G}) \leq \max_{1 \leq i \leq n} \{\min\{n - i + 1, n - d(v_i)\}\},$$

where $v_1, \ldots, v_n$ is a listing of the vertices of G such that $d(v_1) \geq \cdots \geq d(v_n)$.

6.3.8 Using induction on n, the number of vertices of G, prove that $\chi(G) + \chi(\overline{G}) \leq n + 1$.

6.4 Cliques

For any graph G a complete subgraph of G is called a **clique** of G.The number of vertices in a largest clique of G is called the **clique number** of G and denoted by $cl(G)$.

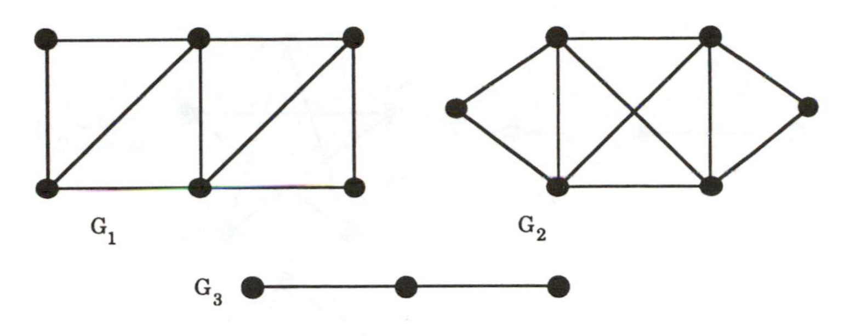

Figure 6.16

For example, in Figure 6.16, $cl(G_1) = 3, cl(G_2) = 4$ and $cl(G_3) = 2$.

It follows immediately from Theorem 6.1(d) that for any graph G,

$$cl(G) \leq \chi(G).$$

In particular if G contains a triangle, i.e., K_3, as a subgraph, then $\chi(G) \geq 3$. To see that the converse does *not* hold, note that if $n \geq 2$ then the cycle C_{2n+1} is triangle-free with chromatic index 3. In fact we now give a theorem which shows that one can construct graphs G with arbitrarily high chromatic index but clique number just 2. The construction is due to Mycielski [48].

Theorem 6.9 (Mycielski, 1955) *For every $k \geq 1$ there is a k-chromatic graph M_k with no triangle subgraphs.*

Proof We use induction on k. The graphs K_1 and K_2 show that the result is true for $k = 1$ and 2 respectively.

Now suppose that $k \geq 2$ and that M_k is a k-chromatic graph containing no triangles. We use M_k to construct a $(k+1)$-chromatic graph M_{k+1} containing no triangles.

First suppose that $v_1, v_2, \ldots, v_n$ are the vertices of M_k. The vertices of the new graph M_{k+1} are defined to be those of M_k together with an extra $n + 1$ vertices denoted by $u_1, u_2, \ldots, u_n, v$. The edge set of M_{k+1} is defined to consist of all the edges of M_k together with an edge from v to each of the u_i's and an edge from each u_i to each of the neighbours of v_i.

Starting with $M_2 = K_2$, Figure 6.17 shows the construction of M_3 and M_4.

We now show that M_{k+1} has no triangles. Assume this is false, i.e., that there is a triangle in M_{k+1}. Since M_k, by assumption, contains no triangles, such a triangle must have either v or at least one of the u_i's as a vertex. Since no two of the u_i's are adjacent and v is only adjacent to the u_i's this forces the triangle to be of the form $v_j v_k u_i v_j$. However, since u_i is adjacent to v_j and v_k, these latter vertices must be neighbours of v_i and we get the triangle $v_j v_k v_i v_j$, impossible since this lies in M_k. Thus M_{k+1} has no triangles, as required.

Now we show that $\chi(M_{k+1}) = k + 1$. Given a k-colouring of M_k we can extend this to a $(k+1)$-colouring of M_{k+1} by colouring each u_i by the colour assigned to v_i and then colouring v with a $(k+1)$st colour. (It is easily checked that this is a

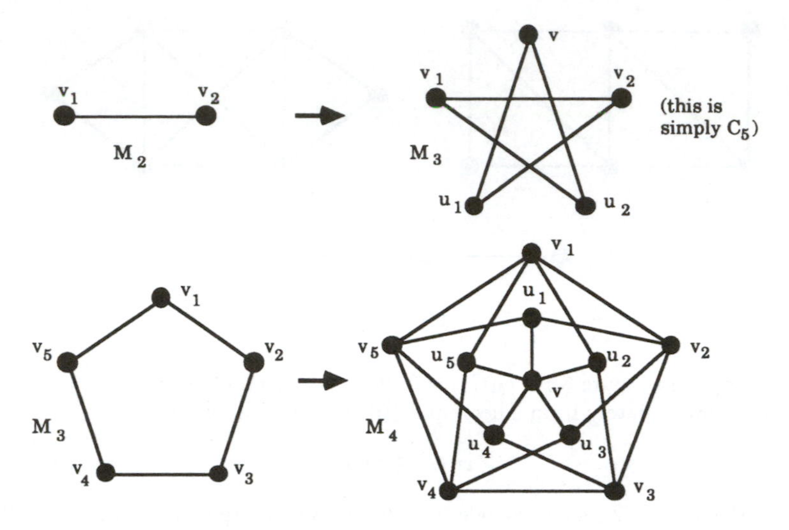

Figure 6.17: The Mycielski graphs M_2, M_3 and M_4.

$(k+1)$-colouring of M_{k+1}, because of the way the edge set of M_{k+1} is defined.) Thus $\chi(M_{k+1}) \leq k+1$ and so we must now show that M_{k+1} can not be k-coloured.

Suppose, to the contrary, that M_{k+1} has a k-colouring using the colours $1, 2, \ldots, k$. Suppose that the vertex v is coloured k. Then each vertex u_i can not be coloured k, since v and u_i are adjacent. Also, since $\chi(M_k) = k$, the colour k must be used to colour some vertices in M_k. Recolour those vertices v_i coloured k in M_k by the colour assigned to their corresponding u_i. Then, since u_i and v_i have exactly the same neighbours in M_k, this has produced a $(k-1)$-colouring of M_k, which is a contradiction since $\chi(M_k) = k$. It now follows that $\chi(M_{k+1}) = k+1$, finishing the proof. □

We note in passing that the graph M_4 shown in Figure 6.17 is know as the **Grötzsch graph**.

Although Mycielski's construction shows that in general the clique number may be quite different fom the chromatic index there is a large class of graphs where the numbers are equal. To describe this class we first need a result on the complement $\overline{G}$ of a graph G.

Theorem 6.10 *Let G be a graph. Then the following two statements are equivalent.*

(a) For every nonempty subset U of $V(G)$, either the induced subgraph $G[U]$ of G is disconnected or the induced subgraph $\overline{G}[U]$ of $\overline{G}$ is disconnected.

(b) G has no vertex induced subgraph isomorphic to P_4, the path of length 3.

Proof Suppose that there is a subset U of G such that the induced subgraph $G[U]$ of G is isomorphic to P_4. Since P_4 is its own complement, the induced subgraph $\overline{G}[U]$ of $\overline{G}$ is also isomorphic to P_4. In particular, $G[U]$ and $\overline{G}[U]$ are both connected. This argument shows that (a) implies (b).

Now suppose that G does not contain a vertex-induced subgraph isomorphic to P_4. We adopt a proof by contradiction to show that (a) holds. Thus we assume that G does contain a nonempty subset U such that both $G[U]$ and $\overline{G}[U]$ are connected. Clearly we can choose such a subset U to have as small a number of vertices as possible. Thus if T is any proper subset of U then either $G[T]$ or $\overline{G}[T]$ is connected. Also a moment's thought shows that our minimum subset U has at least 4 vertices.

Let u_1 be a vertex in U and set $U_1 = U - \{u_1\}$. Then, by the minimality of U, either $G[U_1]$ or $\overline{G}[U_1]$ is disconnected. Suppose that $G[U_1]$ is disconnected. (It suffices to do this since, once our argument is finished, we may replace G by $\overline{G}$ because P_4 is the complement of itself.) Then, since $\overline{G}[U]$ is connected, there is a vertex u_2 in U_1 such that $u_1u_2 \notin E(G)$, the edge set of G.

Denote the vertex set of the component of $G[U_1]$ containing u_2 by U_2. Then any vertex u_3 in U_2 has no neighbours in the set $U_1 - U_2$. Moreover, since $G[U]$ is connected, there is such a vertex u_3 adjacent to u_1 in G and also a vertex $u_4 \in U_1 - U_2$ also adjacent to u_1 in G. So thus far we have $u_1u_2 \in E(\overline{G})$ and $u_1u_3, u_1u_4 \in E(G)$.

Now let

$$X = \{x \in U_2 : x \text{ is a neighbour of } u_1 \text{ in } G\} \text{ and}$$

$$Y = \{y \in U_2 : y \text{ is a neighbour of } u_1 \text{ in } \overline{G}\}.$$

Then X and Y are both nonempty since $u_3 \in X$ and $u_2 \in Y$. Also clearly $U_2 = X \cup Y$ and $X \cap Y = \vee$. Moreover, since $G[U_2]$ is connected (by the definition of U_2) there are vertices $x \in X$ and $y \in Y$ such that $xy \in E(G)$.

Finally let $Z = \{u_4, u_1, x, y\}$. Then we have $u_4u_1, u_1x, xy \in E(G)$ but $u_1y \notin E(G)$, by the definition of Y, and $u_4x, u_4y \notin E(G)$, since u_4 is not in the component $G[U_2]$ of $G[U_1]$. It follows that $G[Z]$ is isomorphic to P_4, in contradiction to our initial assumption. Thus (b) does imply (a), as required. □

We now use Theorem 6.10 to prove our final result of this section. Due to Seinsche [56], it displays a class of graphs where the clique number does equal the chromatic index.

Theorem 6.11 (Seinsche, 1974) *Let G be a graph in which no set of four vertices induces P_4 as a subgraph. Then $\chi(G) = cl(G)$.*

Proof We use induction on n, the number of vertices of G. For $n = 1$ the result is true since $\chi(K_1) = cl(K_1) = 1$.

Now suppose that the result is true for all graphs satisfying the hypothesis and with at most n vertices where $n \geq 1$ is fixed. Let G be a graph with $n+1$ vertices in which no subset of four vertices induces P_4 as a subgraph. Then, by Theorem 6.10, for any nonempty subset U of $V(G)$ either $G[U]$ or $\overline{G}[U]$ is disconnected. In particular, taking $U = V(G)$, we see that either G or its complement $\overline{G}$ is disconnected.

First suppose that G is disconnected and denote its components by $G_1, G_2, \ldots, G_t$, so that $t \geq 2$. Then clearly no G_i contains P_4 as an induced subgraph. Thus, since each G_i has less vertices than G, our induction assumption gives $\chi(G_i) = cl(G_i)$ for each $i = 1, \ldots, t$. Since $\chi(G) = \max\{\chi(G_i) : 1 \leq i \leq t\}$, by Theorem 6.1(e), and $cl(G) = \max\{cl(G_i) : 1 \leq i \leq t\}$ (see Exercise 6.4.1), it follows that $\chi(G) = cl(G)$, as required.

Now suppose that $\overline{G}$ is disconnected and denote the complements of its components by $H_1, H_2, \ldots, H_s$, so that $s \geq 2$. Since G does not contain P_4 as an induced subgraph neither do each of the H_i's. Thus, again by our induction assumption, $\chi(H_i) = cl(H_i)$ for $i = 1, \ldots, s$. Moreover, by Exercise 1.5.5 (d), G is the join

$$G = H_1 + H_2 + \cdots + H_s.$$

Hence, by Exercise 6.3.4 (a), $\chi(G) = \sum_{i=1}^{s} \chi(H_i)$. Since $cl(G) = \sum_{i=1}^{s} cl(H_i)$ (by Exercise 6.4.3) it follows that $\chi(G) = cl(G)$, as required. The result now follows by induction. □

It is quite easy to see that the converse of Theorem 6.11 is false. Just take G to be P_4 itself.

Exercises for Section 6.4

6.4.1 Let the graph G have connected components $G_1, \ldots, G_t$. Prove that $cl(G) = \max\{cl(G_i) : 1 \leq i \leq t\}$.

6.4.2 Determine the clique number of a nonempty bipartite graph, the wheel W_n and the cycle C_n for each $n \geq 4$.

6.4.3 Let $G = G_1 + G_2$, the join of the two graphs G_1 and G_2. Prove that $cl(G) = cl(G_1) + cl(G_2)$.

6.4.4 Let G be a graph with n vertices. If $\chi(G) = n - 1$, what is $cl(G)$?

6.4.5 Let M_k be the kth Mycielski graph as constructed in the proof of Theorem 6.9.

(a) Show that if we take M_2 as K_2, M_k has $3.2^{k-2} - 1$ vertices for each $k \geq 2$. How many edges does M_k have?

(b) Prove, without using Theorem 6.11, that each M_k does not have any set of four vertices which induces P_4 as a subgraph.

(c) Using induction on k, show that each M_k is k-critical.

6.4.6 Prove that a graph G has a k-colouring if and only if it is a subgraph of some graph H such that $cl(H) \leq k$ and H does not contain P_4 as an induced subgraph.

6.5 Edge Colouring

In this section we turn our attention to colouring edges of a graph instead of vertices. Recall that two edges of a graph are called **adjacent** if they have an end vertex in common. As in the vertex colouring case we will assume that all *our graphs are simple.*

Let G be a nonempty graph. An **edge colouring** of G assigns colours, usually denoted by 1, 2, 3, ..., to the edges of G, one colour per edge, so that adjacent edges are assigned different colours.

A **k-edge colouring** of G is a colouring of G which consists of k different colours and in this case G is said to be **k-edge colourable.**

The minimum number n for which there is an n-colouring of G is called the **edge chromatic number** (or **edge chromatic index**) of G and is denoted by $\chi_1(G)$. If $\chi_1(G) = k$ we say that G is **k-edge chromatic.**

We begin our study of edge colouring by noting two elementary properties. Firstly, if H is a subgraph of G, then

$$\chi_1(H) \leq \chi_1(G).$$

Secondly, letting $\Delta(G)$ denote the maximum vertex degree of G as usual, we have

$$\Delta(G) \leq \chi_1(G), \tag{6.1}$$

since if v is any vertex of G with $d(v) = \Delta(G)$ then the $\Delta(G)$ edges incident with v must each have a different colour in any edge colouring of G.

As we will see later the inequality (6.1) is almost an equality in that $\chi_1(G)$ is either $\Delta(G)$ or $\Delta(G)+1$. However, before we prove this we will establish the edge chromatic index for two of your favourite classes of graphs, namely the bipartite graphs and the complete graphs. We begin by describing the edge analogue of the Kempe chains defined in Section 6.1. This will prove useful in proofs to come.

Let G be a graph with an edge colouring involving at least two different colours, denoted by i and j. Let $H(i,j)$ denote the subgraph of G induced by all the edges coloured either i or j. Let K be a connected component of this subgraph. Then, as the reader can easily check, K is just a path whose edges are alternately coloured by i and j and if we interchange the colours on these edges, but leave the colours on all the other edges of G unchanged, the result is a new colouring of G, involving the same initial colours. As in the vertex colouring situation we refer to such a component K as a **Kempe chain** and this recolouring technique as the **Kempe chain argument**.

Given an edge colouring of the graph G involving the colour i we say that i is **present** at a vertex v of G if there is an edge coloured i incident with v. If there is no such edge incident with v then we say that i is **absent** from v.

We are now ready for our bipartite result.

Theorem 6.12 *Let G be a nonempty bipartite graph. Then*

$$\chi_1(G) = \Delta(G).$$

Proof The proof is by induction on the number of edges of G. The result is clearly true if G has just one edge.

Now let G have more than one edge and assume that the result is true for all nonempty bipartite graphs with less edges than G has.

Since $\Delta(G) \leq \chi_1(G)$ it suffices to prove that G has a $\Delta(G)$-edge colouring. To simplify the notation let $\Delta(G) = k$.

Let e be some fixed edge of G. Then the edge-deleted subgraph $G - e$ is bipartite with less edges than G and so, by the induction assumption, has a $\Delta(G - e)$-edge colouring and so a k-colouring since $\Delta(G - e) \leq \Delta(G) = k$. We will show that the same k colours can be used to colour G.

Let the uncoloured edge e have end vertices u and v. Since $d(u) \leq k$ in G and e is uncoloured there is at least one of the k colours absent from u. Similarly at least one of these colours is absent from v. If there is a colour absent from both u and v simply use it to colour e and we get a k-edge colouring of G, as required. Thus we are left to deal with the case where there is a colour i present at u but absent from v and a colour j present at v but absent from u.

Let K be the Kempe chain containing u in the subgraph $H(i, j)$ induced by the edges coloured i or j. Now suppose that v is also in K. Then there is a path P in K from u to v. Since u and v are adjacent they do not belong to the same bipartition subset of the bipartite graph G and so the path P must have odd length. Moreover, since the colour i is present at u, the first edge of P is coloured i. Since the edges of P are alternately coloured i and j and P is of odd length this implies that the last edge of P, that incident with v, is also coloured i. This is a contradiction since i is absent from v. Hence v does not belong to the Kempe chain K.

We now use the Kempe chain argument on K. This interchanging of colours makes i now absent from u, but does not affect the colours of the edges incident with v. Thus i is absent from both u and v in this new k-edge colouring. As before we can safely colour edge e by i to produce a k-edge colouring of G. □

Before we go on to look at the edge chromatic index of complete graphs, let us briefly describe an application of Theorem 6.12 to the construction of Latin squares.

A **Latin square** (of order n) is an $n \times n$ matrix having the numbers $1, 2, \ldots, n$ as entries such that no single number appears in more than one row and in more than one column.

For example

$$\begin{bmatrix} 4 & 1 & 2 & 3 \\ 3 & 4 & 1 & 2 \\ 2 & 3 & 4 & 1 \\ 1 & 2 & 3 & 4 \end{bmatrix} \quad \text{and} \quad \begin{bmatrix} 3 & 4 & 5 & 1 & 2 \\ 5 & 1 & 2 & 3 & 4 \\ 2 & 3 & 4 & 5 & 1 \\ 4 & 5 & 1 & 2 & 3 \\ 1 & 2 & 3 & 4 & 5 \end{bmatrix}$$

are Latin squares of orders 4 and 5 respectively. Latin squares are used frequently by statisticians and quality control analysts in experimental design. We now show how we can construct a Latin square of order n using an n-edge colouring of the complete bipartite graph $K_{n,n}$. (Note that $\Delta(K_{n,n}) = n$ and so, by Theorem 6.12, $K_{n,n}$ does have an n-edge colouring but no edge colouring with less than n colours.)

Let $K_{n,n}$ have bipartition $V = X \cup Y$, where $X = \{x_1, \ldots, x_n\}$ and $Y = \{y_1, \ldots, y_n\}$, and denote the colours of the edge colouring by $1, \ldots, n$. We define the $n \times n$ matrix

$A = (a_{ij})$ by

$$a_{ij} = k \text{ where } x_i y_k \text{ is the edge coloured } j \text{ (incident with } x_i\text{)}.$$

Then, for each distinct pair of indices j_1 and j_2, $a_{ij_1} \neq a_{ij_2}$ (since otherwise there would be a pair of parallel edges).

This shows that each row of A has n different entries. Moreover, if for indices $i_1 \neq i_2$ we have $a_{i_1 j} = a_{i_2 j}$, say with the common value k, then the vertex y_k would be the end vertex of two different edges coloured j, which is of course impossible. Thus each column of A has n different entries. Hence the matrix A we have constructed is a Latin square.

We now turn our attention to the edge chromatic index of the complete graphs.

Theorem 6.13 *Let $G = K_n$, the complete graph on n vertices, $n \geq 2$. Then*

$$\chi_1(G) = \begin{cases} \Delta(G) & (= n-1) \text{ if } n \text{ is even} \\ \Delta(G)+1 & (= n) \text{ if } n \text{ is odd.} \end{cases}$$

Proof We first suppose that n is odd. Draw G as usual so that its vertices form a regular polygon (with the n edges on the perimeter having the same length). Colour the edges around the perimeter using a different colour for each edge. Now each of the remaining "internal" edges of G is parallel to exactly one on the perimeter and we assign it the same colour as we have assigned to this perimeter edge. Figure 6.18 shows K_7 partially edge coloured in this way.

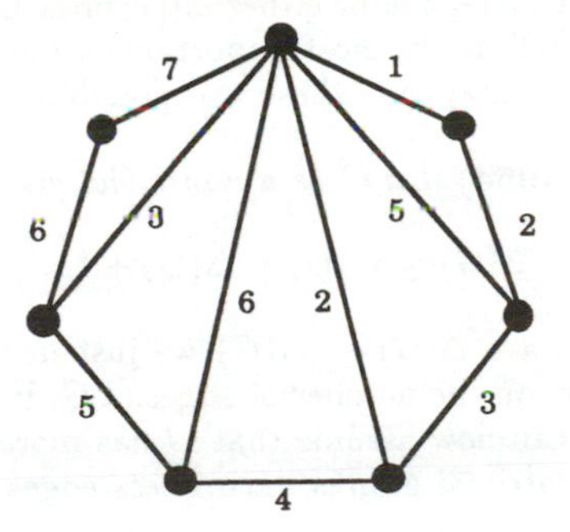

Figure 6.18: The beginnings of an edge colouring of K_7.

Then two edges have the same colour only if they are parallel and from this it follows that we do have an edge colouring of G. Since it involves n colours we have shown that $\chi_1(G) \leq n$ ($= \Delta(G)+1$).

Now suppose G has an $(n-1)$-colouring. Now, from the definition of an edge colouring, the edges of one particular colour form a matching in G and so, since n is odd, the maximum possible number of these is $\frac{1}{2}(n-1)$. This implies that there are at most $(n-1) \times \frac{1}{2}(n-1)$ edges in G, a contradiction since K_n has $\frac{1}{2}n(n-1)$ edges. Hence G does not have an $(n-1)$-colouring and so $\chi_1(G) = n$, as required.

We now deal with the case when n is even. Let v be some fixed vertex of G. Then $G - v$ is complete, with $n - 1$ vertices. Since $n - 1$ is odd we can give it an $(n - 1)$-colouring, as described above. With this colouring there is a colour absent from each vertex, with different vertices having different absentees. G is reformed from $G - v$ by joining each vertex w of $G - v$ to v by an edge. Colour each such edge with the colour absent from w. This gives an $(n - 1)$-colouring of G and since $\Delta(G) = n - 1$ we get $\chi_1(G) = \Delta(G) = n - 1$, as required. □

The last part of the proof of Theorem 6.13 is illustrated in Figure 6.19 for $n = 6$.

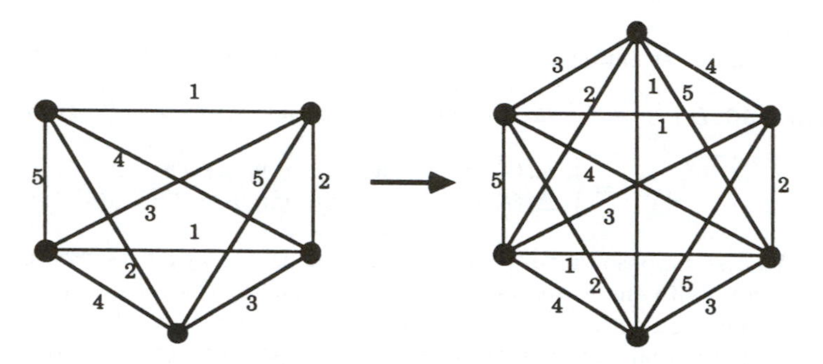

Figure 6.19: The edge colouring of K_6 induced by that of K_5.

Theorem 6.13 shows that $\chi_1(G)$ can be either $\Delta(G)$ or $\Delta(G)+1$ for complete G. The next result, due to Vizing [62], is the most important in the theory of edge colouring. It shows that for *all* (simple) graphs G these two possibilities are the only ones.

Theorem 6.14 (Vizing, 1964) *Let G be a nontrivial graph. Then*

$$\Delta(G) \leq \chi_1(G) \leq \Delta(G) + 1.$$

Proof Since we always have $\Delta(G) \leq \chi_1(G)$ we just need to show that $\chi_1(G) \leq \Delta(G) + 1$. We use induction on the number of edges of G. If G has just one edge then $\Delta(G) = 1 = \chi_1(G)$ so we can now assume that G has more than one edge and that the result is true for all nontrivial graphs having less edges than G has. To simplify the notation let $\Delta(G) = k$.

Let e be a fixed edge of G, with end vertices v_1 and v_2. Then, by our assumption, the edge-deleted subgraph $G - e$ has a $(k + 1)$-edge colouring, using the colours $1, 2, \ldots, k + 1$, say. Since $d(v_1)$ and $d(v_2)$ are both at most k there is at least one of these k colours absent from v_1 and at least one absent from v_2. If there is a common colour absent from both then we may use this to colour e and so produce a $(k + 1)$-colouring of G. So in this case $\chi_1(G) \leq k + 1$, as required. Hence we now assume that there is a colour, say 1, absent from v_1 but present at v_2 and there is a colour, say 2, absent from v_2 but present at v_1.

Starting with v_1 and v_2 we now construct a sequence of distinct vertices $v_1, v_2, \ldots, v_j$ where each v_i, for $i \geq 2$, is adjacent to v_1. The vertex v_3 is chosen so that $v_1 v_3$ is coloured 2 — v_3 does exist since 2 is present at v_1 but absent from v_2. For the next vertex, first note that not all the $k+1$ colours are present at v_3. If possible, choose a new colour 3, absent from v_3 but present at v_1, and let v_1v_4 be the edge coloured 3. Proceed now in this fashion by, if possible, choosing a new colour i absent from v_i, but present at v_1, and let v_1v_{i+1} be the edge coloured i. In this way we get a sequence of vertices $v_1, v_2, \ldots, v_j$ such that

(a) v_i is adjacent to v_1 for each $i > 1$,

(b) the colour i is absent from v_i for each $i = 1, \ldots, j-1$ and

(c) the edge v_1v_{i+1} is coloured i for each $i = 2, \ldots, j-1$.

We illustrate this partly in Figure 6.20.

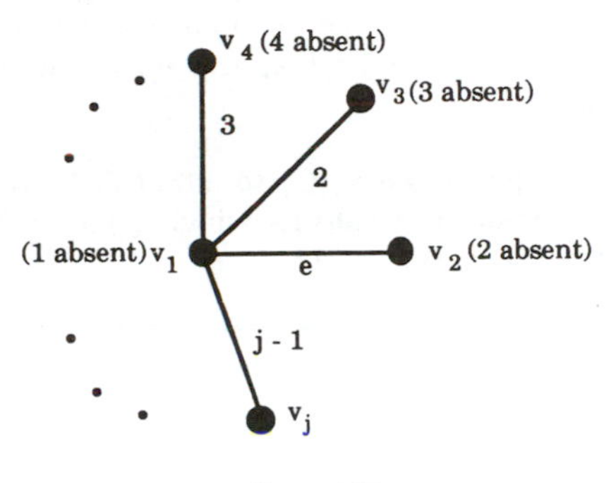

Figure 6.20

Since $d(v_1) \leq k$ it follows from (a) that such a sequence has at most $k+1$ terms, i.e., $j \leq k+1$. We now investigate a longest such sequence $v_1, \ldots, v_j$, i.e., one for which it is not possible to find a new colour j, absent from v_j, together with a new neighbour v_{j+1} of v such that vv_{j+1} is coloured j.

Let us first suppose that for some colour j absent from v_j there is no edge of that colour present at v_1. Now colour edge $e = v_1v_2$ with colour 2 and recolour the edges v_1v_i with colour i for $i = 3, \ldots, j-1$. Since i was absent from v_i for each $i = 2, \ldots, j-1$, this gives a $(k+1)$-colouring of the subgraph $G - v_1v_j$. However, since the colour j absent from v_j is also absent from v_1, we may use j to recolour v_1v_j and this produces a $(k+1)$-colouring of G, as required.

Thus we may now assume that no matter how we choose j as a colour absent from v_j then this colour is present at v_1. If v_{j+1} is a new neighbour of v_1 coloured by such a j then we have extended our sequence to $v_1, v_2, \ldots, v_j, v_{j+1}$, in contradiction to the assumption that $v_1, v_2, \ldots, v_j$ can not be extended further. Thus, choosing j to be a colour absent from v_j, one of the edges $v_1v_3, \ldots, v_1v_{j-1}$ must be coloured j, say v_1v_l (so that $j = l-1$). Now colour edge $e = v_1v_2$ by 2 and recolour each of the edges v_1v_i for $i = 3, \ldots, l-1$ by i. Remove the colour j from v_1v_l. Then we have a $(k+1)$-colouring of the edge-deleted subgraph $G - v_1v_l$ which we wish to extend to G.

Let $H(1,j)$ denote the subgraph of G induced by the edges coloured 1 or j in this partial colouring of G. Since the degree of every vertex in $H(1,j)$ is either 1 or 2, each component of $H(1,j)$ is either a path or a cycle (see Exercise 4.1.12). Since 1 is absent from v_1 and j is absent from both v_j and v_l it follows that these 3 vertices do not all belong to the same connected component of $H(1, j)$. Thus, if we let K and L denote the corresponding Kempe chains containing v_j and v_l respectively, either $v_1 \notin K$ or $v_1 \notin L$.

Suppose that $v_1 \notin L$. The Kempe chain argument used on L, interchanging the colours, gives a $(k+1)$-colouring of $G - v_1v_l$ in which 1 is missing from both v_1 and v_l. Colouring v_1v_l by 1 gives us a $(k+1)$-colouring of G.

Finally, if $v_1 \notin K$, colour the edge v_1v_l by l and recolour the edges v_1v_i for $i = l+1, \ldots, j-1$ by i but remove the colour $j-1$ from v_1v_j. Then, from the definition of the sequence $v_1, v_2, \ldots, v_j$, we obtain a $(k+1)$-colouring of $G - v_1v_j$ which has not affected the two-coloured subgraph $H(1, j)$. Now use the Kempe chain argument to interchange the colours in the component k and we get a $(k+1)$-colouring of $G - v_1v_j$ in which 1 is absent from both v_1 and v_j. Colouring v_1v_j by 1 now gives the desired colouring. □

While Vizing's Theorem pins down $\chi_1(G)$ to either $\Delta(G)$ or $\Delta(G)+1$ there is as yet no good characterisation of the graphs for which $\chi_1(G) = \Delta(G)$ (or, equivalently, of the graphs for which $\chi_1(G) = \Delta(G)+1$).

Exercises for Section 6.5

6.5.1 Let G be the complete bipartite graph $K_{n,n}$ with bipartition $V = X \cup Y$ where $X = \{x_1, \ldots, x_n\}, Y = \{y_1, \ldots, y_n\}$. Let $c_1, \ldots, c_n$ denote n different colours. If $1 \leq i \leq j \leq n$, assign the colour c_{j-i+1} to the edge x_iy_j while if $1 \leq j < i \leq n$, assign the colour $c_{n+j-i+1}$ to the edge x_iy_j. Prove that this gives an n-edge colouring for G and describe the corresponding Latin square of order n.

6.5.2 In an international yacht race, involving $2n$ teams, the preliminary round takes the form of a round-robin tournament in which every team has to compete against every other team. No team competes more than once each day and so this round must take at least $2n-1$ days by scheduling the races to finish. Prove that it is possible to finish in just $2n-1$ days by scheduling the races as follows.

Using the technique described in the proof of Theorem 6.13, edge colour the complete graph K_{2n} with $2n-1$ colours, denoted by $1, 2, \ldots, 2n-1$. Also label the vertices of K_{2n} by 1,2,...,n. Then schedule team i to race against team j on day k if the edge joining vertices i and j is coloured k.

6.5.3 Prove that the Petersen graph (see Exercise 1.6.7) has edge chromatic index 4.

6.5.4 Show that if G is a 3-regular Hamiltonian graph then $\chi_1(G) = 3$. (This together with the previous exercise proves that the Petersen graph is not Hamiltonian.)

6.5.5 A class of 30 school pupils take 12 different subjects although the maximum number of subjects taken by any individual is 7 and the largest subject has 15 pupils. The parents of the pupils are invited to the school one evening to discuss, in private, their child's progress with each of the child's teachers. Discussion times of 10 minutes each are planned. Allowing for 5 minutes between discussion, use edge colouring to determine the minimum total time required to complete the discussions.

6.6 Map Colouring

In this section we look at what was once one of the most famous problems in mathematics, namely the *Four Colour Conjecture.*

This stated that if the plane was divided up into regions and the regions coloured so that no two regions with a common edge had the same colour, then at most four colours were required. We illustrate this in Figure 6.21.

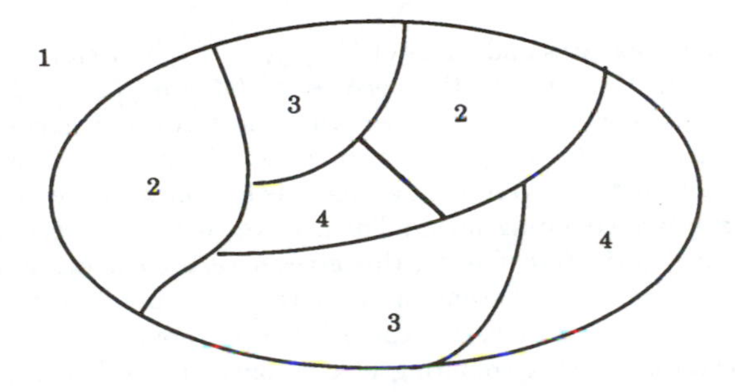

Figure 6.21: A map with its regions coloured.

This problem first came to light in 1852 when a student was supposed to be doing his Geography homework. Instead he spent his time colouring the counties of England in such a way that no two counties with a common boundary had the same colour. To his surprise he found he never needed more than four colours. Of course, in graph theory parlance, the map of the counties of England forms a plane graph with each county corresponding to a face of the graph. Thus the general problem can be expressed as whether one can assign just four colours to the faces or regions of any plane graph G so that adjacent faces of G have different colours. If you are interested in finding out more of the history of this problem, have a look at the book by N. Biggs, E.K. Lloyd and R.J. Wilson [6].

To put our discussion on a more formal footing, we introduce some terminology. We wish to colour the regions of a plane graph G in such a way that, on crossing an edge e, we go from one coloured face to a face with a different colour. This is clearly impossible if the edge e is a bridge, since e lies on the boundary of only one face. We therefore restrict our discussion to graphs without bridges — a very reasonable restriction in the context of map colourings.

A **map** is defined to be a plane conected graph with no bridges.

A map G is said to be **k-face colourable** if we can colour its faces with at most k colours in such a way that no two adjacent faces, i.e., two faces sharing a common boundary edge, have the same colour.

From this we can now give a formal statement of the Conjecture:

The Four Colour Conjecture: Every map is 4-face colourable.

We have now three ways of colouring a plane graph — we may colour its vertices, its edges or its faces. However we will now see that we can change a face colouring problem into a vertex colouring problem.

Theorem 6.15 *(a) A map G is k-face colourable if and only if its dual G^* is k-vertex colourable.*

(b) Let G be a plane connected graph without loops. Then G has a vertex colouring of k colours if and only if its dual G^* has a k-face colouring.

Proof (a) Denote the faces and edges of G by $f_1, \ldots, f_t$ and $e_1, \ldots, e_m$ respectively. Then, as detailed in Section 5.6, the vertices of G^* are $f_1^*, \ldots, f_t^*$ and $e_1^*, \ldots, e_m^*$ respectively, in one-to-one correspondence with the faces and edges of G, and two such vertices f^* and g^* in G^* are joined by an edge e^* if and only if the corresponding faces f and g in G have the corresponding edge e as a common edge on their boundary.

Now choose a k-face colouring for G. Then, if we colour the vertex f^* in G^* with the colour assigned to the face f in G, this gives a vertex colouring of G^* (since the vertices f^* and g^* are only adjacent in G^* if the corresponding faces f and g are adjacent in G). Hence G^* has a vertex colouring of k colours.

Conversely, choose a vertex colouring of G^* consisting of k colours. Then, if we colour the face f in G with the colour assigned to the vertex f^* in G^*, this gives a k-face colouring of G (since the faces f and g are only adjacent in G if the corresponding vertices f^* and g^* are adjacent in G^*).

(b) Since G has no loops, its dual G^* has no bridges (see Section 5.6) and so is a map. Thus, by part (a), G^* is k-face colourable if and only if the double dual G^{**} has a k-face colouring. However, by Theorem 5.17, since G is connected it is isomorphic to G^{**} and so the result follows. □

As an application of Theorem 6.15, we can quickly determine which maps are 2-face colourable.

Theorem 6.16 *A map G is 2-face colourable if and only if it is an Euler graph.*

Proof Let G have a 2-face colouring. Then, by Theorem 6.15 (a), $\chi(G^*) = 2$ and so, by Theorem 6.2, G^* is bipartite. Thus, by Exercise 5.6.4, the double dual G^{**} is an Euler graph. Since G and G^{**} are isomorphic, G is Euler, as required.

Conversely, suppose that G is an Euler graph. Then its double dual G^{**} is Euler and so, again by Exercise 5.6.4, G^* is bipartite. Thus, by Theorem 6.2, $\chi(G^*) = 2$ and so, by Theorem 6.15 (a), G has a 2-face colouring. □

As a further application of the duality principle provided by Theorem 6.15 we have the following result:

Theorem 6.17 *Let G be a cubic map, i.e., a map in which each vertex has degree 3. Then G has a 3-face colouring if and only if each of its faces has even degree, i.e., each of its faces has an even number of edges on its boundary.*

Proof First suppose that G has a 3-face colouring, using colours α, β and γ. Let f be any interior face of G coloured α. Then the faces surrounding f must be coloured β or γ. Looking at these faces in turn as we go clockwise round f, since no two faces of the same colour can be adjacent, those coloured β must alternate with those coloured γ and they must be even in number. Since each of these faces corresponds to an edge on the boundary of f, it follows that f has even degree. A similar argument applies to the exterior face of G.

For the converse, we first prove a "dual" result. Let H be a plane connected Euler graph in which every face has degree three, i.e., each face is a triangle. We will show that H has a 3-colouring, i.e., a vertex colouring of 3 colours. First, it is straightforward to see that every edge of H is part of a cycle and so H has no bridges. Hence H is an Euler map and so, by Theorem 6.16, H is 2-face colourable. Let us choose red and blue to colour the faces of H.

Now choose a red face f of H. Starting at a particular vertex of f and visiting the other two travelling clockwise, colour the first vertex a, the second vertex b and the third c. Any face g adjacent to f is coloured blue and now has two of its vertices coloured. Colour the remaining vertex of g with the third unused colour. This results in the three colours a, b and c being assigned, in that order, to the vertices of g in *anticlockwise* fashion.

We can now extend this vertex colouring to all the vertices of H, resulting in a clockwise allocation of a, b and c to red faces and an anticlockwise allocation to blue faces, as shown in Figure 6.22. Thus we have shown that if H is any plane connected Euler graph in which every face has degree 3 then H has a 3-colouring.

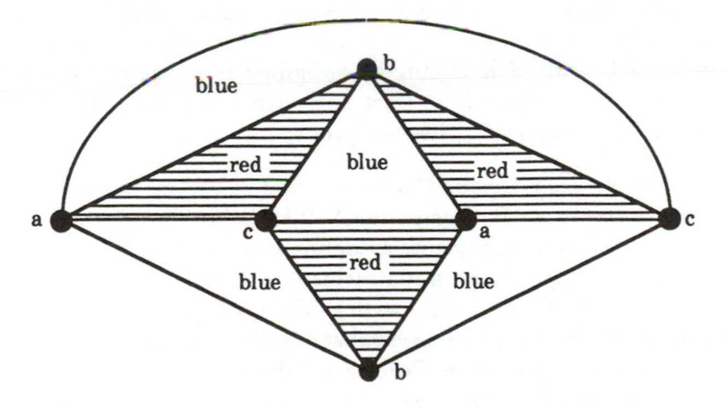

Figure 6.22: The colouring of H starting from a red face.

Now let G be a cubic map in which each face has even degree. Then in the dual map G^* each vertex has even degree while each face has degree 3. Hence, by our arguments of the previous two paragraphs, G^* has a 3-colouring. Thus, by Theorem 6.15 (a), G has a 3-face colouring, as required. □

It follows from Theorem 6.17 that not all maps are 3-face colourable. For example, the cubic map of Figure 6.23 has faces of odd degree and so, by the Theorem, can not be 3-face colourable.

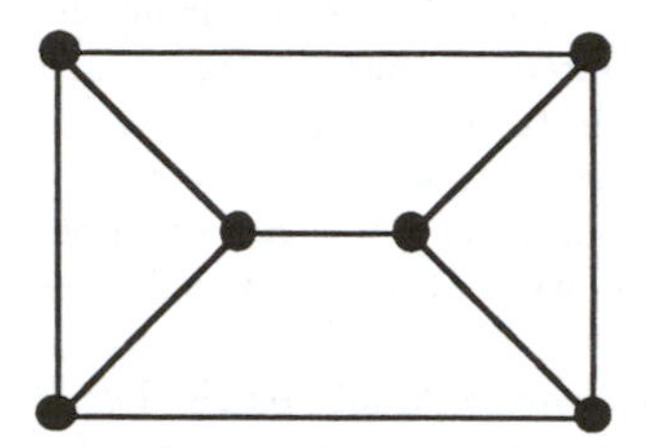

Figure 6.23: A map which is not 3-face colourable — remember the exterior face.

Of course the Four Colour Conjecture says that just one more colour is needed to be able to colour *all* maps. We will prove shortly that five colours are sufficient but first we give an easy proof that six colours suffice.

Theorem 6.18 *Every map G has a face colouring of at most six colours.*

Proof Let H denote the dual of G. By Theorem 6.15 (a), it suffices to prove that H has a vertex colouring of at most six colours. Slightly more generally, we prove that *any* plane graph H is 6-colourable.

We use induction on n, the number of vertices of H. Of course, if H has at most six vertices then it is trivially 6-colourable. Thus we may suppose that $n > 7$ and that all plane graphs on less than n vertices are 6-colourable. Further there is no harm in assuming that H is a simple graph. Then, by Corollary 5.7, H has a vertex v with $d(v) \leq 5$. The vertex deleted subgraph $H - v$ is then a plane graph with $n - 1$ vertices and so, by our induction assumption, is 6-colourable. Such a 6-colouring of $H - v$ can now be extended to all of H simply by assigning to v any of the six colours not given to its neighbours — there is at least one such free colour since v has at most five neighbours. Induction now finishes the proof. □

In the history of the Four Colour Conjecture one of the more significant attempts at its proof was made by A. B. Kempe in 1879 [36]. His "proof" used what is now known as the Kempe chain argument, as defined in Section 6.2. Unfortunately, in 1890, P. J. Heawood [33] produced a map which, although it did not show the Four Colour Conjecture to be false, did show that Kempe's proof failed. However, on a more positive note, Heawood also showed that Kempe's arguments could be used to solve an easier problem — the **Five Colour Problem**.

We now prove Heawood's Five Colour Theorem. The proof involves a particular form of plane graph which we now define.

A plane graph in which every face has degree 3 is called a **triangulation.**

Theorem 6.19 (The Five Colour Theorem, Heawood) *The faces of a map can always be coloured with five or fewer colours.*

Proof We will present the proof in four steps.

Step 1. Look at the part of a map shown in Figure 6.24 (a). We see that five faces meet at the vertex P. Replace P by a face (in a patch-like fashion as shown in Figure 6.24 (b)). Then only three faces meet.

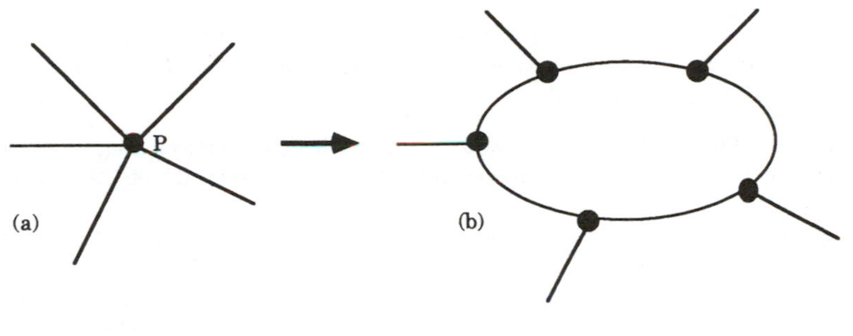

Figure 6.24

Further, if we can colour the map of Figure 6.24 (b) with five colours, we can do the same for the map of Figure 6.24 (a). Thus, from now on we may assume that, in the maps we are considering, at most three faces meet anywhere. (We leave the reader to give a formal argument for this in Exercise 6.6.2.)

Step 2. We now change our map problem into a plane connected graph problem by taking the dual graph of the map, using Theorem 6.15. Because of Step 1, every face of the dual graph is a triangle and so the dual is a triangulation. Thus, by Theorem 6.15, it suffices to show that the vertices of any triangulation can be coloured using at most five colours.

Step 3. We will now show that in every triangulation there is a vertex of degree 2, 3, 4, or 5. The argument is similar to the one we used to prove Theorem 5.10, the cornerstone of our proof that there are only five Platonic bodies.

Let $v_1, \ldots, v_n$ denote the vertices in a particular triangulation G and let n_i be the number of vertices of degree i in G. Clearly $n_1 = 0$. Also, if $\Delta(G) = k$, we have

$$n = n_2 + n_3 + \ldots + n_k = \sum_{i=2}^{k} n_i.$$

Further, if e denotes the number of edges in G,

$$2e = d(v_1) + d(v_2) + \ldots + d(v_n) = 2n_2 + 3n_3 + \ldots + kn_k = \sum_{i=2}^{k} in_i.$$

Now let f denote the number of faces of G and go round the boundary of each face counting up the edges. Then, since each boundary has three edges and in this counting process each edge is counted twice, we get

$$3f = 2e = \sum_{i=2}^{k} i n_i.$$

Substituting this into Euler's Formula $n - e + f = 2$ and tidying up we get

$$4n_2 + 3n_3 + 2n_4 + n_5 = 12 + \sum_{i \geq 6}(i-6)n_i.$$

But, since the right hand side of this equation is positive, the left hand side must be positive. Hence at least one of n_2, n_3, n_4, n_5 is positive and so G must have a vertex of degree 2,3,4 or 5, as claimed.

Step 4. The smallest triangulation is one with three vertices, i.e., a triangle. Its vertices can be coloured with three colours so five will certainly do. We proceed by induction on n.

Suppose first that the triangulation contains a vertex v of degree 4, as in Figure 6.25. In this case, we remove v from the graph and, as shown in the figure, insert the edge u_2u_4. Now we have a triangulation with fewer vertices than the original so we must be able to five colour it. Hence at most four colours are used on the vertices u_1, u_2, u_3, u_4. This leaves a free fifth colour which we can use on u in our original triangulation, thereby colouring all its vertices with at most five colours, as required.

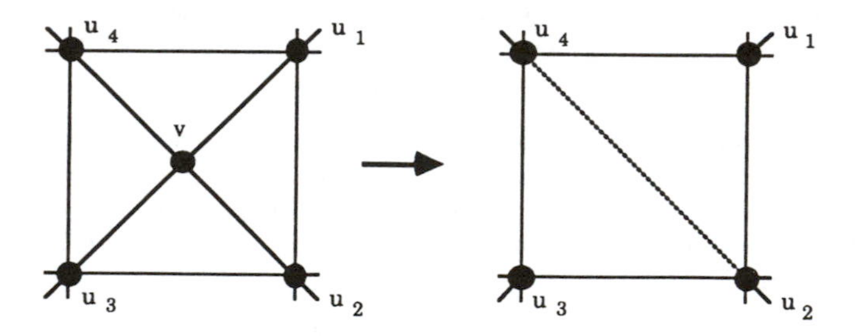

Figure 6.25: Reducing the vertex number of a triangulation having a vertex v of degree 4.

So if the original triangulation has a vertex of degree 4 then it is 5-colourable.

Similar arguments apply to a triangulation with a vertex of degree 2 or a vertex of degree 3 — we leave the details as Exercise 6.6.9. Thus from now on we may suppose that the triangulation contains a vertex v of degree 5.

Again we remove v and set up a smaller triangulation. This time the reduction process converts five faces into two, as shown in Figure 6.26. By our induction hypothesis, we can colour the vertices of this smaller triangulation with five or fewer colours. If the vertices u_1, u_2, u_3, u_4, u_5 (see Figure 6.26) take up no more than four of these colours then, just as before, there is a free fifth colour which we can use on v allowing us to produce a 5-colouring of the original graph.

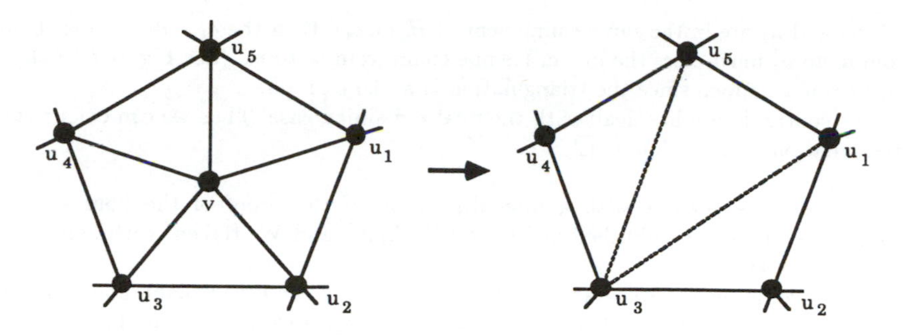

Figure 6.26: Reducing the vertex number of a triangulation having a vertex v of degree 5.

This leaves us to deal with the case in which all five colours are used on $\{u_1, u_2, u_3, u_4, u_5\}$? Suppose that, for each $i = 1, \ldots, 5$, vertex u_i is coloured by colour c_i, as shown in Figure 6.27.

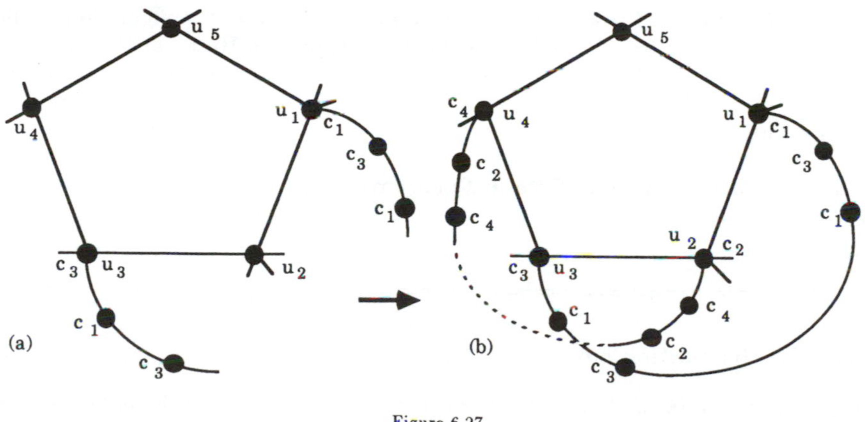

Figure 6.27

Now have a look at the vertices coloured c_1 and c_3. Suppose that u_1 and u_3 are in different components of the Kempe subgraph $H(c_1, c_3)$ induced by those vertices coloured c_1 and c_3. In that case, we can use the Kempe chain argument to interchange the colours c_1 and c_3 on the component containing u_1. This leaves c_1 unused on the set $\{u_1, u_2, u_3, u_4, u_5\}$ and so we have a spare colour to give to v in the original triangulation.

So our only problem now is if u_1 and u_3 are in the same component of the Kempe subgraph $H(c_1, c_3)$. In that case have a look at the Kempe subgraph $H(c_2, c_4)$ induced by the vertices coloured c_2 and c_4. If u_2 and u_4 are in different components of $H(c_2, c_4)$ then, again by the Kempe chain argument, we can swap colours in one component to give a spare colour for vertex v in the original triangulation.

If u_2 and u_4 are in the same component of $H(c_2, c_4)$, then the $c_1 - c_3$ Kempe chain from u_1 to u_3 must cross the $c_2 - c_4$ Kempe chain from u_2 to u_4. (See Figure 6.27 (b).) This cannot happen since the triangulation is a plane graph.

This contradiction has dealt with the final remaining case. Thus we can colour any triangulation in five colours. □

We close this chapter with a brief discussion of the proof of the Four Colour Conjecture. This was published in 1977 by K. Appel and W. Haken, with help from J. Koch ([2], [3]).

Now it would be nice to be able to say that a proof of the Five Colour Theorem can be easily adapted to prove that the Four Colour Conjecture is true, i.e., to prove the Four Colour *Theorem*. The fact that it took over 120 years to prove the Four Colour Theorem suggests that this not the case.

However, the method of proof *is* very similar. Appel and Haken used the triangulation approach of the Five Colour Theorem's proof and showed that all triangulations contain certain configurations. In the Four Colour Theorem there are four configurations — vertices of degree 2, 3, 4 or 5. Appel and Haken had to have nearly 2000 configurations.

In the Five Colour Theorem we had to show that each of the configurations could be coloured starting from a similar triangulation. Appel and Haken did the same.

It should be clear that the computer was an essential tool in Appel and Haken's proof. In fact, they used many weeks of computer time before they settled the Four Colour Theorem.

Theorem 6.20 (The Four Colour Theorem) *Every map can be coloured in four or fewer colours.*

Exercises for Section 6.6

6.6.1 Colour the faces of K_4 with four colours. Can only three colours be used (so that no two neighbouring faces have a common colour)?

6.6.2 Show that the reduction described in Figure 6.24 is good. That is, show that if the map of Figure 6.24 (b) can be coloured in five colours, so can the map in Figure 6.24 (a). Can "five" be replaced by "four"?

6.6.3 Is Step 2 of the proof of the Five Colour Theorem true if we replace "five" by "four"?

6.6.4 Is Step 3 of the proof of the Five Colour Theorem true if we replace "five" by "four"?

6.6.5 In Step 3 of the proof of the Five Colour Theorem we claimed that $n_1 = 0$ in a triangulation. Justify this.

6.6.6 For the triangulation shown in Figure 6.28, show that $2e = \sum_{i=2}^{k} n_i$ and that $3f = 2e$, where e is the number of edges, f is the number of faces, k is the maximum vertex degree and, for each $i = 1, \ldots, k$, n_i denotes the number of vertices of degree i.

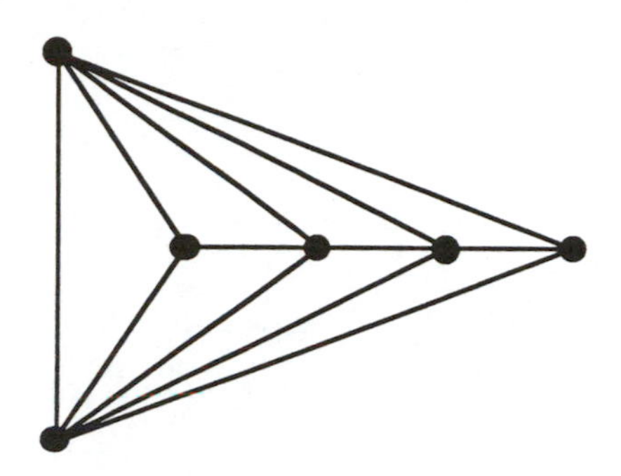

Figure 6.28: A triangulation.

6.6.7 Prove that in any triangulation $2e = \sum_{i=2}^{k} n_i$.

6.6.8 Show that, for a triangulation G,

$$4n_2 + 3n_3 + 2n_4 + n_5 = 12 + \sum_{i \geq 6}(i - 6)n_i,$$

where n_i is the number of vertices of degree i in G, just as in the proof of Theorem 6.19.

6.6.9 Show that a triangulation with a vertex of degree 2 or 3 can be coloured with five colours.

6.6.10 Where does the proof of the Five Colour Theorem break down for four colours?

Chapter 7

Directed Graphs

7.1 Definitions (and More Definitions)

Consider the set of boys Alan, Bill, Charlie and Don and the set of girls Ethel, Florence and Gail. Alan cares only for Ethel, Bill and Charlie are both keen on Florence, but Charlie is also interested in Gail. Don is a confirmed bachelor and has nothing to do with any of the girls. Ethel's heart belongs to Alan, Gail fancies Don and Florence doesn't care for any of them. We can represent this by the diagram in Figure 7.1. Here we have drawn an arrowed line **from** a point x **to** a point y, i.e., the arrow points from x to y, if x "is fond of" y.

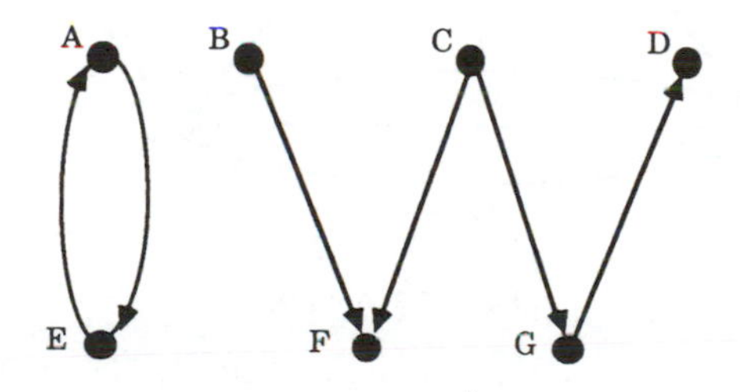

Figure 7.1: A fondness diagram.

Another way of describing this "being fond of" relationship is to list it as a set of **ordered** pairs:

$$(A, F),\ (B, F),\ (C, F),\ (C, G),\ (E, A),\ (G, D)$$

where (x, y) is in the list to signify that x is fond of y. The order in the pair (x, y) is important—X first, Y second, so that (x, y) is different from (y, x) if x is different from y. Hence the term **ordered** pair. In Figure 7.1 this ordering was represented by the direction of the arrow head.

This brings us to the concept of a directed graph.

A **directed graph** $D = (V, A)$ consists of two finite sets

V, the **vertex set**, a nonempty set of elements called the **vertices** of D and
A, the **arc set**, a (possibly empty) set of elements called the **arcs** of D,

such that each arc a in A is assigned an **ordered** pair of vertices (u, v).

If a is an arc, in the directed graph D, with associated ordered pair of vertices (u, v), then a is said to **join** u **to** v, u is called the **origin** or the **initial vertex** or the **tail of** a, and v is called the **terminus** or the **terminal vertex** or the **head of** a.

We represent directed graphs diagrammatically just as for graphs except that arcs are represented by *arrowed* lines. Thus, for example, Figure 7.2 represents the directed graph D with vertex set $V = \{u, v, w, x\}$ and arc set $A = \{a_1, a_2, a_3, a_4, a_5, a_6, a_7\}$ where

a_1 joins u to v and so has associated ordered pair (u, v),
a_2 joins v to v and so has associated ordered pair (v, v),
a_3 joins v to w and so has associated ordered pair (v, w),
a_5 joins w to v and so has associated ordered pair (w, v), and so on.

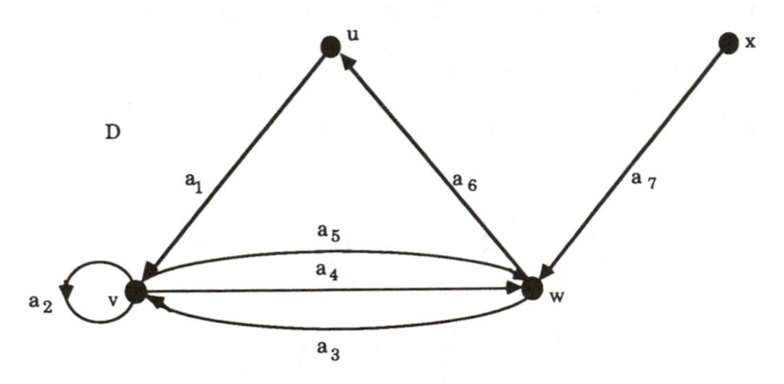

Figure 7.2: A digraph.

We shall abbreviate the term directed graph to **digraph** and also sometimes speak of arcs as **directed edges** (or simply **edges**).

If at some time more than one digraph is being considered we may denote the vertex set of a digraph D by $V(D)$ and similarly the arc set by $A(D)$.

Given a digraph D we can obtain a **graph** G from D by "removing all the arrows" from the arcs. More formally this graph G has the same vertex set as D and corresponding to each arc a in D with associated ordered pair of vertices (u, v) there is an edge e in G with associated *un*ordered pair (u, v). G is then called the **underlying graph** of D.

Thus Figure 7.3 shows the underlying graph of the digraph D of Figure 7.2.

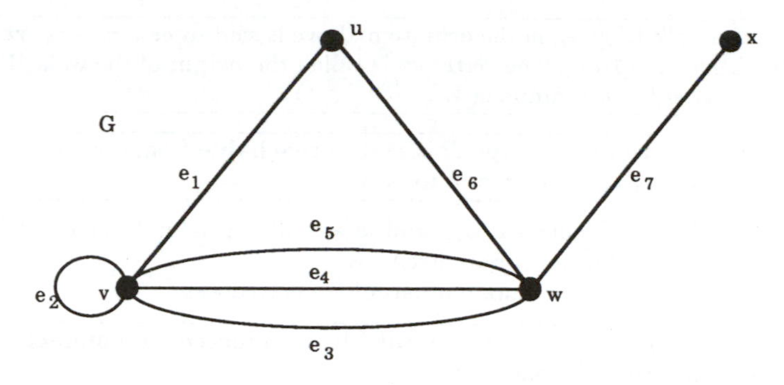

Figure 7.3: The underlying graph of the digraph of Figure 7.2.

Many of the definitions that we gave for graphs have analogues for digraphs. We now give some of these.

> Let D be a digraph. Then a **directed walk** in D is a finite sequence
>
> $$W = v_0 a_1 v_1 \ldots a_k v_k,$$
>
> whose terms are alternately vertices and arcs such that for $i = 1, 2, \ldots, k$, the arc a_i has origin v_{i-1} and terminus v_i.

As in graphs, this directed walk W is often written simply as its sequence of vertices

$$W = v_0 v_1 \ldots v_k,$$

the number k of arcs in W is called the **length** of W, and we can have directed walks of length 0 where the sequence consists solely of one vertex (and no arcs), for example $W = v_0$.

There are similar definitions for **directed trails, directed paths, directed cycles** and **directed tours**.

For example, in the digraph D of Figure 7.2,

$$W = u\ a_1\ v\ a_2\ v\ a_3\ w\ a_5\ v\ a_3\ w$$

is a directed walk of length 5. It is not a directed trail since the arc a_3 occurs twice. Also

$$T = x\ a_7\ w\ a_5\ v\ a_2\ v\ a_3\ w$$

is a directed trail (of length 4). It is not a directed path since the vertex v occurs twice (as does the vertex w). Similarly

$$P = x\ a_7\ w\ a_6\ u\ a_1\ v$$

is a directed path, while

$$C = v\ a_4\ w\ a_6\ u\ a_1\ v$$

is a directed cycle.

The walk W given in the definition above is said to be a $v_0 - v_k$ **walk** or a **walk from v_0 to v_k**. The vertex v_0 is called the **origin** of the walk W, while v_k is called the **terminus** of W.

A vertex v of the digraph D is said to **reachable** from a vertex u if there is a directed path in D from u to v.

In the digraph D of Figure 7.2, u, v and w are all reachable from x but x is not reachable from any vertex apart from itself.

In digraphs there are two natural notions of connectedness.

A digraph D is said to be **weakly connected** or **connected** if its underlying graph is connected.

Thus our example D of Figure 7.2 is weakly connected.

A digraph D is said to be **strongly connected** (or **diconnected**) if for any pair of vertices u and v in D there is a directed path from u to v, i.e., given any pair of vertices in D, each is reachable from the other.

Since x is not reachable from any other vertex, the digraph of Figure 7.2 is not strongly connected. However the digraph of Figure 7.4 *is* strongly connected.

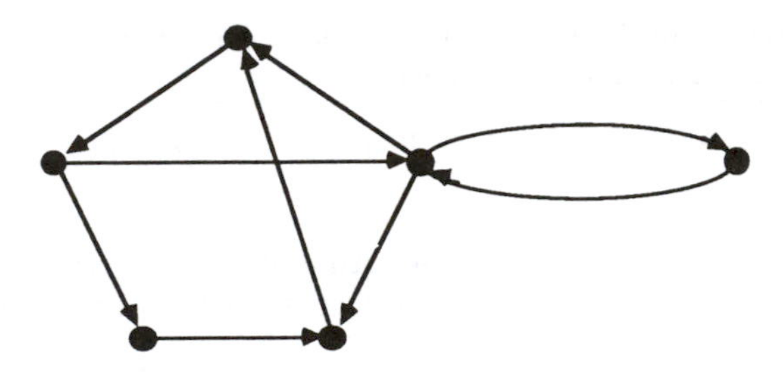

Figure 7.4: A strongly connected digraph.

Given a *graph* G we can obtain a digraph from G by specifying for each edge in G an order to its end vertices. Such a digraph D is called an **orientation** of G.

The graph K_3 has eight different orientations, shown as $D_1, \ldots, D_8$ in Figure 7.5, although, of these, only two are strongly connected, namely D_1 and D_8.

Two digraphs D_1 and D_2 are said to be **isomorphic** if there is a one-to-one and onto correspondence between $V(D_1)$ and $V(D_2)$ and also a one-to-one and onto correspondence between $E(D_1)$ and $E(D_2)$ such that if arc e_1 in D_1 goes from vertex u_1 to v_1 then the corresponding arc e_2 in D_2 goes from vertex u_2 to vertex v_2 where u_2 and v_2 are the vertices in D_2 corresponding to u_1 and v_1 respectively. Of course, such a pair of correspondences is called a (digraph) **isomorphism**.

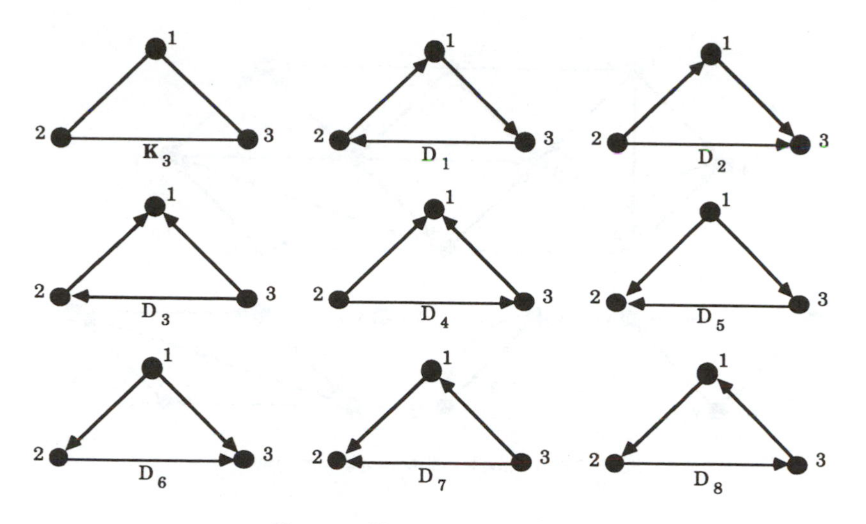

Figure 7.5: The eight orientations of K_3.

It can be easily seen that, among the orientations of K_3 shown in Figure 7.5, D_1 and D_8 are isomorphic with a suitable vertex correspondence given by

$$1 \sim 1,\ 3 \sim 2,\ 2 \sim 3,$$

while $D_2, D_3, \ldots, D_7$ are all isomorphic to each other. For example, for D_2 and D_6 the vertex correspondence is

$$1 \sim 2,\ 2 \sim 1,\ 3 \sim 3.$$

> A **strong component** S of a digraph D is a subdigraph of D which is strongly connected and is not a proper subdigraph of any other strongly connected subdigraph of D.

This notion corresponds to that of connected components in graphs. Using arguments similar to that used for graphs we can show that if the vertex u belongs to the strong component S of the digraph D then S consists of all those vertices v of D such that v is reachable from u and u is reachable from v together with all the arcs in D which join such vertices. (We realise that we have not formally defined the term **subdigraph** used in the definition but feel confident in leaving this to the reader.) Figure 7.6 shows a digraph D and its four strong components.

> A digraph D is called **simple** if, for any pair of vertices u and v of D, there is at most one arc from u to v and there is no arc from u to itself.

For example, the digraph D of Figure 7.2 is not simple but that of Figure 7.4 is.

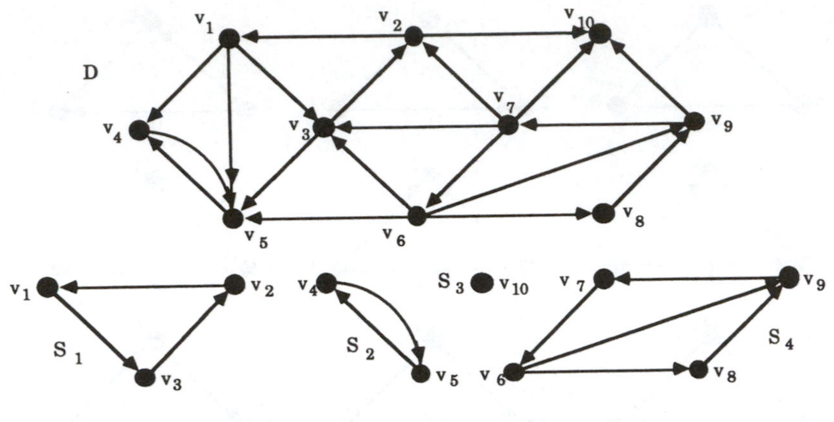

Figure 7.6: A digraph and its strong components.

Exercises for Section 7.1

7.1.1 Let D be the digraph of Figure 7.7.

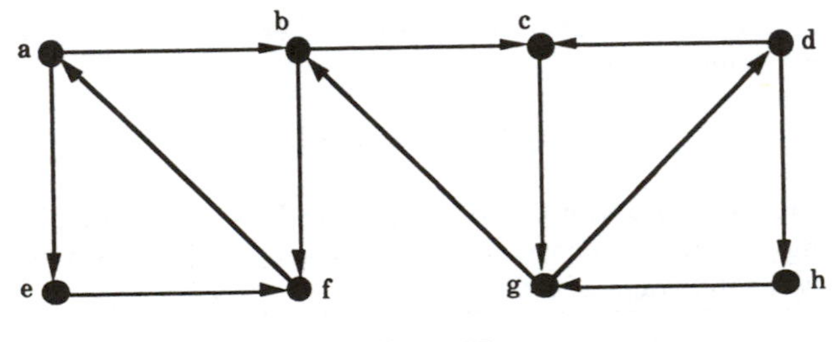

Figure 7.7

(a) Find a directed walk in D of length 8. Is this walk a directed path?

(b) Find a directed trail in D of length 10.

(c) Find a directed path in D of longest possible length.

(d) Find a directed cycle in D of longest possible length.

(e) Is D weakly connected?

(f) Is D strongly connected? If so, give an example of a directed path from u to v for each pair of vertices u and v of D. If D is not strongly connected, find a pair of vertices u and v such that u is not reachable from v.

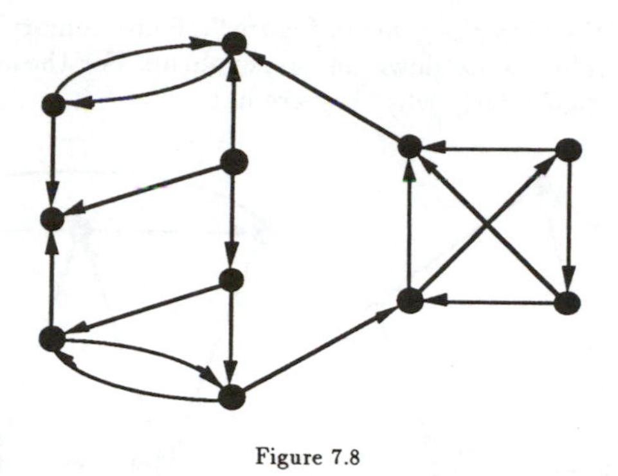

Figure 7.8

7.1.2 Find the strong components of the digraph of Figure 7.8.

7.1.3 A digraph D is said to be **unilaterally connected** if, given any pair of vertices u and v of D, either u is reachable from v or v is reachable from u, but not necessarily both.

(a) Show that the digraph of Figure 7.9 is unilaterally connected but not strongly connected.

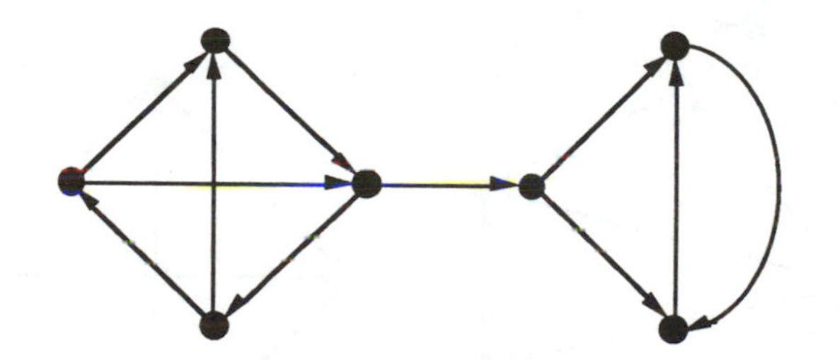

Figure 7.9: A unilaterally connected digraph which is not strongly connected.

(b) Determine whether or not the digraph of Figure 7.8 is unilaterally connected.

(c) Prove that a digraph D is unilaterally connected if and only if there is a directed walk in D involving all the vertices of D, i.e., D has a directed **spanning** walk.

7.1.4 Let u and v be distinct vertices of the digraph D. Prove that every directed $u - v$ walk in D contains a directed $u - v$ path. (This is the digraph analogue of Theorem 1.3.)

7.1.5 Prove that a digraph D is strongly connected if and only if it has a closed directed walk going through all its vertices. (Compare this result with Exercise 7.1.3 (c).)

7.1.6 Which pairs of the digraphs given in Figure 7.10 are isomorphic? For those pairs that are isomorphic, write down an isomorphism. For those pairs that are not isomorphic, explain clearly why they are not.

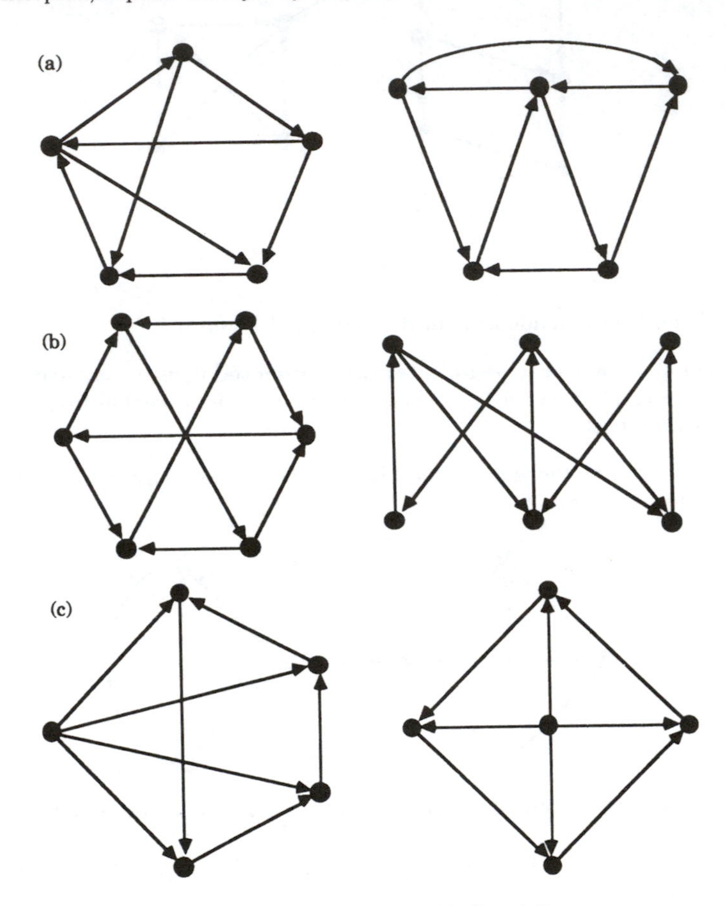

Figure 7.10: Pairs of isomorphic digraphs?

7.1.7 Let D be a simple digraph. We define the **complement** of D to be the simple digraph $\overline{D}$ with the same vertex as D and where there is an arc from a vertex u to a vertex v if and only if there is no such arc in D. Figure 7.11 illustrates a digraph and its complement.

Give an example of a simple weakly connected digraph D which is not unilaterally connected but its complement is

(a) strongly connected,

(b) unilaterally connected but not strongly connected.

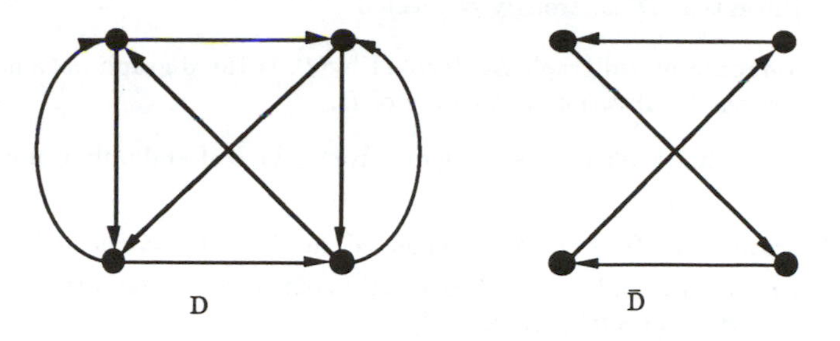

Figure 7.11: A simple digraph and its complement.

7.1.8 Given a digraph D we define its **condensation** to be the simple digraph D^* whose vertices $v_1, \ldots, v_n$ are in one-to-one correspondence with the strong components $S_1, \ldots, S_n$ of D and such that there is an arc a from u_i to u_j (for $i \neq j$) if and only if there is an arc from some vertex of S_i to some vertex of S_j. Figure 7.12 shows a digraph D and its condensation D^*.

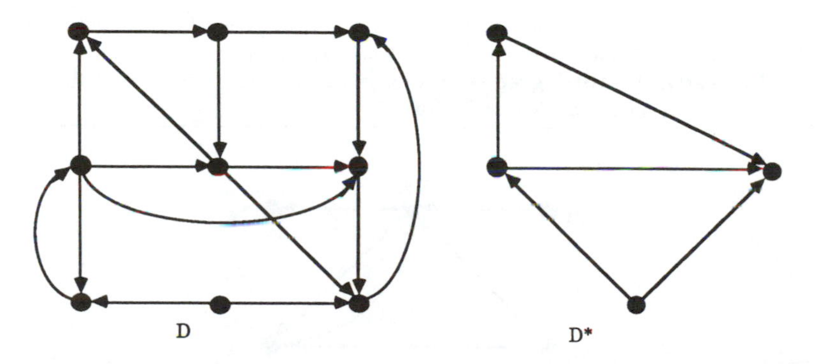

Figure 7.12: A digraph and its condensation.

(a) Find the condensation D^* of the digraph D of Figure 7.8.

(b) Prove that the condensation D^* of any digraph D has no directed cycles.

(c) Prove that the condensation D^* of any digraph D is strongly connected, unilaterally connected or weakly connected if and only if D is strongly connected, unilaterally connected or weakly connected, respectively. (This shows that the condensation construction produces a digraph D^*, usually much simpler than the given digraph D, with the same connectedness attributes as D.)

7.1.9 Construct the digraph D' from a given digraph D by introducing a new vertex v and an arc joining v to each vertex of D and an arc joining each vertex of D to v. Prove that D' is strongly connected.

7.1.10 The **converse** of a digraph D, denoted by $\overleftrightarrow{T}$, is the digraph obtained from D by reversing the direction of each arc of D.

(a) Prove that a vertex v is reachable from u in D if and only if u is reachable from v in $\overleftrightarrow{T}$.

(b) Prove that D is weakly connected, unilaterally connected or strongly connected if and only if $\overleftrightarrow{T}$ is weakly connected, unilaterally connected or strongly connected, respectively.

(c) Show that for every $n \geq 1$ there is a digraph D on n vertices which is isomorphic to its converse.

7.2 Indegree and Outdegree

Let v be a vertex in the digraph D. The **indegree** $id(v)$ of v is the number of arcs of D that have v as its head, i.e., the number of arcs that "go to" v. Similarly, the **outdegree** $od(v)$ of v is the number of arcs of D that have v as its tail, i.e., that "go out" of v.

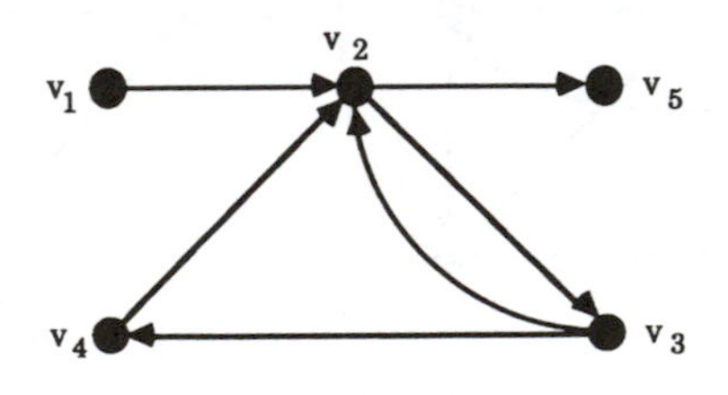

Figure 7.13

Thus in the digraph of Figure 7.13 we have $id(v_1) = 0$, $id(v_2) = 3$, $id(v_3) = 1$, $id(v_4) = 1$, $id(v_5) = 1$, while $od(v_1) = 1$, $od(v_2) = 2$, $od(v_3) = 2$, $od(v_4) = 1$, $od(v_5) = 0$.

Our first theorem on digraphs is the analogue of Theorem 1.1 on graphs.

Theorem 7.1 (The First Theorem of Digraph Theory) *Let D be a digraph with n vertices and q arcs. If $\{v_1, \ldots, v_n\}$ is the set of vertices of D then*

$$\sum_{i=1}^{n} id(v_i) = \sum_{i=1}^{n} od(v_i) = q.$$

Proof When the indegrees of the vertices are summed, each arc is counted exactly once since every arc goes to exactly one vertex. Thus

$$\sum_{i=1}^{n} id(v_i) = q.$$

Similarly, when the outdegrees are summed, each arc is counted exactly once since every arc goes out of exactly one vertex and this gives the other equality. □

We continue with our graph theory analogies.

> Let D be a weakly connected digraph. Then a **directed Euler trail** in D is a directed open trail of D containing all the arcs of D (once and only once). A **directed Euler tour** of D is a directed closed trail of D containing all the arcs of D (once and only once). A digraph D containing a directed Euler tour is called an **Euler digraph**.

For example, in Figure 7.14 the digraph D_1 is Euler with directed Euler tour

$$a_1\ a_2\ a_3\ a_4\ a_5\ a_6\ a_7\ a_8\ a_9\ a_{10}\ a_{11}$$

while the digraph D_2, although it is not Euler, has a directed Euler trail, e.g.,

$$a_1\ a_2\ a_3\ a_4\ a_5.$$

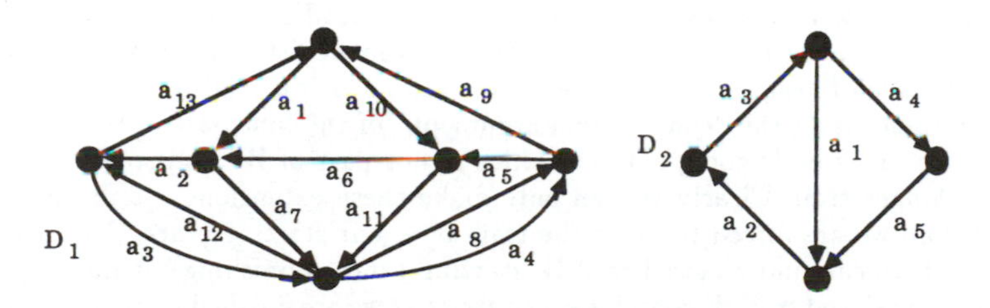

Figure 7.14: D_1 is an Euler digraph and D_2 has a directed Euler trail.

We now give a characterisation of Euler digraphs which is very similar to that of Euler graphs given in Theorem 3.2.

Theorem 7.2 *Let D be a weakly connected digraph with at least one arc. Then D is Euler if and only if $od(v) = id(v)$ for every vertex v of D.*

Proof Suppose first that D is Euler and let T denote a directed Euler tour of D, beginning (and so finishing) at the vertex v. Then, if u is a vertex different from v, each time that u is encountered on T it is entered by an arc and left by an arc and so each occurrence of u in T represents a contribution of 1 to the indegree of u and 1 to the outdegree of u. Since every arc incident with u occurs in T it follows that

$od(u) = id(u)$. It is left to consider the initial vertex v of T. Since T begins and ends at v the first arc of T contributes 1 to the outdegree of v while the last arc contributes 1 to its indegree. Since every other occurrence of v on the tour T contributes 1 each to its outdegree and indegree, it follows once again that $od(v) = id(v)$.

For the converse, suppose that D is a weakly connected digraph with $od(v) = id(v)$ for every vertex v of D. We use induction on q, the number of arcs in D. To begin with, if D has only one arc, a, say, with head u and tail v, then $u = v$ since otherwise either $od(v) = 1$ but $id(v) = 0$ or $od(u) = 1$ but $id(u) = 0$, in contradiction to the hypothesis on D. Thus a is a directed loop and by itself gives a directed Euler tour of D.

Although it is not necessary for us to do this, let us now suppose that $q = 2$. Then, again by the hypothesis, either both of the two arcs are loops incident with the same vertex or they are both non-loops with opposite heads and tails as shown in Figure 7.15. Clearly both the resulting digraphs have a directed Euler tour.

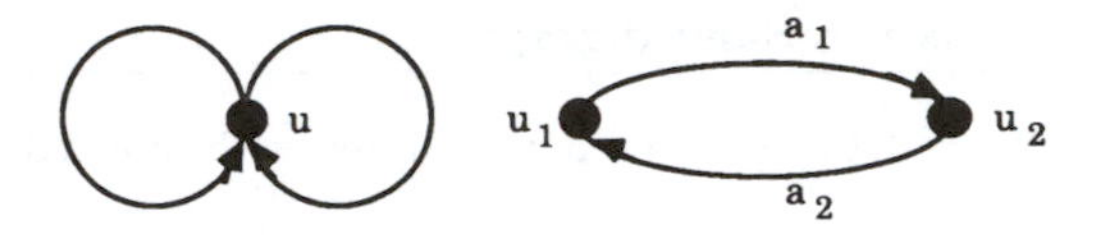

Figure 7.15: The only Euler digraphs with two arcs.

Now assume that q is a fixed number with $q \geq 3$, and that all weakly connected digraphs with less than q arcs in which every vertex has equal outdegree and indegree are Euler. Since our digraph D is weakly connected and $od(v) = id(v)$ for every vertex v, every vertex of D has positive outdegree, i.e., its outdegree is not zero.

Now select any vertex u in D. Since $od(u) > 0$ there exists a trail W' in D starting at u. If W' also finishes at u then we have a closed $u - u$ trail in D. If W' finishes at $v \neq u$ then, using the "contribution argument" of the first part of the proof there must be an arc a in D going out of v which is not part of W' and so we can extend W' to a longer trail. Clearly we can only make these extensions a finite number of times before we are forced to finish the trail at u, our starting place. This argument shows that we can find a closed trail W starting (and so finishing) at the vertex u.

Now if this closed trail W contains every arc of D we are finished since it is a directed Euler tour and so D is Euler. Otherwise there are arcs of D that do not belong to W. Remove from D all those arcs in W together with any resulting isolated vertices to obtain a new digraph F. Since for every vertex v of W we have $od_W(v) = id_W(v)$ (where $od_W(v)$ and $od_W(v)$ denote the outdegree and indegree, respectively, of v *in* W) it follows that *in* F $od(v) = id(v)$ for every v in F. Now the connected components of the underlying graph G of F produce weakly connected subdigraphs of F each having less than q edges and with every vertex having equal indegree and outdegree. Thus, by our induction hypothesis, each such subdigraph is Euler. Moreover, since D is weakly connected, each of these subgraphs has a vertex in common with W. A directed Euler tour can now be constructed by attaching to W at each of these common vertices the Euler tour of the subdigraph. □

The following is the digraph analogue of Theorem 3.3.

Theorem 7.3 *Let D be a weakly connected digraph with at least two vertices. Then D has a directed Euler trail if and only if D has two vertices u and v such that*

$$od(u) = id(u) + 1 \quad \text{and} \quad id(v) = od(v) + 1$$

and, for all other vertices w of D, $od(w) = id(w)$. Furthermore, in this case the trail begins at u and ends at v.

Proof Suppose that D contains an Euler trail W that begins at u and ends at v. Then, as in the first part of the proof of Theorem 7.2, for every vertex w different from both u and v we get $od(w) = id(w)$. Moreover the first arc of W contributes 1 to the outdegree of u while every other occurrence of u in W contributes 1 each to the outdegree of u and the indegree of u. Therefore $od(u) = id(u) + 1$. Similarly $id(v) = od(v) + 1$.

Conversely, let D be a weakly connected digraph containing vertices u, v as described in the statement and with $od(w) = id(w)$ for all other vertices w. Add a new arc a to D joining v to u. This produces a new digraph F in which the new outdegree of v is one more than its old outdegree so now $id(v) = od(v)$, similarly $id(u) = od(u)$, and for every other vertex w we still have $id(w) = od(w)$. Moreover F is weakly connected and so since $id(x) = od(x)$ for every vertex x in F it follows from Theorem 7.2 that F is Euler. Let T then be a directed Euler tour. Then T contains all the arcs of D together with the added arc a. Deleting this arc a produces a directed Euler trail back in our digraph D, and this trail must start at u and finish at v, as required. □

We can apply Theorems 7.2 and 7.3 to the digraphs D_1 and D_2 respectively of Figure 7.14 to verify, just by looking at the outdegrees and indegrees of the vertices, that they have a directed Euler tour and trail respectively (and in D_2 the trail must start at the tail of the arc a_1 and finish at the head of the arc a_2).

We now discuss an application of directed Euler tours to a problem in coding theory.

Let $\Sigma = \{0, \ldots, n-1\}$ be an alphabet of n letters. We can form precisely n^k different sequences of length k using these letters. Such a sequence is called a **word** of length k from Σ.

An (n, k) **de Bruijn sequence** is a sequence

$$a_0 a_1 \ldots a_{t-1}$$

of letters from $\Sigma = \{0, \ldots, n-1\}$ such that every word w of length k from Σ can be written in the form

$$w = a_i a_{i+1} \ldots a_{i+k-1}$$

for a unique $i \in \{0, \ldots, t-1\}$. (In the case where $i \geq t - 1 - k$ this form is interpretted to be

$$a_i a_{i+1} \ldots a_{t-1} a_0 \ldots a_{i+k-t-1},$$

so that, in effect, the sequence $a_0 a_1 \ldots a_{t-1}$ is cyclical.)

For example, consider the case alphabet $n = 2$, so that $\Sigma = \{0, 1\}$. Then the sequence

$$1\,0\,0\,0\,1\,1\,1\,0$$

is a $(2,3)$ de Bruijn sequence. To see this, take sections of three consecutive letters $a_j a_{j+1} a_{j+2}$ from the sequence, starting at its beginning (and running over to the beginning again for the last two words). This produces all of the $2^3 = 8$ different words of length 3 from Σ namely:

$$100, 000, 001, 011, 111, 110, 101, 010.$$

In general the process of producing all of the n^k words from a given (n,k) de Bruijn sequence is to take the first k letters from the sequence for the first word, then "shift" this along by one place to get the second word and continue this shifting procedure until we come full circle back to the first word. Since there are n^k shifts in all, we see that any (n,k) de Bruijn sequence has precisely n^k terms.

De Bruijn sequences are very important in coding theory and are used by shift registers in the case where $\Sigma = \{0, 1\}$. For this reason they are also called **shift register sequences**.

We now associate to any given (n,k) de Bruijn sequence a digraph $D_{n,k}$, called a **de Bruijn diagram** or **Good diagram**. The vertex set V of $D_{n,k}$ is defined to be the set of all words of length $k-1$ from the alphabet $\Sigma = \{0, \ldots, n-1\}$, so that $D_{n,k}$ has n^{k-1} vertices. We now introduce an arc from each such vertex (word) $b_1 b_2 \ldots b_{k-1}$ to each vertex (word) of the form $b_2 b_3 \ldots b_k$, (so that we cancel off the first term of the tail of the arc and add on a new last term to get the head of the arc). We label this arc unambiguously by $b_1 b_2 \ldots b_n$. Note that there are n^k such arcs, and that each corresponds uniquely to an (n,k) de Bruijn sequence. Figures 7.16 and 7.17 illustrate the two digraphs $D_{2,3}$ and $D_{3,3}$.

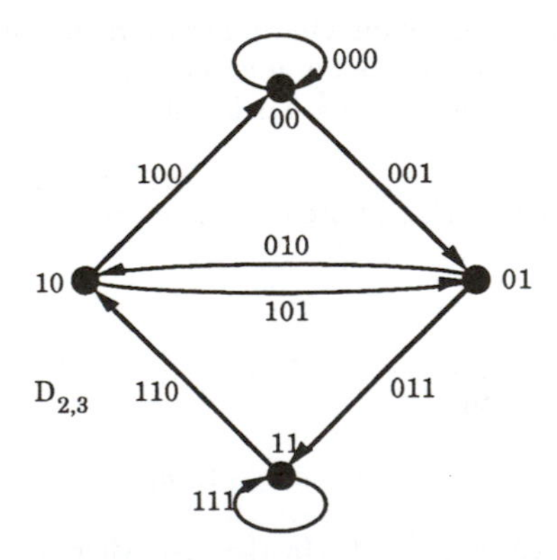

Figure 7.16: The de Bruijn diagram $D_{2,3}$.

We now use the digraphs $D_{n,k}$ to show that, for every pair of positive integers n and k, both greater than 1, there is an (n,k) de Bruijn sequence.

First suppose that the digraph $D_{n,k}$ has a directed Euler tour T say. Now choose the first term of each arc of T in turn to form a sequence σ of n^k terms. We claim that σ is a de Bruijn sequence. To see this, let $w = b_1b_2b_3\ldots b_n$ be any word of length n from Σ. Then this corresponds to the unique arc a from the vertex $b_1b_2b_3\ldots b_{n-1}$ to the vertex $b_2b_3b_4\ldots b_n$. On the tour T the next arc after a must begin $b_2b_3b_4\ldots$, while the next one after that must begin $b_3b_4\ldots$, and so on. Thus, in our construction of the sequence σ, we must encounter $w = b_1b_2b_3\ldots b_n$ as a subsequence of n consecutive terms. This shows that each of the n^k words is generated by σ in the desired way and so, since σ has n^k terms, it follows that σ is a de Bruijn sequence, as claimed.

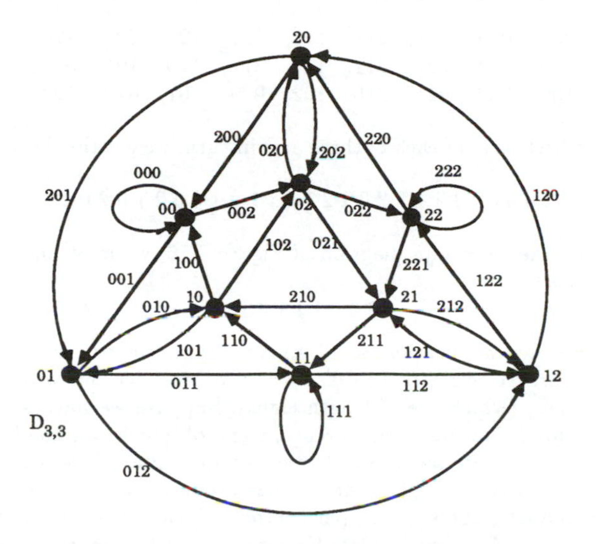

Figure 7.17: The de Bruijn diagram $D_{3,3}$.

We have proved that an (n, k) de Bruijn sequence exists provided the corresponding digraph $D_{n,k}$ has a directed Euler tour. The following theorem guarantees such tours. It was proved by N. G. de Bruijn [15] for the case $k = 2$ and for arbitrary k by I. J. Good [27].

Theorem 7.4 (de Bruijn, 1946; Good, 1946) *For each pair of positive integer n and k, both greater than one, the de Bruijn diagram $D_{n,k}$ has a directed Euler tour.*

Proof By Theorem 7.2 it suffices to show that $D_{n,k}$ is weakly connected and that $\text{id}\,(v) = \text{od}\,(v)$ for each of its vertices v.

Let x and y be two vertices of $D_{n,k}$, say $x = b_1b_2b_3\ldots b_{n-1}$ and $y = c_1c_2c_3\ldots c_{n-1}$. Then there is a directed path from x to y given by the sequence of vertices

$$b_1b_2b_3\ldots b_{n-1},\ b_2b_3\ldots b_{n-1}c_1,\ b_3b_4\ldots b_{n-1}c_1c_2,\ \ldots\ ,c_1c_2c_3\ldots c_{n-1}.$$

This shows that $D_{n,k}$ is weakly connected, in fact, strongly connected.

Consider again our vertex $x = b_1b_2 \ldots b_{n-1}$. By the definition of the arcs in $D_{n,k}$, any arc a having x as its tail is of the form $y = b_1b_2b_3 \ldots b_{n-1}c_1$. It follows that x has outdegree n. Similarly, any arc b having x as its head is of the form $y = a_1b_1b_2b_3 \ldots b_{n-2}$ and so x has also indegree n. Thus, since $od(x) = id(x)$ for all vertices x, $D_{n,k}$ has a directed Euler tour, as required. □

To illustrate the above discussion, we construct a $(3,3)$ de Bruijn sequence using the de Bruijn diagram of Figure 7.17. As the reader may easily check from the Figure, a directed Euler tour of $D_{3,3}$ is given by the sequence of arcs

$$\begin{array}{ccccccccc} 200, & 000, & 001, & 011, & 111, & 112, & 122, & 222, & 220, \\ 202, & 022, & 221, & 212, & 121, & 211, & 110, & 101, & 010, \\ 100, & 002, & 021, & 210, & 102, & 020, & 201, & 012, & 120. \end{array}$$

Thus, taking the first term of each of these arcs in turn, we get the de Bruijn sequence

$$2\,0\,0\,0\,1\,1\,1\,2\,2\,2\,0\,2\,2\,1\,2\,1\,1\,0\,1\,0\,0\,2\,1\,0\,2\,0\,1.$$

Similarly, using the de Bruijn diagram of Figure 7.16, we may construct the $(2,3)$ de Bruijn sequence

$$0\,0\,0\,1\,1\,1\,0\,1.$$

We now discuss an application of $(2,k)$ de Bruijn sequences to a problem in telecommunications. To simplify the discussion we restrict our attention to $(2,3)$ de Bruijn sequences, (which are of length eight). Suppose we have a rotating drum with eight segments round its circumference, some of which can conduct an electric current while the others are insulated. Three electric contacts are placed against the drum so that after any rotation of the drum there are three consecutive drum segments touching these contacts, one segment per contact. Then, for example, if we denote a conducting segment by 1 and an insulated one by 0, we get a binary sequence of length three for each position of the drum. The question arises as to whether we can arrange the segments of the drum so that each position gives a different sequence.

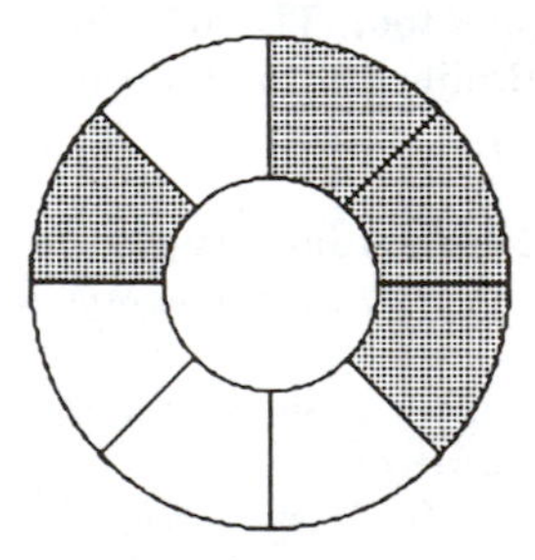

Figure 7.18: A drum design constructed using a $(2,3)$ de Bruijn sequence.

Clearly such an arrangement amounts to finding a $(2,3)$ de Bruijn sequence. Using the one above, we get the drum design shown in Figure 7.18 where the dark segments, corresponding to the 0's in the sequence, are insulated and the positioning starts at the top and procedes clockwise. Finally we note that this problem occurs in a various forms, ranging from the generation of codes in cryptography to the design of washing machine dials, and is known as the **teleprinter's problem.**

Exercises for Section 7.2

7.2.1 Find $od(v)$ and $id(v)$ for each vertex of the digraph of Figure 7.7.

7.2.2 A digraph D is called k-**regular** if $od(v) = id(v) = k$ for each vertex v of D.

(a) Give an example of a 1-regular digraph with n vertices for each $n \geq 2$.

(b) Give an example of a 2-regular digraph with five vertices.

(c) Prove that given any $n \geq 1$ and any k with $0 \leq k < n$ there is a simple k-regular digraph D with n vertices.

7.2.3 Let D be a digraph with an odd number of vertices. Prove that if each vertex of D has an odd outdegree then there is an odd number of vertices of D with odd indegree.

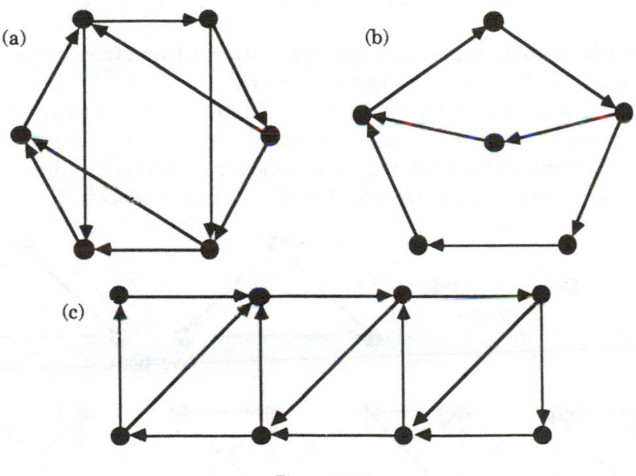

Figure 7.19

7.2.4 Which digraphs of Figure 7.19 are Euler and which have a directed Euler trail?

7.2.5 Let D be an Euler digraph. Prove that D is strongly connected.

7.2.6 Let D be a digraph with a directed Euler trail. Prove that D is unilaterally connected.

7.2.7 Draw the $(3,2)$ de Bruijn diagram and use it to construct a $(3,2)$ de Bruijn sequence.

7.2.8 Draw the $(2,4)$ de Bruijn diagram and use it to construct a $(2,4)$ de Bruijn sequence. Use this sequence to design a rotating drum, having the same properties as that described in the text but with 16 segments round its circumference and four electric contacts.

7.2.9 Prove by induction on n that for each $n \geq 1$ there is a simple digraph D with n vertices $v_1, \ldots, v_n$ such that $od(v_i) = i-1$ and $id(v_i) = n-i$ for each $i = 1, \ldots, n$.

7.2.10 Let D be a digraph such that either every vertex of D has positive outdegree or every vertex of D has positive indegree. Prove that D has a directed cycle.

7.2.11 Let D be a digraph such that $id(v) \geq k$ for every vertex v, where k is some fixed positive integer. Prove that D has a directed cycle of length at least $k+1$.

7.3 Tournaments

A **tournament** is an orientation of a complete graph.

In other words, a tournament is a digraph with no (directed) loops in which any two distinct vertices are joined by exactly one arc.

The number of non-isomorphic tournaments increases sharply with the number of vertices. For example, there is only one tournament with exactly 1 vertex and only one with exactly 2 vertices. There are two tournaments on 3 vertices, four on 4 vertices, 12 on 5 vertices. However there are over 9 million on 10 vertices.

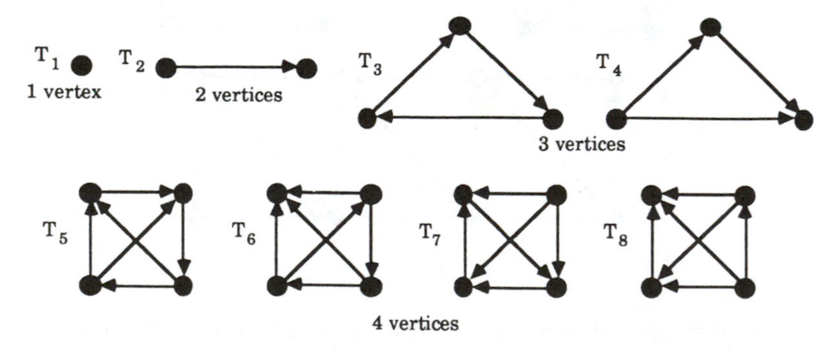

Figure 7.20: The tournaments on at most four vertices.

The tournaments on less than five vertices are shown in Figure 7.20. Here, in the tournaments on four vertices, T_5 is the only one with a directed cycle of length 4, T_6

is the only one with a directed cycle of length 3 and a vertex of indegree 3, T_7 is the only one with a directed cycle of length 3 and a vertex of outdegree 3 and T_8 is the only one with no directed cycle of length 3.

The reason for the name "tournament" is that the digraph can be used to record the results of games in a round-robin tournament in any game in which draws are not allowed, such as tennis. The arc from a to b then indicates that a has beaten b.

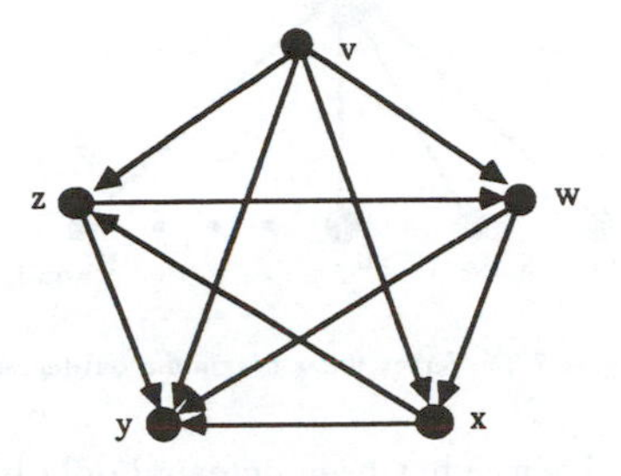

Figure 7.21: A tournament on five vertices.

For example, in the digraph of Figure 7.21 representing a round-robin tournament, v has beaten every other competitor, y has lost each match, each of the other competitors have won two and lost two matches. The vertex v has maximum outdegree 4. Every other vertex can be reached from v by a directed path of length at most 2. This illustrates the following general result:

Theorem 7.5 *Let v be any vertex having maximum outdegree in the tournament T. Then for every vertex w of T there is a directed path from v to w of length at most 2.*

Proof Let $od(v) = m$ and let the vertices joined by an arc *from* v be $v_1, v_2, \dots, v_m$. If T has n vertices then each of the remaining $n - m - 1$ vertices $u_1, u_2, \dots, u_{n-m-1}$ are adjacent to v, since T is a tournament, i.e., for these remaining vertices u_j, $1 \le j \le n - m - 1$, there are arcs *from* u_j *to* v. (See Figure 7.22.)

Then for each i, $1 \le i \le m$, the arc from v to v_i gives a directed path of length 1 from v to v_i. It remains to show that there is a directed path of length 2 from v to u_j for each j, $1 \le j \le n - m - 1$.

Given such a vertex u_j, if there is an arc from v_i to u_j for some i then vv_iu_j gives a directed path of the desired type. However, now suppose there is a u_k, $1 \le k \le n - m - 1$, such that *no* vertex v_i, $1 \le i \le m$, has an arc *from* v_i *to* u_k. Then, because T is a tournament, there must be an arc *from* u_k *to* each of the m vertices v_i. Since we also have an arc *from* u_k *to* v this gives $od(u_k) \ge m + 1$. This contradicts the fact that v has maximum outdegree with $od(v) = m$. Thus each u_j must have an arc joining it from some v_i and so the proof is complete by using the directed path vv_iu_j. □

Theorem 7.5 has the following interpretation in round-robin tournaments. Let w be a winner in such a tournament, i.e., any player with the most victories — there may

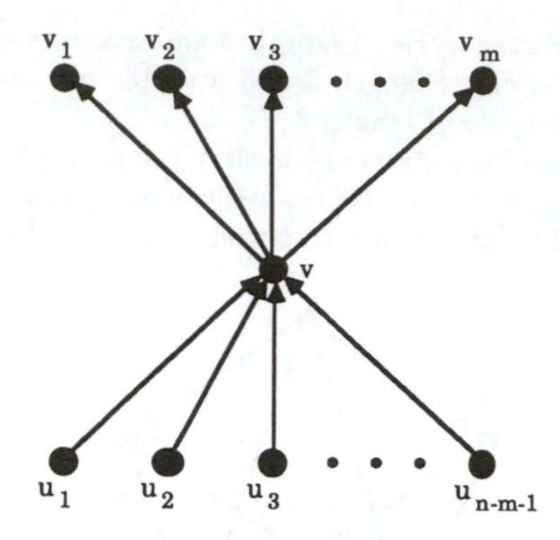

Figure 7.22: Vertex v has maximum outdegree.

be more than one winner. Then w has been defeated only by players who themselves have lost to players defeated by w.

Suppose T is a tournament on n vertices and let v be any vertex of T. Then by $T-v$ we mean the directed graph obtained from T by removing v and all arcs incident with v. Now any two vertices of $T-v$ are joined by exactly one arc, since these two vertices are joined by exactly one arc in T. Hence $T-v$ is also a tournament. We use this property of tournaments in the proof of the next theorem, due to Rédei [53].

> A **directed Hamiltonian path** of a digraph D is a directed path in D that includes every vertex of D (once and only once).

Theorem 7.6 (Rédei, 1934) *Every tournament T has a directed Hamiltonian path.*

Proof Assume that T has n vertices. If $n = 1$, 2 or 3 we can easily check from Figure 7.20 that T has a directed Hamiltonian path. Thus we may assume $n \geq 4$.

Fix such an n and assume that the result is true for all tournaments on $n-1$ vertices. Let v be a vertex of T. Then $T-v$ has $n-1$ vertices and so, since by the remarks above $T-v$ is a tournament, there is, by our assumption, a directed Hamiltonian path in $T-v$. Let $P = v_1v_2\ldots v_{n-1}$ be such a path.

Now, if there is an arc from v to v_1, then

$$P' = vv_1v_2\ldots v_{n-1}$$

is a directed Hamiltonian path in T. Similarly, if there is an arc from v_{n-1} to v then

$$P'' = v_1v_2\ldots v_{n-1}v$$

is a directed Hamiltonian path in T. Thus, in both of these cases, we are finished.

Hence we may now suppose that there is no arc from v to v_1 and no arc from v_{n-1} to v. Then there is at least one vertex w on the path P with the property that there is an arc from w to v and w is not v_{n-1} (since v_1 has this property). Let v_i be the last vertex on P having this property, so that the next vertex v_{i+1} does not have this property. Then, in particular, there is an arc *from* v_i *to* v and an arc *from* v *to* v_{i+1}, as illustrated in Figure 7.23. But then $Q = v_1 v_2 \ldots v_i v v_{i+1} v_{i+2} \ldots v_{n-1}$ gives us a directed Hamiltonian path in D. Our proof is now complete by induction. □

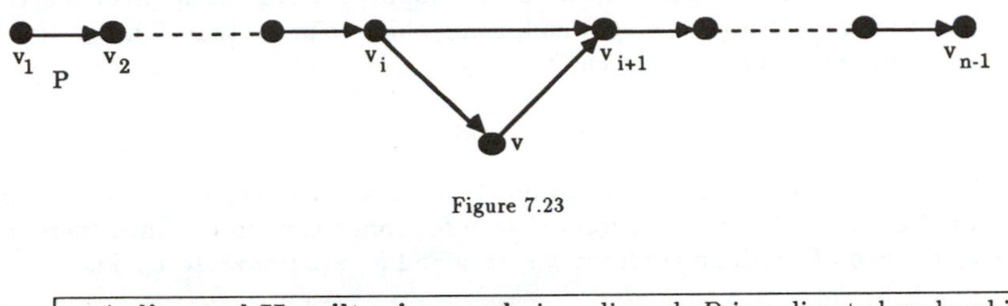

Figure 7.23

A **directed Hamiltonian cycle** in a digraph D is a directed cycle which includes every vertex of D. If D contains such a cycle then D is called **Hamiltonian**.

The previous result shows that every tournament is "nearly Hamiltonian". The next two, due to Camion [11], completely determine when a tournament is Hamiltonian.

Theorem 7.7 *A strongly connected tournament T on n vertices contains directed cycles of length* $3, 4, \ldots, n$.

Proof We first show that T contains a directed cycle of length 3. Let v be any vertex of T. Let W denote the set of all vertices w of T for which there is an arc from v to w. Let Z denote the set of all vertices z of T for which there is an arc from z to v. (Note that since T is a tournament $W \cap Z = \emptyset$.)

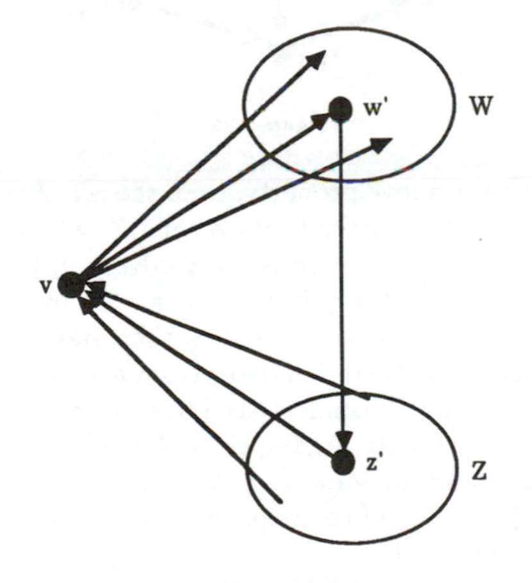

Figure 7.24

Then, since T is strongly connected, W and Z must both be nonempty. (For example, if W were empty, then there would be no arc going out of v, impossible because T is strongly connected). Moreover, again because T is strongly connected, there must be an arc in T going from some w' in W to some z' in Z. This gives the directed cycle $vw'z'v$ of length 3. (See Figure 7.24.)

We now use induction to finish the proof. We suppose that T has a directed cycle C of length k where $k < n$ (and $k \geq 3$) and, using this, we prove that T has a directed cycle of length $k+1$. Let C be given by

$$v_1 v_2 \ldots v_k v_1.$$

Suppose that there is a vertex v, not on the cycle C, with the property that there is an arc from v to v_i and an arc from v_j to v for some v_i, v_j on C. Then there must be a vertex v_i on C with an arc from v_{i-1} to v and an arc from v to v_i. Then

$$C' = v_1 v_2 \ldots v_{i-1} v v_i v_{i+1} \ldots v_k v_1$$

is a directed cycle of length k+1, i.e., of the desired length. (See Figure 7.25.)

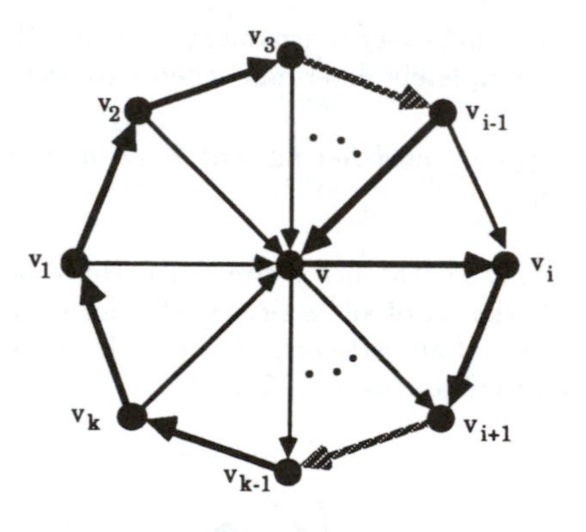

Figure 7.25

If no vertex exists with the above property, then the set of vertices not contained in the cycle can be divided into two distinct sets W and Z, where W is the set of vertices w such that for each i, $1 \leq i \leq k$, there is an arc from v_i to w, and Z is the set of vertices z such that for each i, $1 \leq i \leq k$, there is an arc from z to v_i. If W is empty then the vertices of C and the vertices of Z together make up all the vertices in T. However, by the definition of Z there is no arc from a vertex on C to a vertex in Z, a contradiction since T is strongly connected. Thus W must be nonempty. A similar argument shows that Z is nonempty. Again, since T is strongly connected, there must be an arc from some w' in W to some z' in Z. Then $C' = v_1 w' z' v_3 v_4 \ldots v_k v_1$ is a directed cycle of length $k+1$, i.e., of the required length. (See Figure 7.26.) The proof is now complete by induction. □

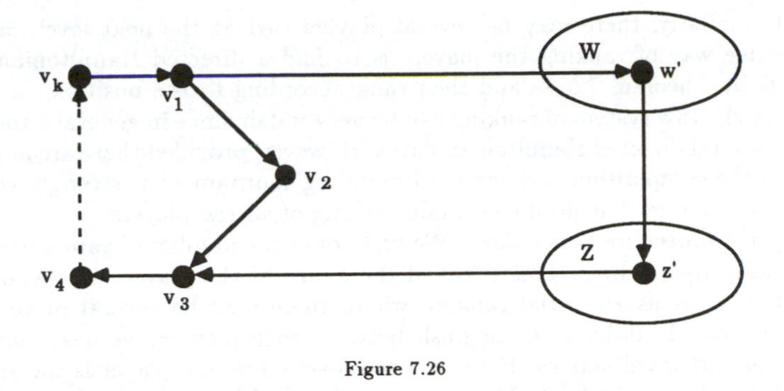

Figure 7.26

Corollary 7.8 (Camion, 1959) *A tournament T is Hamiltonian if and only if it is strongly connected.*

Proof Suppose that T has n vertices. If T is strongly connected then, by the theorem, T must have a directed cycle of length n. Such a cycle is a directed Hamiltonian cycle since it includes every vertex of T. Hence T is Hamiltonian.

Conversely if T is Hamiltonian with directed Hamiltonian cycle $C = v_1v_2\dots v_nv_1$, then given any v_i, v_j in the vertex set of T, if $i \geq j$ then $v_jv_{j+1}\dots v_i$ is a directed path P_1 from v_j to v_i while $v_iv_{i+1}\dots v_{n-1}v_nv_1\dots v_{j-1}v_j$ is a directed path P_2 from v_i to v_j. (See Figure 7.27.) Thus each vertex is reachable from any other vertex and so T is strongly connected, as required. □

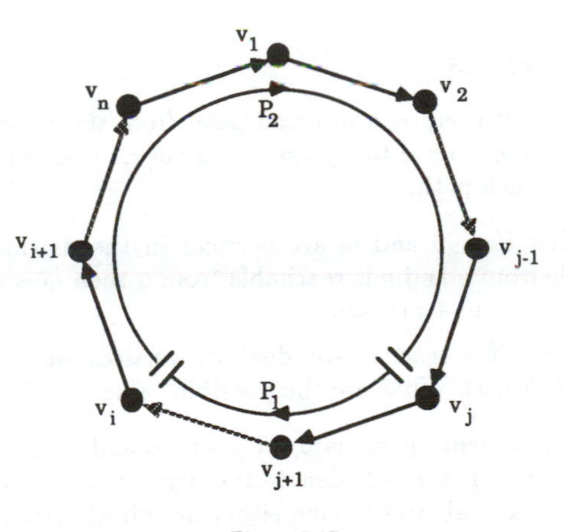

Figure 7.27

We finish this section by noting that strongly connected tournaments may be used to give a ranking of participants in a round-robin competition. In such a competition, there may be several players tied on first place (with the same maximum number of wins). Similarly, there may be several players tied at the next level, and so on. One possible way of ranking the players is to find a directed Hamiltonian path — one exists by Theorem 7.6 — and then rank according to the position on the path. Unfortunately, this system of ranking can be very unfair since in general a tournament can have several directed Hamiltonian paths. However, provided there are at least four players in the competition and the corresponding tournament is strongly connected, there is a procedure that produces a fair ranking of all the players.

Briefly, the procedure is as follows. We first count the number of games won by each player and compare them. This is called the **score** of the player. (See also Exercise 7.3.6.) This gives us an initial ranking where there may be several players on any particular score. In order to distinguish between such players, we next consider the players' **second level scores**. Here a second level score of a player is the sum of the scores of the players she beat. The sequence of second level scores may still result in ties so we compute each player's third level score, i.e., the sum of the second level scores of the players she beat. Even if no ties occur among the second level scores there may appear to be an unfair rearranging of the players' relative positions and so the second level score sequence should be computed regardless.

We now continue the process, finding the fourth level scores, then the fifth level scores, and so on. But when do we stop? The answer to this question lies in matrix theory. Using convergence of matrices one can show (but we won't) that the nth level scores settle down to a fixed pattern after taking a high enough level provided the tournament in question is *strongly connected and has at least four vertices.* For further details and a worked example, we refer the reader to Section 10.7 of Bondy and Murty [7].

Exercises for Section 7.3

7.3.1 In a digraph D, if there is a directed path from the vertex u to the vertex v, then the **distance** from u to v, denoted by $d(u,v)$, is defined to be the length of the shortest such path.

(a) Prove that if u, v and w are vertices in the digraph D such that w is reachable from v and v is reachable from u then w is reachable from u and $d(u,w) \leq d(u,v) + d(v,w)$.

(b) Prove that if u and v are distinct vertices of a tournament T then $d(u,v) \neq d(v,u)$. What are the possible values of $d(u,v)$?

7.3.2 Let T be a tournament on at least two vertices and let U be a proper subset of the vertex set of T. Let $T - U$ denote the digraph obtained from D by deleting all vertices in U and all arcs having either an initial vertex or a terminal vertex in U. Show that $T - U$ is a tournament.

7.3.3 (a) Prove that if five teams play in a round robin tournament then it is possible that all five teams tie for first place, i.e., all have the same number of wins (and losses).

(b) Prove that if six teams play in a round robin tournament then it is *not* possible that all six teams tie for first place.

(c) Prove that if n teams play in a round robin tournament (where $n \geq 3$) then it is possible that all teams tie for first place if and only if n is odd.

7.3.4 Let T be any tournament. Prove that $\overleftrightarrow{T}$, the converse of T, and $\overline{T}$, the complement of T, are isomorphic. (See Exercises 7.1.7 and 7.1.10 for definitions.)

7.3.5 A simple digraph D is called **transitive** if, whenever there is an arc in D from vertex u to vertex v and there is an arc from v to vertex w, then there is an arc from u to w.

(a) Prove that a tournament T is transitive if and only if it has a unique directed Hamiltonian path.

(b) Give an example of a tournament T on four vertices which is not transitive. Justify your example by showing that the transitive condition is not satisfied and also by showing that T has more than one directed Hamiltonian path.

(c) Give an example of a tournament T on four vertices which *is* transitive. What is the unique directed Hamiltonian path in your example T?

(d) Prove that if a simple digraph D has a directed cycle of length three then it is not transitive. Is the converse to this true?

(e) Prove that a tournament T is transitive if and only if it has no directed cycles.

7.3.6 The **score** of a vertex v in a tournament T is defined to be its outdegree. (If T represents a round robin tournament and v a player in this tournament then v's score is the number of games v has won.) If T has vertex set $\{v_1, v_2, \ldots, v_n\}$ where $od(v_1) \leq od(v_2) \leq \cdots od(v_n)$ then the sequence $(od(v_1), od(v_2), \ldots, od(v_n))$ is called a **score sequence** of T.

(a) Find score sequences of the tournaments on four vertices in Figure 7.20 and the tournament on five vertices of Figure 7.21.

(b) Prove that if $(s_1, \ldots, s_n)$ is a score sequence of a tournament T then $\sum_{i=1}^{n} s_i = n(n-1)/2$.

(c) Draw a tournament with score sequence $(0, 1, 2, 3, 4)$.

(d) Is it possible for a tournament to have $(3, 3, 3, 3, 3)$ as its score sequence?

(e) Prove that a tournament T on n vertices is transitive if and only if it has score sequence $(0, 1, 2, \ldots, n-1)$.

7.4 Traffic Flow

One-way street assignment is often used by cities to help alleviate traffic flow problems. Given a street map of a city one may ask whether or not it is possible to make each street on the map a one-way street in such a way that one can still drive from many part of the city to any other part (obeying the one-way rules of course!). We may rephrase this question using a graph and an associated digraph as follows.

First construct a graph G in which each vertex represents a street intersection. Join two vertices x and y of G by an edge if it is possible to travel between x and y without passing through any other intersection. (If it is possible to make such a trip between x and y in several ways then there should be an edge corresponding to each of these ways.) The resulting graph G gives us, in effect, a street map of the city, albeit without the street names and possibly not to scale.

If we now want to make each of the streets a one-way street the this amounts to assigning a direction to each of the edges of G, i.e., **orienting** each edge of G. Doing this creates a digraph D which is an orientation of G. Our initial question now amounts to whether or not we can find such a digraph D in which every vertex x is reachable from every other vertex y.

A graph G is called **orientable** if it has a strongly connected orientation.

Clearly then our question asks if our graph G is orientable. To determine when a graph is orientable we first consider the graph G of Figure 7.28.

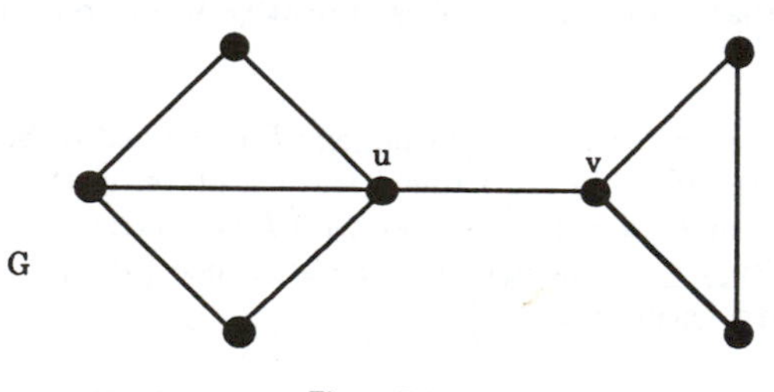

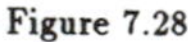
Figure 7.28

In any orientation of G the edge uv is oriented either from u to v or from v to u. If the former then there will be no $v-u$ directed walk in the orientation while if the latter happens then, similarly, v will not be reachable from u. Thus in any event the orientation can not be strongly connected and so G is not orientable.

Of course the reason we are denied a strongly connected orientation here is because the edge uv is a bridge in G. It turns out that bridges are the key to orientability, as the following result, due to Robbins [54], shows.

Theorem 7.9 (Robbins, 1939) *A graph G is orientable if and only if it is connected and has no bridges.*

Proof Clearly if G is not connected then it has no strongly connected orientation. Moreover, if G has a bridge $e = uv$, then, as the argument above shows, in any orientation of G either u is not reachable from v or vice versa. This shows that if G is orientable then G is connected and has no bridges.

Conversely, suppose that G is connected and has no bridges. Any subgraph induced by a single vertex v of G is orientable, since if there are any loops incident with v then changing them to arcs produces a strongly connected digraph with v as the only vertex. Since this shows that G does have vertex induced orientable subgraphs we may now choose a largest possible subset U of the vertex set V of G such that the induced subgraph $H = G[U]$ is orientable. If $H = G$ then, of course, G is orientable and our goal is achieved.

Thus we are left to deal with the case when $H \neq G$, i.e., $U \neq V$. In this case we choose $u \in U, v \notin U$, and, since H is orientable, we may orient the edges of H to get a strongly connected digraph D. Since G is connected and has no bridges it follows from Exercise 7.4.3 (see also Theorem 8.7 of the next chapter) that there are two edge-disjoint paths from u to v in G say

$$P = u_0u_1 \ldots u_{m-1}v \text{ and } P' = v_0v_1 \ldots v_{n-1}v,$$

where $u_0 = v_0 = u$. Since $u \in H$ but $V \notin H$, there is a vertex $u_i \in P$ which is the last vertex in P belonging to H and there is a vertex $v_j \in P'$ which is the first vertex in P' belonging to H. Now let Q and Q' be those parts of P and P' which start at u_i and v_j respectively, i.e.,

$$Q = u_iu_{i+1} \ldots u_{m-1}v \text{ and } Q' = v_jv_{j+1} \ldots v_{n-1}v.$$

Orient the edges in Q from u_i to u_{i+1}, from $ui + 1$ to u_{i+2}, and so on and orient the edges in Q' from v to v_{n-1}, from v_{n-1} to v_{n-2} and so on (so that the edges of Q' are oriented in reverse to the direction of the path). We illustrate this in Figure 7.29.

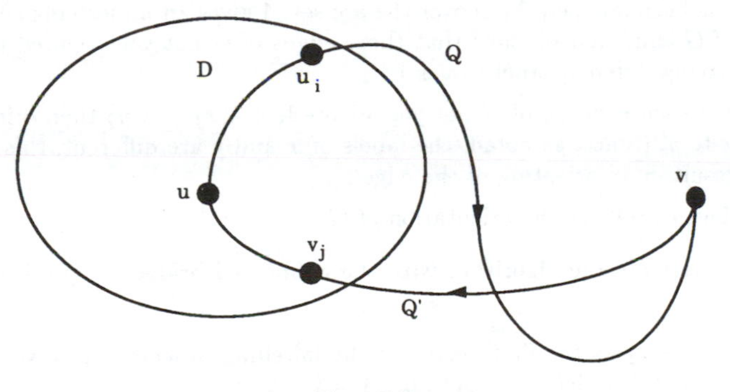

Figure 7.29

Now let D' be the digraph with vertex set U together with all the vertices of Q and Q' and with the arc set of D together with those arcs we have just defined on Q and Q'. Then, since D is strongly connected and the directed walk Q concatenates with the reverse of Q' to give a directed walk from u_i to v_j it follows easily that D' is strongly connected.

However, D' has at least one more vertex than D has, namely v, and so this contradicts our choice of U as a largest possible subset of vertices of G which induces an orientable subgraph. This contradiction shows that $H = G$ after all. □

We now present an algorithm due to J. E. Hopcroft and R. Tarjan [35] which produces a strongly connected orientation for a connected bridgeless graph. The algorithm is not based on the proof above and we leave its justification to Exercise 7.4.5. It uses a labelling technique.

The Hopcroft and Tarjan algorithm

Step 1. Let G be a connected graph with no bridges.

Let x be an arbitrary vertex of G and label x by setting $\lambda(x) = 1$.

Set $L = \{x\}$ and $U = V(G) - \{x\}$. (Here L denotes the set of labelled vertices while U denotesthe set of unlabelled vertices.)

Set $A = \emptyset$. (A denotes the set of arcs produced by orienting edges of G.)

Step 2. Let v be a vertex in L, of highest possible label value, which is adjacent to some vertex $u \in U$. Set $\lambda(u) = \lambda(v) + 1$.

Replace L by $L \cup \{u\}$ and U by $U - \{u\}$ (since u has just been labelled).

Orient the edge vu from v to u and replace A by $A \cup \{(v, u)\}$ (since there is now a new arc from v to u). (Note that, in this process, each new arc of A goes from a labelled vertex to one with a higher label.)

Step 3. If $L \neq V(G)$, repeat Step 2.

Step 4. (When we reach this step we must have $L = V(G)$, i.e., every vertex of G has been labelled. Moreover the arc set A gives an underlying spanning tree of G and one can show that those edges of G not yet oriented join vertices having different label values.)

For each edge xy of G not yet oriented, if $\lambda(x) > \lambda(y)$ then orient xy from x to y. (Since, as noted, the labels of x and y are different, this will always result in an orienting of the edge xy.)

This completes the orientation of G.

We now illustrate the algorithm with the connected bridgeless graph G of Figure 7.30.

Step 1. Choose v_1 as the first vertex in the labelling process. Thus we set $\lambda(v_1) = 1, L = \{v_1\}, U = V(G) - \{v_1\}$ and $A = \emptyset$.

Step 2. Choose $v_2 \in U \cap N(v_1)$. (Another possibility is v_7.) Set $\lambda(v_2) = \lambda(v_1) + 1 = 2$.

L becomes $\{v_1, v_2\}$ and U becomes $\{v_3, \ldots, v_8\}$.

Orient the edge v_1v_2 from v_1 to v_2 so that A becomes $\{(v_1, v_2)\}$.

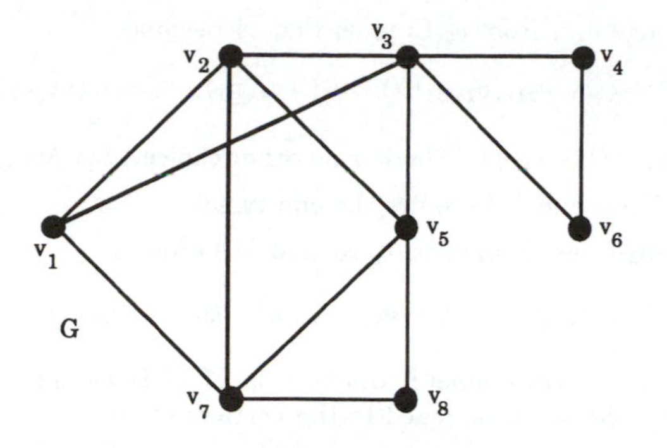

Figure 7.30: A connected graph with no bridges.

Step 3. $L \neq V(G)$ so we return to step 2. (For the sake of brevity, we will omit all but the last of the subsequent step 3's.)

Step 2. Choose $v_3 \in U \cap N(v_2)$. (Other possible choices are v_5 and v_7.) Set $\lambda(v_3) = \lambda(v_2) + 1 = 3$.

L becomes $\{v_1, v_2, v_3\}$ and U becomes $\{v_4, \ldots, v_8\}$.

Orient the edge v_2v_3 from v_2 to v_3 so that A becomes $\{(v_1, v_2), (v_2, v_3)\}$.

Step 2. Choose $v_4 \in U \cap N(v_3)$. (Other possible choices are v_5 and v_6.) Set $\lambda(v_4) = \lambda(v_3) + 1 = 4$.

L becomes $\{v_1, \ldots, v_4\}$ and U becomes $\{v_5, \ldots, v_8\}$.

Orient the edge v_3v_4 from v_3 to v_4 so that A becomes $\{(v_1, v_2), (v_2, v_3), (v_3, v_4)\}$.

Step 2. Choose $v_6 \in U \cap N(v_4)$. (There is no other choice.) Set $\lambda(v_6) = \lambda(v_4) + 1 = 5$.

L becomes $\{v_1, \ldots, v_4, v_6\}$ and U becomes $\{v_5, v_7, v_8\}$.

Orient the edge v_4v_6 from v_4 to v_6 so that A becomes

$$\{(v_1, v_2), (v_2, v_3), (v_3, v_4), (v_4, v_6)\}.$$

Step 2. Now v_3 is the highest labelled vertex with an unlabelled neighbour. Choose $v_5 \in U \cap N(v_3)$. (There is no other choice.) Set $\lambda(v_5) = \lambda(v_3) + 1 = 4$.

L becomes $\{v_1, \ldots, v_6\}$ and U becomes $\{v_7, v_8\}$.

Orient the edge v_3v_5 from v_3 to v_5 so that A becomes

$$\{(v_1, v_2), (v_2, v_3), (v_3, v_4), (v_4, v_6), (v_3, v_5)\}.$$

Step 2. Now v_5 is the highest labelled vertex with an unlabelled neighbour. Choose $v_7 \in U \cap N(v_5)$. (Another possible choice is v_8.) Set $\lambda(v_7) = \lambda(v_5) + 1 = 5$.

L becomes $\{v_1, \ldots, v_7\}$ and U becomes $\{v_8\}$.

Orient the edge v_5v_7 from v_5 to v_7 so that A becomes

$$\{(v_1, v_2), (v_2, v_3), (v_3, v_4), (v_4, v_6), (v_3, v_5), (v_5, v_7)\}.$$

Step 2. Choose $v_8 \in U \cap N(v_7)$. (There is no other choice.) Set $\lambda(v_8) = \lambda(v_7) + 1 = 5$.
L becomes $V(G)$ and U becomes the empty set.
Orient the edge v_7v_8 from v_7 to v_8 so that A becomes

$$\{(v_1, v_2), (v_2, v_3), (v_3, v_4), (v_4, v_6), (v_3, v_5), (v_5, v_7), (v_7, v_8)\}.$$

At this stage our (incomplete) orientation of G is as shown in Figure 7.31, together with the labels assigned to the vertices of G.

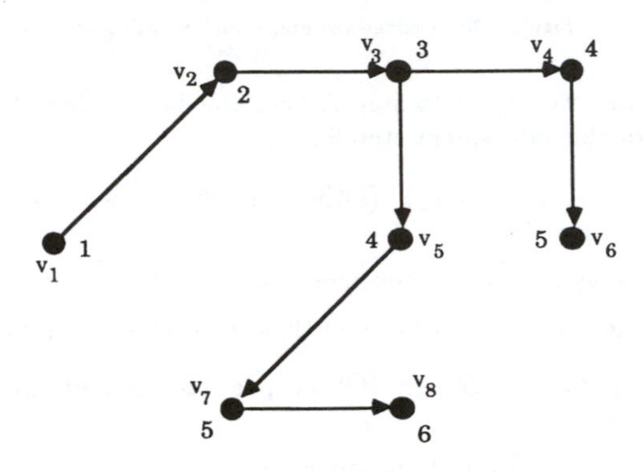

Figure 7.31: A partial orientation of the graph G of Figure 7.28.

Step 3. We now have $L = V(G)$.

Step 4. The edges of G not yet oriented are $v_1v_3, v_1v_7, v_2v_5, v_2v_7, v_3v_6$ and v_5v_8. These are each oriented from the vertex with the higher label to the vertex with the lower label.

This completes the orientation of G, giving the strongly connected digraph shown in Figure 7.32.

Exercises for Section 7.4

7.4.1 Let G be a Hamiltonian graph. Prove that G is orientable without using Theorem 7. 9.

7.4.2 Give a simple construction of a strongly connected orientation for each complete graph with at least three vertices and also for each complete bipartite graph which is not a star graph.

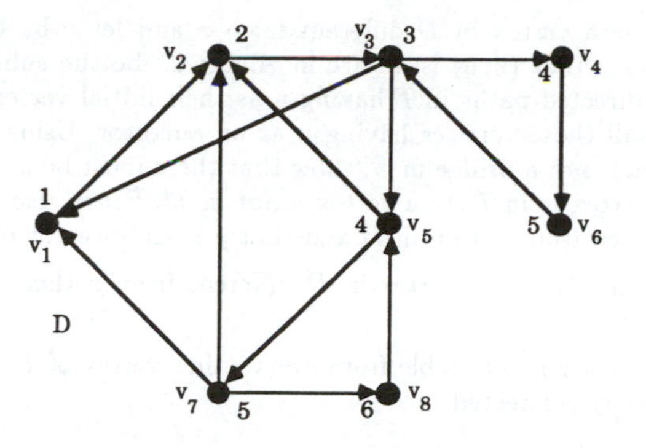

Figure 7.32: A strongly connected orientation of G.

7.4.3 Prove that if G is a connected graph with no bridges then between any two distinct vertices u and v of G there are two edge-disjoint paths.

7.4.4 Using the Hopcroft and Tarjan algorithm, find a strongly connected orientation for the graphs of Figure 7.33.

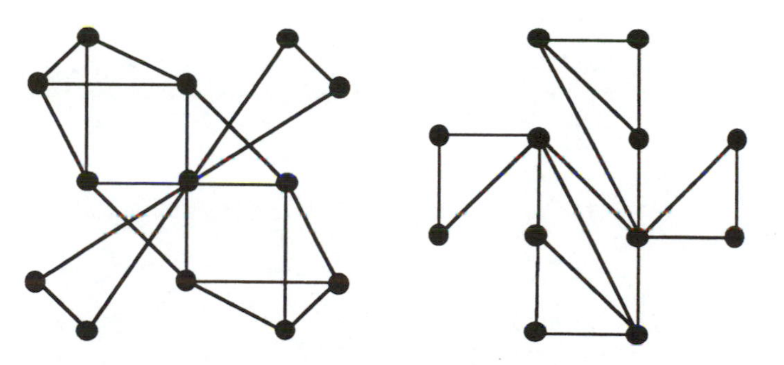

Figure 7.33: Orientable graphs.

7.4.5 This exercise sketches a proof that the Hopcroft and Tarjan algorithm does produce a strongly connected orientation for the connected bridgeless graph G.

(a) Prove that, once all the vertices of G have been labelled in the algorithm, the arc set A gives an underlying spanning tree T of G.

(b) Prove that no arc in the set A joins two vertices having the same label value.

(c) We will say that a vertex a is an **ancestor** of a vertex b in the digraph D finally produced by the algorithm if there is a directed path from a to b involving only arcs from the set A. Prove that x, the first vertex to be labelled, is an ancestor of every vertex.

(d) Let u be a vertex in D different from x and let v be the unique vertex in D such that (v, u) is an arc in A. Let T' be the subtree of T induced by all directed paths in T having u as their initial vertex, i.e., induced by u and all those vertices having u as an ancestor. Using the fact that the edge vu is not a bridge in G, show that there must be an edge in G joining some vertex w in T' to a vertex y not in T'. Prove also that this edge wy is oriented from w to y in D and that y is an ancestor of u.

(e) Prove that if u is a vertex in D different from x then it has a reachable ancestor.

(f) Prove that x is reachable from every other vertex of D and hence that D is strongly connected.

Chapter 8

Networks

8.1 Flows and Cuts

A manufacturer in New Zealand wants to export several boxes of one of his products, clockwork kiwifruit, to a department store in Taiwan. There are various channels through which the boxes can be sent and the digraph of Figure 8.1 represents these, with vertex s as the manufacturer and t the department store. The numbers assigned to each arc represent the maximum loads which each of the corresponding channels can handle.

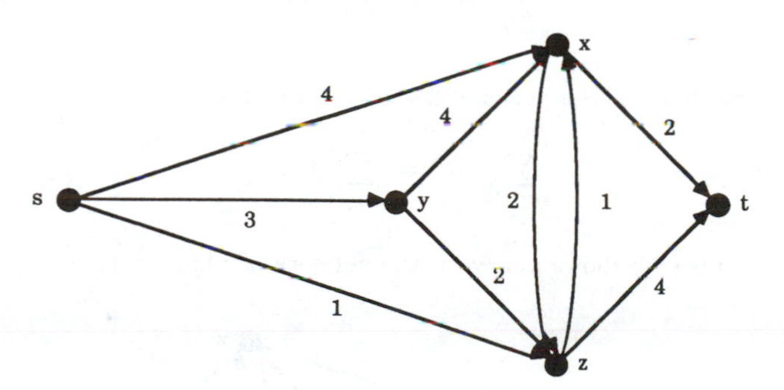

Figure 8.1: A clockwork kiwifruit export network.

The manufacturer wishes to find the maximum number of boxes he can send through the network of channels and "middle men" x, y and z to t so that he never exceeds the permitted capacity of any channel.

Using this example as motivation we now define the concept of a network.

> A **network** N is a weakly connected simple digraph in which every arc a of N has been assigned a non-negative integer $c(a)$, called the **capacity** of a.
> A vertex s of a network N is called a **source** if it has indegree 0 while a vertex t of N is called a **sink** if it has outdegree 0. Any other vertex of N is called an **intermediate** vertex.

It follows that any arc incident with a source s goes *from* s *to* another vertex and any arc incident with a sink t goes *from* some vertex *to* t.

We will assume from now on that any network N we consider *has exactly one source and exactly one sink.*

Roughly speaking, the source of any network can be thought of as the manufacturer and the sink as the market. The intermediate vertices represent middle men, while the capacity on each arc is the maximum amount of goods that can be sent from the tail of the arc to its head. In Figure 8.1, s, being the only vertex with zero indegree, is the source and similarly t is the sink.

Given any vertex u of the network N we let denote the set of arcs going *into* u and going *out of* u by $I(u)$ and $O(u)$ respectively.

A **flow** in a network N from the source s to the sink t is a function f which assigns a non-negative integer to each of the arcs a in N such that
(i) (**capacity constraint**) $f(a) \leq c(a)$ for each arc a,
(ii) the total flow into the sink t equals the total flow out of the source s and
(iii) (**flow conservation**) for any intermediate vertex x, the total flow into x equals the total flow out of x.

To be more specific, (ii) means that for the source s and the sink t,

$$\sum_{a \in O(s)} f(a) = \sum_{a \in I(t)} f(a),$$

while (iii) means that if x is a vertex different from s and t then

$$\sum_{a \in O(x)} f(a) = \sum_{a \in I(x)} f(a).$$

For example, Figure 8.2 shows a flow for the network of Figure 8.1.

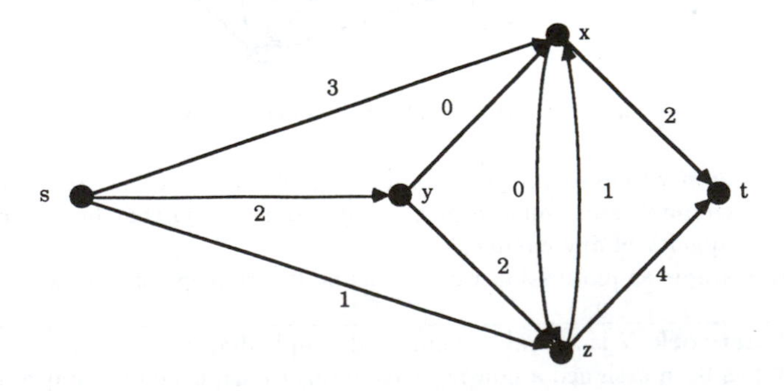

Figure 8.2: A flow for the network of Figure 8.1.

In this example we have

$$\begin{array}{rcccccccc}
\sum_{a\in O(s)} f(a) & = & 3+2+1 & = & 6 & = & 2+4 & = & \sum_{a\in I(t)} f(a), \\
\sum_{a\in O(x)} f(a) & = & 2+1 & = & 3 & = & 3+0 & = & \sum_{a\in I(x)} f(a), \\
\sum_{a\in O(y)} f(a) & = & 0+2 & = & 2 & = & & & \sum_{a\in I(y)} f(a), \\
\sum_{a\in O(z)} f(a) & = & 0+4 & = & 4 & = & 1+2+1 & = & \sum_{a\in I(z)} f(a),
\end{array}$$

and for each arc a of the network, $0 \leq f(a) \leq c(a)$.

The number

$$d = \sum_{a\in O(s)} f(a) = \sum_{a\in I(t)} f(a),$$

where s and t are the source and sink of the network N, is called the **value** of the flow f.

Thus the flow in our example above has value 6.

Let f be a flow on the network $N = (V, A)$ and for any proper subset X of the vertex set V of N let $\overline{X}$ denote the complement of X in V, i.e., $\overline{X} = V - X$.

If X contains the source s but not the sink t, then intuitively we would expect the *net* flow from the vertices in X to those not in X, i.e., to $\overline{X}$, to equal d, the value of the flow. We illustrate this with the network in Figure 8.3 where the first number assigned to each arc a is its capacity $c(a)$, while the second is $f(a)$, the flow across a.

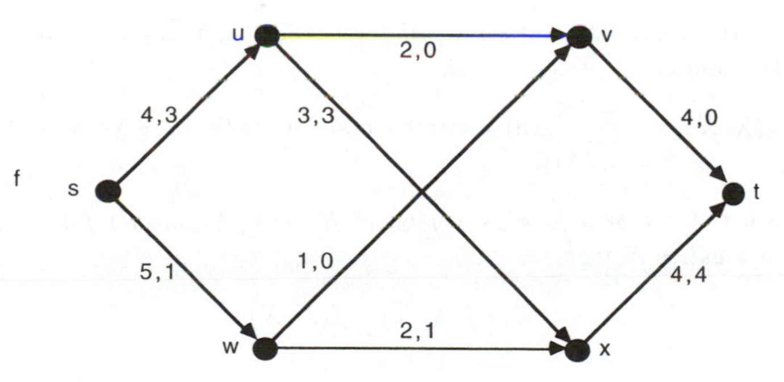

Figure 8.3: A network with a flow.

Here if we take $X = \{s, u, x\}$, so that $\overline{X} = \{v, w, t\}$, then the arcs from vertices in X to vertices in $\overline{X}$ are uv, sw and xt while there is only one arc from vertices in $\overline{X}$ to vertices in X, namely wx. Thus the *net* flow from vertices in X to vertices in $\overline{X}$ is

$$f(uv) + f(sw) + f(xt) - f(wx) = 0 + 1 + 4 - 1 = 4,$$

which is the value of the flow f, as expected. (The value of the flow is given by $f(su) + f(sw) = 3 + 1$, or, alternatively, by $f(vt) + f(xt) = 0 + 4$.)

Also, intuitively we would expect the net flow from any such subset X to $\overline{X}$ not to exceed the capacity of the arcs from X to $\overline{X}$, where by this we mean the sum of the capacities of each arc going from a vertex in X to a vertex in $\overline{X}$. For example for our subset X in the network of Figure 8.3, the capacity of the arcs from X to $\overline{X}$ is

$$c(uv) + c(sw) + c(xt) = 2 + 5 + 4 = 11,$$

which is certainly larger than the value of the flow, namely 4.

The preceding ideas are generalized in our first result below. We first fix some notation.

If X and Y are any two subsets of vertices of the network N we let $A(X,Y)$ denote the set of arcs from vertices in X to vertices in Y.

If g is any function which assigns non-negative integers to the arcs of the network N (for example, g could be the capacity function c or a flow f), then for any two subsets of vertices X, Y of N we define

$$g(X,Y) = \sum_{a \in A(X,Y)} g(a).$$

In other words, $g(X,Y)$ is the sum of the values of the function g on each arc from a vertex in X to a vertex in Y.

A **cut** is a set of arcs $A(X,\overline{X})$ where the source s is in X and the sink t is in $\overline{X}$.

Thus, for the subset $X = \{s, u, x\}$ of Figure 8.3, $A(X,\overline{X})$ is a cut and for the capacity function c,

$$c(X,\overline{X}) = \sum_{a \in A(X,\overline{X})} c(a) = c(uv) + c(sw) + c(xt) = 2 + 5 + 4 = 11.$$

Theorem 8.1 *Let f be a flow on a network $N = (V, A)$ and let f have value d. If $A(X,\overline{X})$ is a cut in N then*

$$d = f(X,\overline{X}) - f(\overline{X},X)$$

and

$$d \leq c(X,\overline{X}).$$

In other words, the total flow **out of** *X minus the total flow* **into** *X, i.e., the* **net flow** *out of X, equals d, the value of the flow, and this never exceeds the total capacity of the arcs from X to $\overline{X}$.*

Proof From the definition of flow, for the source s we have

$$f(\{s\}, V) = d \text{ and } f(V, \{s\}) = 0,$$

while, for any vertex u different from both s and the sink t,

$$f(\{u\},V) = \sum_{a\in O(u)} f(a) = \sum_{a\in I(u)} f(a) = f(V,\{u\}),$$

i.e., $f(\{u\},V) - f(V,\{u\}) = 0$ for $u \neq s,t$.

Thus, for our cut $A(X,\overline{X})$, we have

$$\sum_{x\in X}\{f(\{x\},V) - f(V,\{x\})\} = f(\{s\},V) - f(V,\{s\}) + 0 = d - 0 + 0 = d,$$

i.e., $f(X,V) - f(V,X) = d$. However

$$f(X,V) = f(X, X \cup (\overline{X})) = f(X,X) + f(X,\overline{X})$$

and similarly

$$f(V,X) = f(X,X) + f(\overline{X},X).$$

Thus

$$d = f(X,V) - f(V,X) = f(X,X) + f(X,\overline{X}) - f(X,X) - f(\overline{X},X) = f(X,\overline{X}) - f(\overline{X},X),$$

i.e., $d = f(X,\overline{X}) - f(\overline{X},X)$, thus establishing the first part of the Theorem.

Moreover, since for each arc a of N we have $f(a) \leq c(a)$ (from the definition of a flow), we get $f(X,\overline{X}) \leq c(X,\overline{X})$ and so

$$d = f(X,\overline{X}) - f(\overline{X},X) \leq f(X,\overline{X}) \leq c(X,\overline{X})$$

giving $d \leq c(X,\overline{X})$, as required. □

The second part of the theorem tells us that the value of *any* flow is less than or equal to the capacity of the arcs from X to $\overline{X}$ for *any* cut $A(X,\overline{X})$. Thus, if f is a flow with value d, we have

$$d \leq \min\{c(X,\overline{X}) : A(X,\overline{X}) \text{ is any cut}\}.$$

We are particularly interested in flows having values equal to the upper bound imposed by the last inequality.

A flow with value equal to

$$\min\{c(X,\overline{X}) : A(X,\overline{X}) \text{ is any cut}\}$$

is called a **maximal** (or **maximum**) **flow**.

In the network of Figure 8.3, if $A(X,\overline{X})$ is a cut then s must be in X, t in $\overline{X}$ and each of the four intermediate vertices u, v, w and x can be either in X or $\overline{X}$. It follows that there are $2^4 = 16$ possible cuts in this network. (More generally, if there are n intermediate vertices in the network N then N will have 2^n cuts.) If we examine the capacity of each of these cuts in turn it turns out that the cut $A(X,\overline{X})$ where

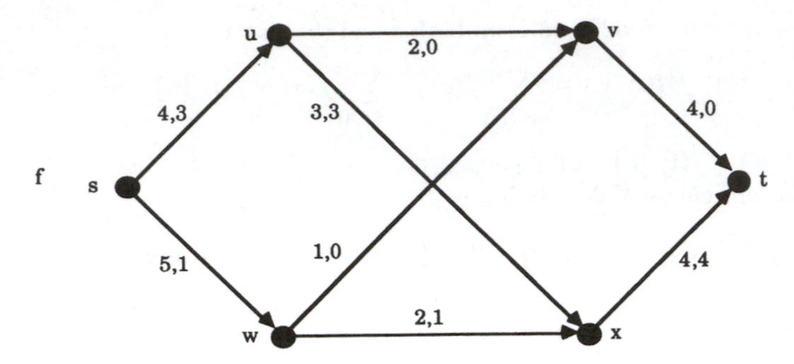

Figure 8.4: This is just Figure 8.3 in a more convenient place.

$X = \{s, u, w, x\}$ (and so $\overline{X} = \{v, t\}$) has capacity $c(uv) + c(wv) + c(xt) = 2+1+4 = 7$ and that this is the smallest capacity of any of the possible cuts. Hence, by the above inequality, any flow on the network must have value at most 7.

We now try to construct a maximal flow on our network. We increase the given f by steps. Figure 8.4 shows f.

For example, the flow on the directed path $s\ u\ v\ t$ is not at a maximum since the flow's value on each of su, uv, vt can be increased by 1 and still remain within the capacity of each arc. This gives a new flow f_1, shown in Figure 8.5.

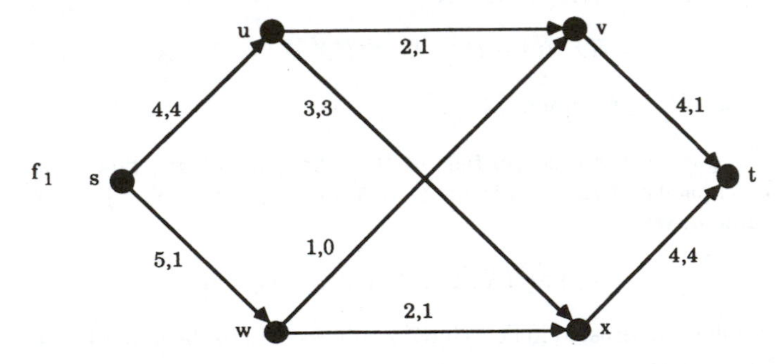

Figure 8.5: Flow f_1 has value 5, an improvement on f's value.

The value of this new flow is $4 + 1 = 5$, one more than the value of f. Likewise f_1 can be increased since the flow on the directed path $s\ w\ v\ t$ is not at a maximum but can be increased by 1 by increasing the flow on each of its arcs sw, wv and vt by 1. This gives a new flow f_2, shown in Figure 8.6, with value $4 + 2 = 6$.

Now each of the four possible directed paths from s to t, namely $s\ u\ v\ t$, $s\ u\ x\ t$, $s\ w\ x\ t$ and $s\ w\ v\ t$, have an arc whose flow equals the capacity of the arc: $s\ u\ v\ t$ has su, as does $s\ u\ x\ t$, while $s\ w\ x\ t$ has xt and $s\ w\ v\ t$ has wv. Thus we may suspect that f_2 is a flow of maximal value, i.e., no flow can have value more than 6. However we must still investigate the possibilty that the flow can be increased by some other kind of adjustment. In fact, if we increase the flow on sw, wx, uv and vt each by 1 and

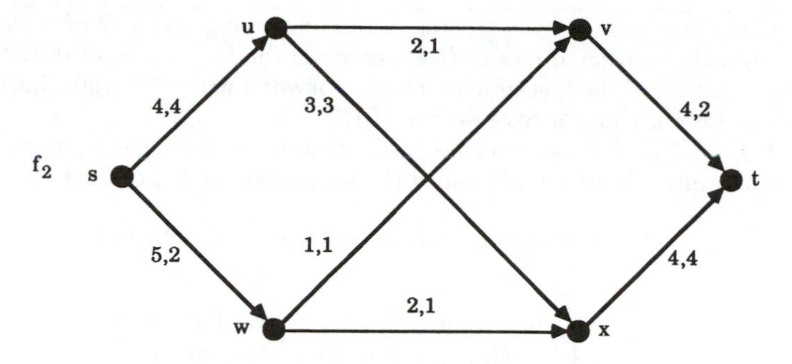

Figure 8.6: Flow f_2 has value 6, an improvement on f_1's value.

decrease the flow on ux by 1 we get a new flow f_3, shown in Figure 8.7, with value $3 + 4 = 7$. Moreover, since there is a cut with capacity 7, f_3 is a maximal flow.

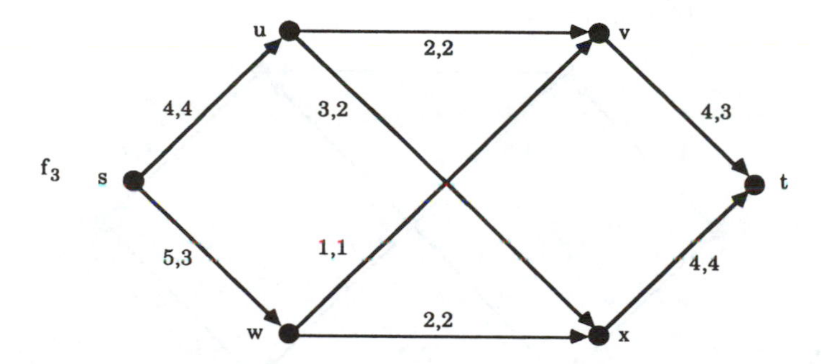

Figure 8.7: Flow f_3 has value 7, an improvement on f_2's value, and is a maximal flow.

We note that each of the adjustments made in the process of obtaining f_3 did in fact give a flow according to our definition: if we increased the flow of an arc *into* an intermediate vertex z then we made a similar increase on an arc *out* of z or a corresponding *decrease* on another arc *into* z, so that flow conservation was maintained, while flow increase out of source s was matched by a flow increase into sink t, so that condition (ii) of the definition was also respected.

The main result of this chapter assures us that for any network there is *always* a maximal flow. Furthermore the proof of the result yields an algorithm which constructs such a maximal flow. Before we state and prove it we introduce some more terminology.

Given any trail $W = v_0v_1\ldots v_k$ in the underlying graph G of a network N, then the associated arcs in N are either of the form $v_{i-1}v_i$ or of the form v_iv_{i-1}. An arc of the first form is called a **forward arc** of W while one of the second form is called a **reverse arc** of W.

If f is a flow in N we associate to the trail W, in the underlying graph G, a non-negative integer $i(W)$, called the **increment** of W, defined by

$$i(W) = \min\{i(a) : a \text{ is an arc associated with } W\},$$

where

$$i(a) = \begin{cases} c(a) - f(a) & \text{if } a \text{ is a forward arc of } W \\ f(a) & \text{if } a \text{ is a reverse arc of } W. \end{cases}$$

For example, for the flow f on the network N of Figure 8.4, shown again in Figure 8.8, the walk $W = s\ w\ x\ u\ v$ has forward arcs sw, wx and uv while the remaining associated arc xu is a reverse arc of W. Thus

$$\begin{array}{lclclcl} i(sw) & = & c(sw) - f(sw) & = 5 - 1 = 4, & i(wx) & = & c(wx) - f(wx) & = 2 - 1 = 1, \\ i(ux) & = & f(ux) & = 3, & i(uv) & = & c(uv) - f(uv) & = 2 - 0 = 2, \end{array}$$

and so $i(W) = \min\{4, 1, 3, 2\} = 1$.

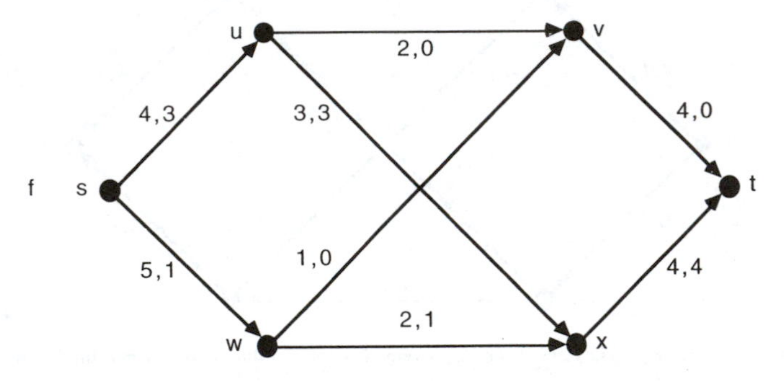

Figure 8.8: This is just Figure 8.4 in a more convenient place.

The amount $i(W)$ is the largest by which the flow f along W can be increased subject to the capacity constraint. For example, for $W = s\ w\ x\ u\ v$ above, we can increase the flow on sw, wx and uv by 1 and make the corresponding decrease by 1 on the arc xu (and this gives the largest increase possible on W). (See Figure 8.9.)

The walk W is said to be f-**saturated** if $i(W) = 0$ and f-**unsaturated** if $i(W) > 0$.

This simply means that f-unsaturated walks are precisely those that are not being used to their full capacity. Thus, for example, our walk $W = s\ w\ x\ u\ v$ above is f-unsaturated.

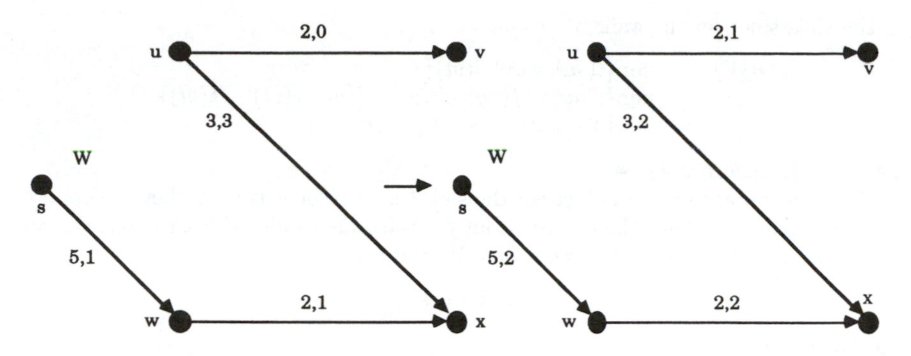

Figure 8.9: The flow is increased along W by $i(W) = 1$.

> An f-**incrementing** walk is an f-unsaturated walk from the source s to the sink t.

We are now ready for our main result, due to Ford and Fulkerson [23]. We shall interrupt its proof at several stages to illustrate it using our favourite network, that of Figure 8.8 (or is it Figure 8.4?). Each of these interruptions begins and ends with the symbol ■.

Theorem 8.2 (The Max-Flow, Min-Cut Theorem) *Let N be a network with capacity function c. Then there exists a maximum flow in N, i.e., there exists a flow f in N with value*

$$\min\{c(X,\overline{X}) : A(X,\overline{X}) \text{ is a cut}\}.$$

Proof We already know from Theorem 8.1 that for any flow f in N with value d we have

$$d \le \min\{c(X,\overline{X}) : A(X,\overline{X}) \text{ is a cut}\}.$$

Given an arbitrary flow f, we let X be the set of vertices z in N such that either $z = s$ or in the underlying graph G of N there is an f-unsaturated walk $W = v_0 \ldots v_k$ from s to z (so that $s = v_0$, $z = v_k$).

■ Thus, for our example N of Figure 8.8, the set X consists of:
s, the source,
u, since, for the walk $W = su$, $i(W) = i(su) = c(su) - f(su) = 4 - 3 > 0$,
w, since, for the walk $W = sw$, $i(W) = i(sw) = c(sw) - f(sw) = 5 - 1 > 0$,
v, since, for the walk $W = suv$,

$$\begin{aligned} i(W) &= \min\{i(su), i(uv)\} = \min\{c(su) - f(su), c(uv) - f(uv)\} \\ &= \min\{4-3, 2-0\} > 0, \end{aligned}$$

x, since, for the walk $W = swx$,

$$\begin{aligned} i(W) &= \min\{i(sw), i(wx)\} \\ &= \min\{c(sw) - f(sw), c(wx) - f(wx)\} \\ &= \min\{5-1, 2-1\} > 0, \end{aligned}$$

t, the sink, since for the walk $W = suvt$,

$$\begin{aligned} i(W) &= \min\{i(su), i(uv), i(vt)\} \\ &= \min\{c(su) - f(su), c(uv) - f(uv), c(vt) - f(vt)\} \\ &= \min\{4-3, 2-0, 4-0\} > 0, \end{aligned}$$

i.e., $X = \{s, u, w, v, x, t\}$. ■

Now, returning to the proof, either the sink t is in X or it is in $\overline{X}$. Let us suppose first that t is in X. Then there must be an f-unsaturated walk W from s to t, i.e., an f-incrementing walk. Choose such a walk W and let

$$i(W) = \epsilon,$$

so that $\epsilon > 0$.

■ Thus, in our example, we take W as $suvt$, so that $i(W) = \epsilon = 1$. ■

We now define a new function f_1 on the arcs a of N by

$$f_1(a) = \begin{cases} f(a) + \epsilon & \text{if } a \text{ is a forward arc in } W \\ f(a) - \epsilon & \text{if } a \text{ is a reverse arc in } W \\ f(a) & \text{if } a \text{ is any other arc in } N. \end{cases}$$

Then f_1 is a flow with value $d+\epsilon$. (We leave the proof of this as Exercise 8.1.2.) Since ϵ is, by its definition, a positive integer, we have increased the flow f, with value d, to a new flow f_1 with value $d + \epsilon$.

This new flow f_1 is called the **revised flow based on (the f-incrementing walk) W**.

■ In our example, f_1 is actually the flow f_1 given in Figure 8.5 since

$$\begin{array}{lllllll} f_1(su) = f(su) + \epsilon = 3+1 = 4, & f_1(sw) = f(sw) = 1, \\ f_1(uv) = f(uv) + \epsilon = 0+1 = 1, & f_1(ux) = f(ux) = 3, \\ f_1(vt) = f(vt) + \epsilon = 0+1 = 1, & f_1(wv) = f(wv) = 0, \\ f_1(wx) = f(wx) = 1, & f_1(xt) = f(xt) = 4, \end{array}$$

and the value of this new flow f_1 is $d + \epsilon = 4 + 1 = 5$. ■

This procedure of increasing the flow is always possible provided the sink t is in the set X. Thus we may repeat the process, progressively revising the flow based on incrementing walks until we reach a stage when t is not in the associated set X. At this stage, there is no longer an incrementing walk available to us. Also, since $t \notin X$, $A(X, \overline{X})$ is a cut.

■ In our example, the set X associated with f_1 (see Figure 8.5) consists of:
s, the source,
w, since, for the walk $W = sw$, $i(W) = i(sw) = c(sw) - f_1(sw) = 5 - 1 > 0$,
x, since, for the walk $W = swx$,

$$\begin{aligned} i(W) &= \min\{i(sw), i(wx)\} = \min\{c(sw) - f_1(sw), c(wx) - f_1(wx)\} \\ &= \min\{5-1, 2-1\} > 0, \end{aligned}$$

v, since, for the walk $W = swv$,

$$\begin{aligned} i(W) &= \min\{i(sw), i(wv)\} = \min\{c(sw) - f_1(sw), c(wv) - f_1(wv)\} \\ &= \min\{5-1, 1-0\} > 0, \end{aligned}$$

u, since, for the walk $W = swvu$,

$$\begin{aligned} i(W) &= \min\{i(sw), i(wv), i(vu)\} \\ &= \min\{c(sw) - f_1(sw), c(wv) - f_1(wv), f_1(uv)\} \\ &= \min\{5-1, 1-0, 1\} > 0, \end{aligned}$$

t, since, for the walk $W = swvt$,

$$\begin{aligned} i(W) &= \min\{i(sw), i(wv), i(vt)\} \\ &= \min\{c(sw) - f_1(sw), c(wv) - f_1(wv), c(vt) - f_1(vt)\} \\ &= \min\{5-1, 1-0, 4-1\} > 0, \end{aligned}$$

i.e., $X = V$, as for f.

Taking the walk $W = swvt$ we get $\epsilon = 1$. This is used to define a new function f_2 on the arcs of N by

$$\begin{aligned} f_2(sw) &= f_1(sw) + \epsilon = 1 + 1 = 2, \quad f_2(wv) = f_1(wv) + \epsilon = 0 + 1 = 1, \\ f_2(vt) &= f_1(vt) + \epsilon = 0 + 1 = 1, \end{aligned}$$

and, for every other arc a, $f_2(a) = f_1(a)$. In fact, f_2 is just the flow f_2 given in Figure 8.6.

We now find the set X associated with f_2. It consists of (see Figure 8.6) the following vertices:
s, the source,
w, since, for the walk $W = sw$, $i(W) = i(sw) = c(sw) - f_2(sw) = 5 - 2 > 0$,
x, since, for the walk $W = swx$,

$$\begin{aligned} i(W) &= \min\{i(sw), i(wx)\} = \min\{c(sw) - f_2(sw), c(wx) - f_2(wx)\} \\ &= \min\{5-2, 2-1\} > 0, \end{aligned}$$

u, since, for the walk $W = swxu$,

$$\begin{aligned} i(W) &= \min\{i(sw), i(wx), i(xu)\} \\ &= \min\{c(sw) - f_2(sw), c(wx) - f_2(wx), f_2(ux)\} \\ &= \min\{5-2, 2-1, 3\} > 0, \end{aligned}$$

v, since, for the walk $W = swxuv$,

$$\begin{aligned} i(W) &= \min\{i(sw), i(wx), i(xu), i(uv)\} \\ &= \min\{c(sw) - f_2(sw), c(wx) - f_2(wx), f_2(ux), c(uv) - f_2(uv)\} \\ &= \min\{5-2, 2-1, 3, 2-1\} > 0, \end{aligned}$$

t, since for the walk $W = swxuvt$,

$$\begin{aligned} i(W) &= \min\{i(sw), i(wx), i(xu), i(uv), i(vt)\} \\ &= \min\{c(sw) - f_1(sw), c(wx) - f_2(wx), f_2(ux), c(uv) - f_2(uv), c(vt) - f_2(vt)\} \\ &= \min\{5-2, 2-1, 3, 2-1, 4-2\} > 0, \end{aligned}$$

i.e., $X = V$ again.

Then taking the walk $W = swxuvt$ from s to t we get $\epsilon = 1$. This is used to define a new function f_3 on the arcs of N by

$$\begin{array}{lllllll} f_3(sw) &=& f_2(sw)+\epsilon &=& 2+1=3, & f_3(wx) = f_2(wx)+\epsilon = 1+1=2,\\ f_3(ux) &=& f_2(ux)-\epsilon &=& 3-1=2, & f_3(uv) = f_2(uv)+\epsilon = 1+1=2,\\ f_3(vt) &=& f_2(vt)+\epsilon &=& 2+1=3, & \end{array}$$

and $f_3(a) = f_2(a)$ for all other arcs a. In fact, f_3 is just the flow f_3 given in Figure 8.7.

We now find the associated set X. It consists of (see Figure 8.7) the vertices:
s, the source,
w, since, for the walk $W = sw$, $i(W) = i(sw) = c(sw) - f_3(sw) = 5 - 3 > 0$,
and no other vertices since all the other arcs incident with s or w go from s or w to another vertex and their current flow value equals their capacity value, i.e.,

$$X = \{s, w\}.$$

Since $t \notin X$, $A(X,\overline{X})$ is a cut. ■

Meanwhile, back at the proof of the Theorem, we have also reached a flow, call it f', which has associated set X with $t \notin X$.

Then $A(X,\overline{X})$ is a cut. Now if the vertex x is in X then, by the definition of X, there is an f'-unsaturated walk $W = v_0 \dots v_k$ from the source s to x, (so that $s = v_0$ and $v_k = x$). Suppose that y is a vertex not in X, i.e., $y \in \overline{X}$. Then, if there is an arc xy from x to y satisfying $f'(xy) < c(xy)$, the walk $W_1 = v_0 \dots v_k y$ from s to y would also be f-unsaturated, implying that y is in X, not $\overline{X}$, a contradiction. Similarly, if there is an arc yx from y to x satisfying $f'(yx) > 0$ then the walk $W_2 = v_0 \dots v_k y$ from s to y would be f-unsaturated, again giving a contradiction. Thus any arc of the form xy where $x \in X$ and $y \in \overline{X}$ must have $f'(xy) = c(xy)$, while any arc of the form yx where $x \in X$ and $y \in \overline{X}$ must have $f'(yx) = 0$. This shows that

$$f'(X,\overline{X}) = c(X,\overline{X}) \text{ while } f(\overline{X},X) = 0.$$

Now, if f' has value d then, since $A(X,\overline{X})$ is a cut, we have, by Theorem 8.1,

$$d = f'(X,\overline{X}) - f'(\overline{X},X),$$

Thus $d = c(X,\overline{X}) - 0 = c(X,\overline{X})$. In other words, the value of our flow f' equals the capacity of cut $A(X,\overline{X})$. It follows that f' is a maximal flow, completing the proof.

■ In our example, $f' = f_3$ and the set X was found to be $\{s,w\}$. The arcs from X to $\overline{X}$, i.e., $A(X,\overline{X})$, are su, wv, wx and so $c(X,\overline{X}) = c(su)+c(wv)+(wx) = 4+1+2 = 7$, which equals the value of the flow f_3, as expected. ■ □

Exercises for Section 8.1

8.1.1 For the two networks of Figure 8.10, list all the cuts and find a minimum cut.

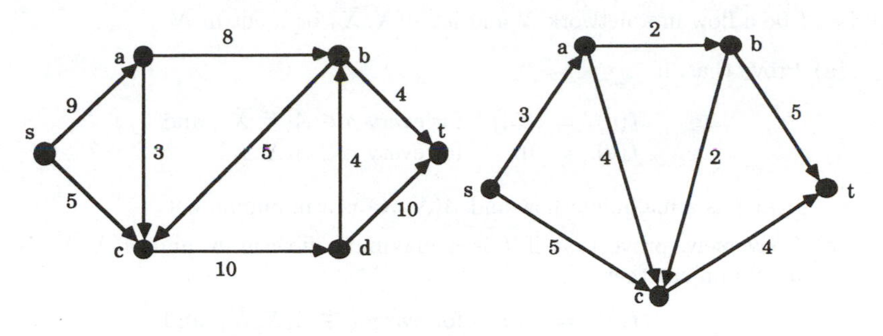

Figure 8.10: Two networks.

8.1.2 Let f be a flow on a network N and let W be an f-incrementing walk. Let f_1 be the revised flow based on W as defined in the proof of Theorem 8.2. Prove that f_1 is indeed a flow on N. (This fills a gap left in the proof of the Theorem.)

8.1.3 In this exercise we consider networks with several sources and several sinks and indicate how to modify these so the the theory of the single source, single sink situation can be applied.

Let N be a network with sources $s_1, \ldots, s_k$ and sinks $t_1, \ldots, t_l$. Then a **flow** in N from these sources to these sinks is a function f which assigns non-negative real numbers to each of the arcs a in N such that

(a) $f(a) \leq c(a)$ for each arc a,

(b) the total flow out of the sources $s_1, \ldots, s_k$ equals the total flow into the sinks $t_1, \ldots, t_l$, and

(c) for any vertex x which is neither a source nor a sink the total flow into x equals the total flow out of x.

The **value** d of f is defined to be the total flow out of the sources and so by (b),

$$d = \sum_{i=1}^{k} \sum_{a \in O(s_i)} f(a) = \sum_{j=1}^{l} \sum_{a \in I(t_j)} f(a).$$

From N we now create a network N^* with a single source s and a single sink t. We do this by simply introducing s and t as two new vertices, joining an arc a_i from s to each s_i and joining an arc b_j from each t_j to t, these new arcs having infinite capacity (or, in practice, a capacity which is the sum of all the capacities of the arcs already in N).

Now let f^* be a flow in our new network N^*, and let f be the function obtained by restricting f^* to the arcs of N.

(a) Show that f is a flow in N with the same value as f^*.

(b) Show that if f^* is a maximal flow in N^* then f is a maximal flow in N.

8.1.4 Let f be a flow in a network N and let $A(X,\overline{X})$ be a cut in N.

(a) Prove that, if

$$\begin{aligned} f(a) &= c(a) \quad \text{for every } a \in A(X,\overline{X}), \text{ and} \\ f(a) &= 0 \quad \text{for every } a \in A(\overline{X},X), \end{aligned}$$

then f is a maximum flow and $A(X,\overline{X})$ is a minimum cut.

(b) Conversely, prove that if f is a maximum flow in N and $A(X,\overline{X})$ is a minimum cut then

$$\begin{aligned} f(a) &= c(a) \quad \text{for every } a \in A(X,\overline{X}), \text{ and} \\ f(a) &= 0 \quad \text{for every } a \in A(\overline{X},X). \end{aligned}$$

8.1.5 Let f_1 and f_2 be flows in a network N and let $A(X,\overline{X})$ be a cut in N.

(a) Show that if f_1 and f_2 are both maximum flows then we need not have $f_1(a) = f_2(a)$ for every $a \in A(X,\overline{X})$ and every $a \in A(\overline{X},X)$.

(b) If f_1 and f_2 agree on both $A(X,\overline{X})$ and $A(\overline{X},X)$, are both f_1 and f_2 maximum flows?

8.2 The Ford and Fulkerson Algorithm

We now present an algorithm, due to Ford and Fulkerson [24], and based on the proof of the Max-Flow Min-Cut Theorem, which constructs a maximal flow. It uses a labelling technique to produce a maximal flow. Starting with a known flow, for example the flow which has 0 assigned to each arc (i.e., the **zero flow**), it recursively constructs a sequence of flows of increasing value, terminating with a maximal flow.

To describe the labelling technique we need the following definition.

> An f-**unsaturated tree** of the network N (with respect to the flow f) is a subtree T of the underlying graph G of N such that
> (i) the source s is a vertex of T,
> (ii) for every vertex v of T the unique $s-v$ path in T is an f-unsaturated path.

In Figure 8.11 an example of an f-unsaturated tree is given by the shaded edges.

The sequence of flows of increasing value is constructed using f-incrementing walks, which are found by "growing" f-unsaturated trees. The growing procedure is as follows.

(i) Initially the f-unsaturated tree T consists of just the source s,

(ii) If X is the set of vertices of T at any given stage, an arc a is adjoined to T in either of the two following ways, *provided this process does not create cycles, (i.e., provided the end result is still a tree)*:

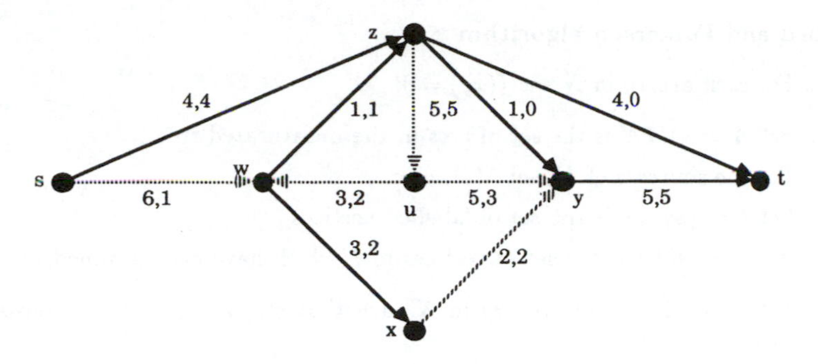

Figure 8.11: An f-unsaturated tree.

(a) if there is an arc a in $A(X, \overline{X})$ such that $f(a) < c(a)$ then adjoin both a and its head to T,
(b) if there is an arc a in $A(\overline{X}, X)$ such that $f(a) > 0$ then adjoin both a and its tail to T.

Under this construction our tree T either eventually grows out as far as the sink t or it stops short of t. If it does grow out as far as t we say we have **breakthrough** and in this case the $s - t$ path in T is an f-incrementing path. On the other hand, if T stops growing before it reaches t then, as the last part of the proof of the theorem shows, our flow f must be a maximal flow.

The actual labelling technique assigns labels to the vertices of an f-unsaturated tree T as it grows. For each vertex v of T the label $\lambda(v) = i(P_v)$ where P_v is the unique path in T from the source s to v. The advantage of this labelling is that if breakthrough does occur then not only do we have an f-incrementing path P_t from the source s to the sink t but also we have calculated $\lambda(t) = i(P_t)$, the increment used to obtain our revised flow based on P_t. The labelling procedure begins by assigning to the source s the label $\lambda(s) = \infty$. It continues according to the following rules:

(1) If a is an f-unsaturated arc whose tail u is already labelled but whose head v is not, then v is labelled

$$\lambda(v) = \min\{\lambda(u), c(a) - f(a)\}.$$

(2) If a is an arc with $f(a) > 0$ whose head u is already labelled but whose tail v is not, then v is labelled

$$\lambda(v) = \min\{\lambda(u), f(a)\}.$$

In both cases we say that v is labelled **based on** u. To **scan** a labelled vertex u is to label all unlabelled vertices that can be labelled based on u. The labelling procedure is continued until either the sink t is labelled (i.e., we have breakthrough) or all labelled vertices have been scanned and no more vertices can be labelled (i.e., the flow f is a maximal flow).

We now give a formal statement of the algorithm. It uses the zero flow as the initial flow on the network N.

The Ford and Fulkerson algorithm

Step 1. For each arc xy in N set $f(xy) = 0$.

Step 2. Set $A' = \emptyset$. (A' is the set of arcs in the unsaturated tree.)

For the source s of N, set $\lambda(s) = \infty$.

Set $L = \{s\}$. (L is the set of labelled vertices.)

Set $S = \emptyset$. (S is the set of vertices in L which have been scanned.)

Step 3. Let F be the set of arcs xy in N such that $c(xy) > f(xy)$. (F consists of forward arcs.)

Let R be the set of arcs xy in N such that $f(xy) > 0$. (R consists of reverse arcs.)

Step 4. If $L - S = \emptyset$ go to Step 10.

Step 5. Choose $x \in L - S$.

Step 6. If there is *no* $y \in \overline{L}$ such that $xy \in F$ and the set of arcs $A' \cup \{xy\}$ induces an underlying tree, then go to Step 7.

If there *is* a $y \in \overline{L}$ such that $xy \in F$ and the set of arcs $A' \cup \{xy\}$ induces an underlying tree then label y by

$$\lambda(y) = \min\{\lambda(x), c(xy) - f(xy)\}.$$

Change A' to $A' \cup \{xy\}$ and L to $L \cup \{y\}$.

Now repeat this step.

Step 7. If there is *no* $y \in \overline{L}$ such that $yx \in R$ and the set of arcs $A' \cup \{yx\}$ induces an underlying tree, then go to Step 8.

If there *is* a $y \in \overline{L}$ such that $yx \in R$ and the set of arcs $A' \cup \{yx\}$ induces an underlying tree then label y by

$$\lambda(y) = \min\{\lambda(x), f(yx)\}.$$

Change A' to $A' \cup \{yx\}$ and L to $L \cup \{y\}$.

Now repeat this step.

Step 8. Change S to $S \cup \{x\}$. If $t \notin L$ return to Step 4.

Step 9. (We reach this step when the sink t has been labelled, i.e., when we have breakthrough.)

Using a backtracking procedure, identify the incrementing walk W from s to t in N using the arcs from A' and for each a in W change $f(a)$ to $f(a) + \lambda(t)$ if $a \in F$ and to $f(a) - \lambda(t)$ if $a \in R$.

Return to Step 2.

Step 10. (We reach this step when all labelled vertices have been scanned and there is no breakthrough.)

The values $f(xy)$ for each arc xy in N give a maximum flow in N and the set of vertices L gives a minimum cut.

We now illustrate the algorithm in Figures 8.12 – 8.16 using the network N of Figure 8.11, starting with the flow f given there (instead of the zero flow as used in the algorithm). At each stage the unsaturated tree is shown having shaded edges, the vertex being scanned is in white and the label assignment is shown below the diagram. To save space, we omit listing the specific steps of the algorithm used for the example and also recording the changes to the sets A', L, and S, but these should be clear to the reader from the diagrams.

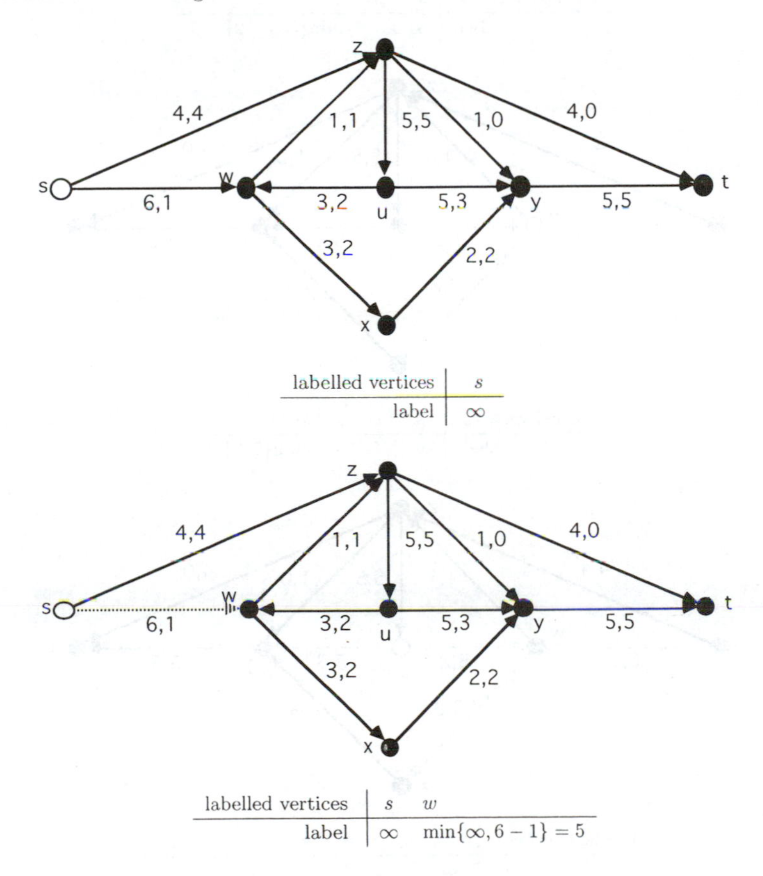

labelled vertices	s
label	∞

labelled vertices	s	w
label	∞	$\min\{\infty, 6-1\} = 5$

Figure 8.12

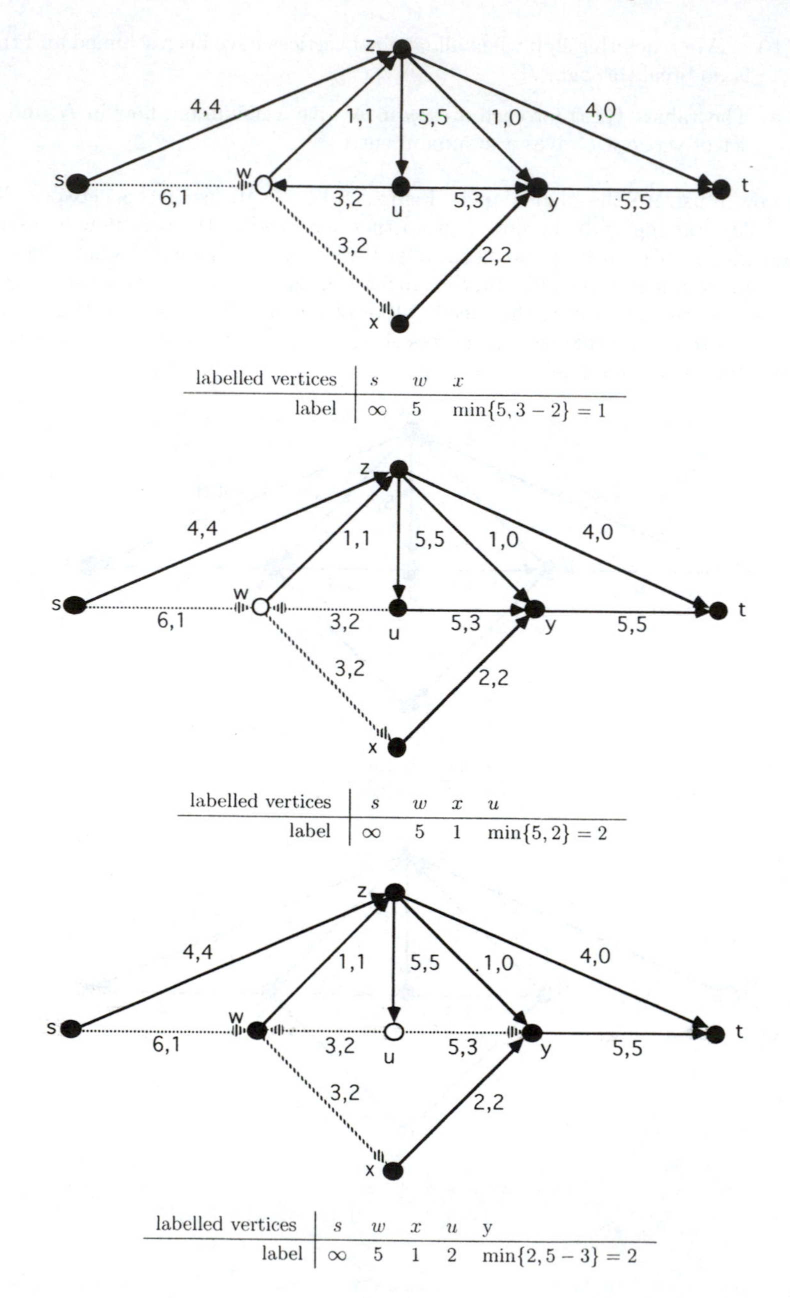

labelled vertices	s	w	x
label	∞	5	$\min\{5, 3-2\} = 1$

labelled vertices	s	w	x	u
label	∞	5	1	$\min\{5, 2\} = 2$

labelled vertices	s	w	x	u	y
label	∞	5	1	2	$\min\{2, 5-3\} = 2$

Figure 8.13

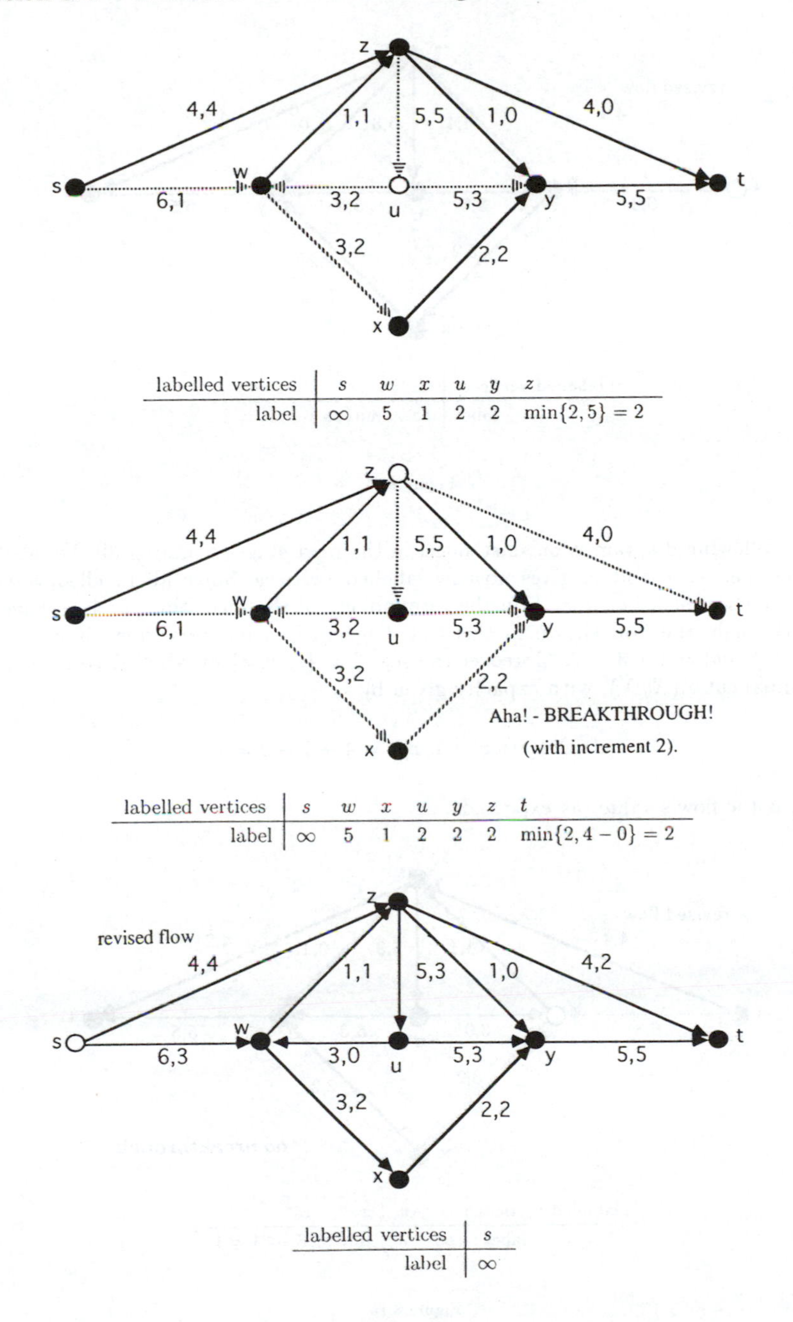

labelled vertices	s	w	x	u	y	z
label	∞	5	1	2	2	$\min\{2,5\}=2$

labelled vertices	s	w	x	u	y	z	t
label	∞	5	1	2	2	2	$\min\{2,4-0\}=2$

labelled vertices	s
label	∞

Figure 8.14

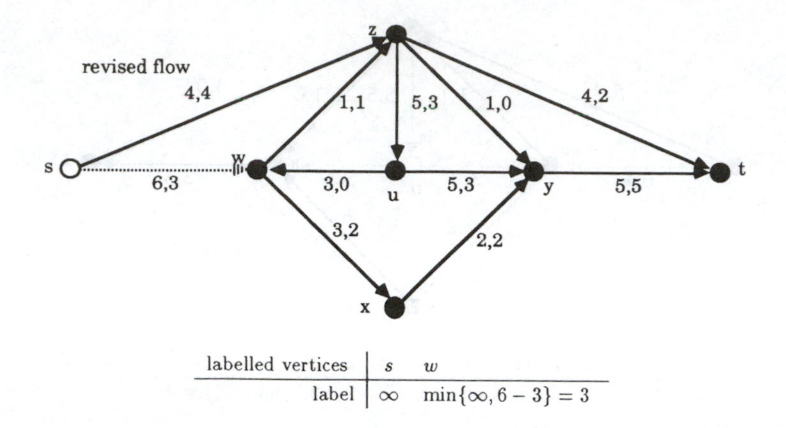

labelled vertices	s	w
label	∞	$\min\{\infty, 6-3\} = 3$

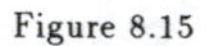

Figure 8.15

The following diagram is our last since, in the next stage, scanning all the labelled vertices, i.e., s, w and x, gives no new labelled vertices. Since all labelled vertices have been scanned, i.e., $L = S$ in the terminology of the algorithm, and we have no breakthrough, the flow shown in this last diagram is a maximal flow. It has value $f(sz) + f(sw) = 4 + 3 = 7$. Moreover the set $X = \{s, w, x\}$ of labelled vertices gives a minimal cut $A(X, \overline{X})$, with capacity given by

$$c(sz) + c(wz) + c(xy) = 4 + 1 + 2 = 7,$$

which is the flow's value (as expected).

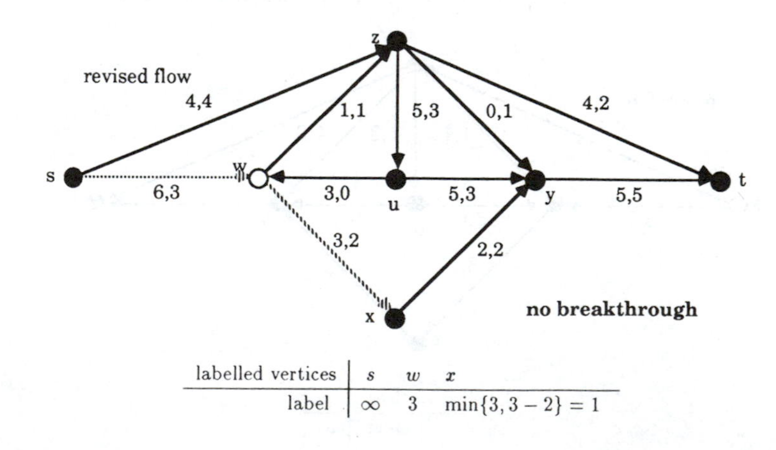

labelled vertices	s	w	x
label	∞	3	$\min\{3, 3-2\} = 1$

Figure 8.16

Exercises for Section 8.2

8.2.1 Use the Ford and Fulkerson algorithm to find a maximal flow and a cut with capacity equal to this flow for both of the networks in Figure 8.10. (Draw the unsaturated trees at each stage of their growth and identify the sets A', L, S, F and R as defined in the step-wise presentation.)

8.2.2 Use the Ford and Fulkerson algorithm to find a maximal flow and a cut with capacity equal to this flow for each of the networks in Figure 8.17.

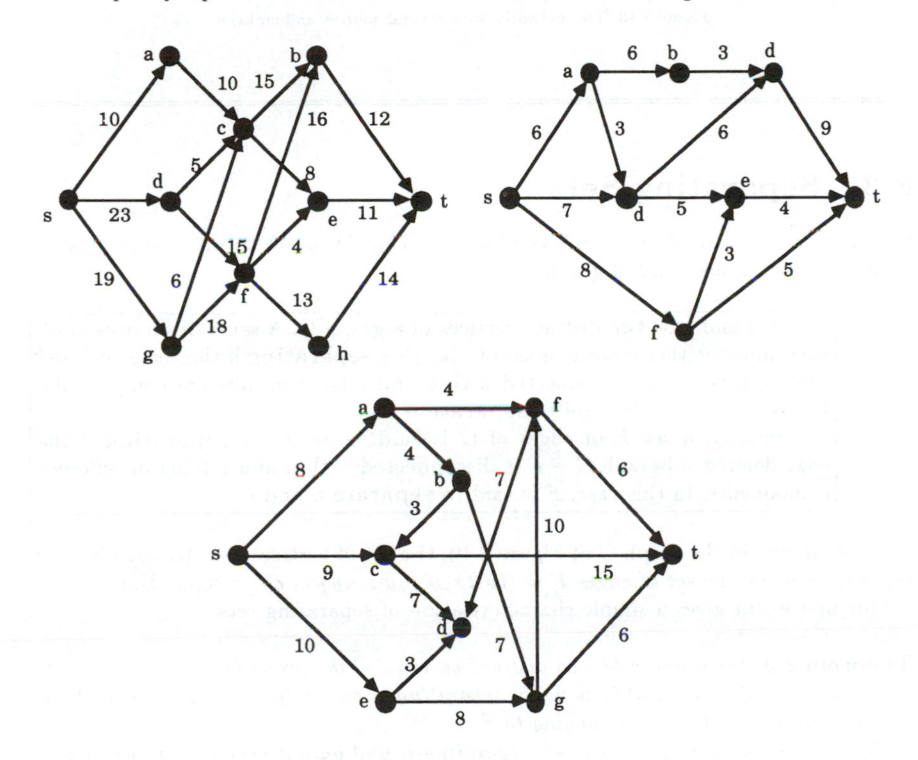

Figure 8.17: Two networks.

8.2.3 The two networks of Figure 8.18 have several sources and sinks. Use the technique described in Exercise 8.1.2 together with the Ford and Fulkerson algorithm to find maximal flows for both networks. What is the value of each of your flows? What is the total value of each of your flows out of each individual source and into each individual sink?

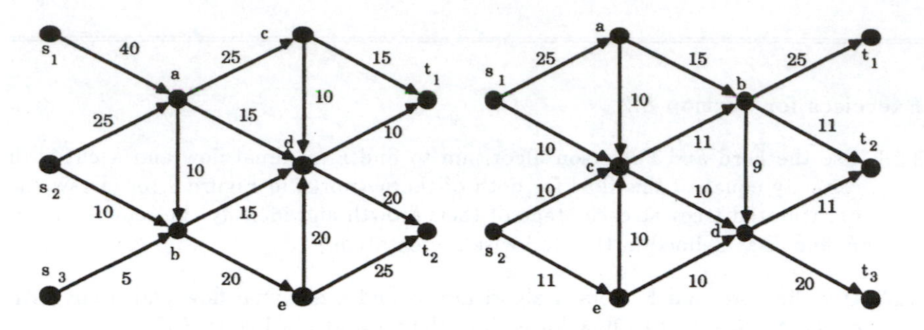

Figure 8.18: Two networks with several sources and sinks.

8.3 Separating Sets

In this section we shall use the Max-Flow Min-Cut Theorem to prove some results related to the connectivity of graphs.

> Let u and v be two distinct vertices of a graph G. A set S of vertices of G, containing neither u nor v, is said to be $u-v$ **separating** if the vertex deleted subgraph $G - S$ is disconnected with u and v lying in different components. In this case, S is also said to **separate** u and v.
>
> Similarly, a set F of edges of G is said to be $u - v$ **separating** if the edge deleted subgraph $G - F$ is disconnected with u and v lying in different components. In this case, F is said to **separate** u and v.

For example, in the graph G of Figure 8.10, the set of vertices $S = \{t, w, y\}$ is $u - v$ separating while the set of edges $F = \{tv, tx, wx, wz, uy\}$ is $u - v$ separating.

Our first result gives a simple characterisation of separating sets.

Theorem 8.3 *Let u and v be two distinct vertices of the graph G.*

(a) A set S of vertices of G is $u - v$ separating if and only if every $u - v$ path has at least one internal vertex belonging to S.

(b) A set F of edges of G is $u - v$ separating if and only if every $u - v$ path has at least one edge belonging to F.

Proof (*a*) Let S be a $u-v$ separating set of vertices in G and let P be a $u-v$ path. If P has no internal vertices belonging to S then the deletion of S from G leaves P intact. But then u and v will be in the same connected component of $G - S$, which is a contradiction since S is $u - v$ separating. Thus every $u - v$ path must have at least one internal vertex belonging to S.

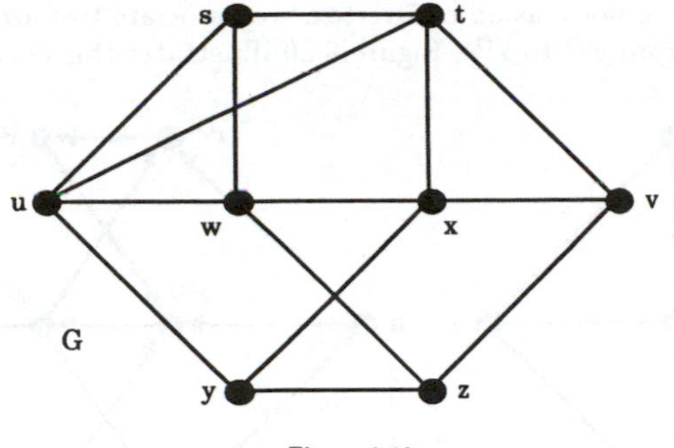

Figure 8.19

Conversely, suppose that every $u - v$ path has an internal vertex belonging to S. Then if we delete S from G there will be no path from u to v in the resulting subgraph $G - S$. In other words, u and v are in different connected components of $G - S$ and so S is $u - v$ separating, as required.

(b) The proof of this is similar to part (a)'s and is left as Exercise 8.3.1. □

Our goal now is to show that, for two nonadjacent vertices u and v, the size of a smallest possible $u - v$ separating set of vertices (or edges) is equal to the largest possible number of internally disjoint (respectively edge disjoint) $u - v$ paths. Since this equates minimums with maximums, perhaps it is not too surprising that we will prove these results using the max-flow min-cut theorem. The vertex version was first proved by K. Menger in 1927 [44], but the proof given here is from a 1956 article by G. B. Dantzig and D. R. Fulkerson [14].

Theorem 8.4 (Menger's Theorem) *Let u and v be two nonadjacent vertices of a graph G. Then the maximum number of internally disjoint $u - v$ paths in G equals the minimum number of vertices in a $u - v$ separating set.*

Proof Let m be the maximum number of internally disjoint $u - v$ paths in G and let n be the minimum number of vertices in a $u - v$ separating set. Let $P_{(1)}, \ldots, P_{(m)}$ be m internally disjoint $u - v$ paths and let S be a $u - v$ separating set of vertices. Then, by Theorem 8.3 (a), each of the paths $P_{(1)}, \ldots, P_{(m)}$ contains at least one member of S. On the other hand, since the paths are internally disjoint, no member of S can occur as an internal vertex in more than one of the paths. From this we see that we have at least as many vertices in S as we have paths, i.e., $|S| \geq m$. Since S was any separating set, it follows that $m \leq n$.

To prove that $m \geq n$ we construct a network N from G, having u as source and v as sink, as follows. The vertex set of N consists of u and v together with a pair of vertices, denoted by $x^{(1)}$ and $x^{(2)}$, for each vertex x in G different from both u and v. Each such pair is joined by an arc from $x^{(1)}$ to $x^{(2)}$. Such an arc will be called an **internal** arc of N while all other arcs (still to be defined) will be called **external**. Now to each edge of the form ux in G we associate an arc from u to $x^{(1)}$ in N and to each edge of the form yv we associate an arc from $y^{(2)}$ to v. Finally, to each edge xy

in G, having neither u nor v as an end vertex, we associate two arcs, one from $x^{(2)}$ to $y^{(1)}$ and the other from $y^{(2)}$ to $x^{(1)}$. Figure 8.20 illustrates this construction.

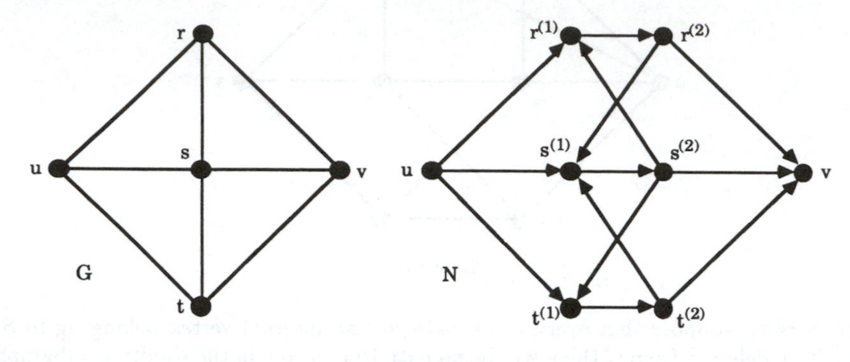

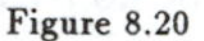
Figure 8.20

Now assign a capacity of one to each internal edge and infinite capacity to each of the external edges. (If the reader is unhappy with infinite capacities then he can change these to a large number, bigger than the total number of arcs in N, say.)

We now prove that m, the maximum number of internally disjoint $u - v$ paths in G, is equal to the value d of a maximum flow f in the network N. First note that in N any vertex of the form $x^{(1)}$ is the tail of only one arc, the associated internal arc, and this has capacity one. Similarly, any vertex of the form $x^{(2)}$ is the head of only one arc, again the associated internal arc with capacity one. It follows from this that the value of a flow into and out of any vertex in N is either one or zero. In particular, the maximum flow f of value d must be obtained using d internally disjoint directed $u - v$ paths in N, each contributing the value one to the total flow value.

Now any $u - v$ path $P = uu_1u_2 \dots u_kv$ in G uniquely induces the $u - v$ directed path

$$Q = uu_1^{(1)}u_1^{(2)}u_2^{(1)}u_2^{(2)} \dots u_k^{(1)}u_k^{(2)}v$$

in N and this path Q permits a flow of value one along it. On the other hand, any directed $u-v$ path Q permitting such a flow must be induced in this way from a $u-v$ path P in G since the internal arcs are the only arcs of capacity one in N. Moreover, if the $u-v$ paths $P_{(1)}, \dots, P_{(t)}$ induce the directed paths $Q_{(1)}, \dots, Q_{(m)}$ in N then it is easy to see that $P_{(1)}, \dots, P_{(t)}$ are internally disjoint in G if and only if $Q_{(1)}, \dots, Q_{(m)}$ are internally disjoint in N. It now follows that the maximal flow value d corresponds to d internally disjoint $u - v$ paths in G and so $d \leq m$. Similarly, since m internally disjoint paths in G can induce a flow of value m in N and so $m \leq d$. Hence $m = d$, as required.

Now the Max-Flow Min-Cut Theorem tells us that there is some cut $A(X, \overline{X})$ in N such that $c(X, \overline{X}) = d$. Since d is finite, each arc from X to $\overline{X}$ must be an internal arc. Moreover, since $A(X, \overline{X})$ is a cut, each directed path from u to v must use at least one of these arcs. It follows that deleting the tails of all the arcs in $A(X, \overline{X})$ removes all possible $u - v$ directed paths in N. This in turn implies that the set of d

vertices in G corresponding to these d tails must be a $u-v$ separating set. Since this has produced a $u-v$ separating set of size d we must have $d \geq n$. Thus, since $d = m$ and $m \leq n$, we have $m = n$, as required. □

As a simple consequence of Theorem 8.3 we have the following characterisation of n-connected graphs. Note that it generalises Whitney's Theorem (Theorem 2.21).

Theorem 8.5 *A simple graph G is n-connected if and only if, given any pair of distinct vertices u and v of G, there are at least n internally disjoint paths from u to v.*

Proof Suppose that G is n-connected and u and v are two distinct vertices of G. Then any $u-v$ separating set must have at least n vertices. Thus, by Theorem 8.4, there must be at least n internally disjoint $u-v$ paths, as required.

Conversely, suppose that, given any pair of distinct vertices u and v of G, there are at least n internally disjoint paths from u to v. Then, by Theorem 8.4, every $u-v$ separating set must have at least n vertices for each pair of vertices u and v. Thus it requires the deletion of at least n vertices from G in order to produce a disconnected graph or K_1. In other words, G is n-connected, as required. □

We now turn our attention to the edge analogues of Theorems 8.4 and 8.5. Their proofs are similar but a little easier. They are due to Ford and Fulkerson.

Theorem 8.6 (The edge version of Menger's Theorem) *Let u and v be two vertices of a graph G. Then the maximum number of edge disjoint $u-v$ paths in G equals the minimum number of edges in a $u-v$ separating set.*

Proof Let m be the maximum number of edge disjoint $u-v$ paths in G and let n be the minimum number of edges in a $u-v$ separating set. Let $P_{(1)}, \ldots, P_{(m)}$ be m edge disjoint $u-v$ paths and let F be a $u-v$ separating set of edges. Then, by Theorem 8.3 (b), each of the paths $P_{(1)}, \ldots, P_{(m)}$ contains at least one member of F. On the other hand, since the paths are edge disjoint, no member of F can occur as an edge in more than one of the paths. From this we see that we have at least as many edges in F as we have paths, i.e., $|F| \geq m$. Since F was any separating set, it follows that$m \leq n$.

To prove that $m \geq n$ we construct a network N from G, having u as source and v as sink, as follows. The vertex set of N is defined to be $V(G)$. To each edge of the form ux in G we associate an arc from u to x in N and to each edge of the form yv we associate an arc from y to v. To complete the arc set of N, for each edge uv in G, having neither u nor v as an end vertex, there are two associated arcs in N, namely one from u to v and one from v to u. Figure 8.21 illustrates this construction. Now assign a capacity of one to each arc in N.

We now prove that m, the maximum number of edge disjoint $u-v$ paths in G, is equal to the value d of a maximum flow f in the network N. First note that any $u-v$ path $P = uu_1u_2 \ldots u_kv$ in G translates simply over to a directed p$u-v$ path P in N and, under this translation, any set of edge disjoint $u-v$ paths retains the property of being edge disjoint. Thus we can choose a set of m edge disjoint directed

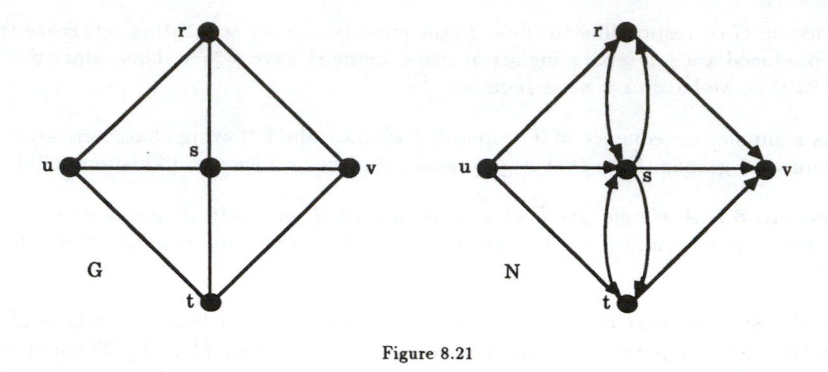

Figure 8.21

$u - v$ paths in N. Clearly each path in this set can contribute one to a flow in N and so there is a flow from u to v with value m. Thus $d \leq m$.

To see the opposite inequality, let f be a maximal flow with value d. Then since the capacity of each arc in N is just one, each arc can only contribute at most once to the flow's value d and so the flow is achieved by d edge disjoint directed $u - v$ trails, (each having flow component of value one). Moreover, given one of these directed trails, if it involves a pair of arcs (x, y) and (y, x), both having flow value one, then, by omitting these arcs, we may prune the trail to obtain a new trail still with flow value one. Indeed, doing this for all such pairs of arcs in each of the d trails, we produce d edge disjoint $u - v$ paths in N. We illustrate this pruning procedure in Figure 8.22. From this we get that $d \leq m$ and so $d = m$.

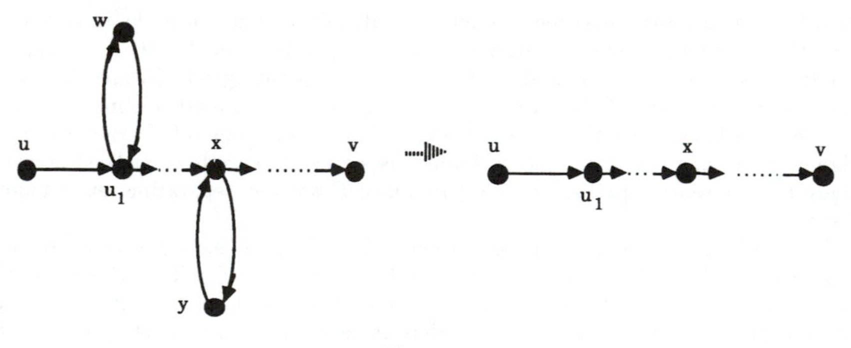

Figure 8.22

The max-flow min-cut theorem now tells us that there is some cut $C = A(X, \overline{X})$ in N such that $c(X, \overline{X}) = d$. Since C is a cut, every directed $u - v$ path in N uses at least one arc of C. Moreover, such a path is induced from a $u - v$ path in G and so each $u - v$ path in G uses at least one edge of the form xy such that the induced arc xy belongs to C. It follows that the set of all such edges in G is a $u - v$ separating set of edges and this set has at most $|C|$ we get $n \leq |C|$. However, $d = c(X, \overline{X}) = |C|$, since the capacity of each arc in C is one, and so $n \leq d$. Thus, since $d = m$ and $m \leq n$, we have $m = n$, as required. □

We finish this chapter with a quick look at edge connectivity.

> Let G be a simple graph. The **edge connectivity** of G, denoted by $\kappa_e(G)$, is the smallest number of edges in G whose deletion from G either leaves a disconnected graph or an empty graph.

For example, any nonempty simple graph G with a bridge has $\kappa_e(G) = 1$. Clearly $\kappa_e(G) = 0$ if and only if either G is disconnected or an empty graph. The graph G of Figure 8.23 has $\kappa_e(G) = 2$.

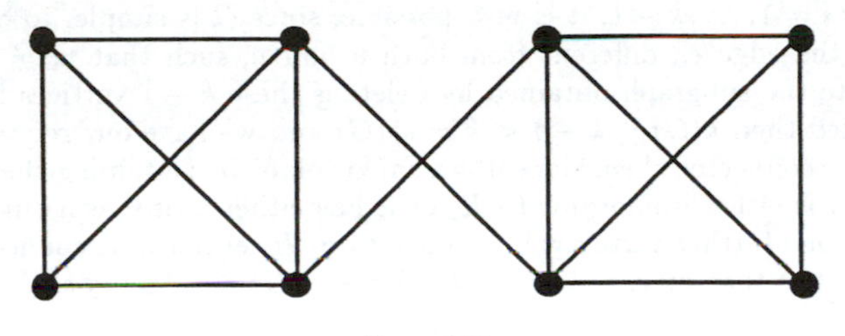

Figure 8.23

> A simple graph G is called n-**edge-connected** (where $n \geq 1$) if $\kappa_e(G) \geq n$.

As a simple consequence of Theorem 8.5 we have the following characterisation of n-connected graphs.

Theorem 8.7 *A simple graph G is n-edge-connected if and only if, given any pair of distinct vertices u and v of G, there are at least n edge disjoint paths from u to v.*

Proof Suppose that G is n-edge-connected and u and v are two distinct vertices of G. Then any $u - v$ separating set of edges must have at least n members. Thus, by Theorem 8.6, there must be at least n edge disjoint $u - v$ paths, as required.

Conversely, suppose that, given any pair of distinct vertices u and v of G, there are at least n edge disjoint paths from u to v. Then, by Theorem 8.6, for each pair of vertices u and v, every $u - v$ separating set of edges must have at least n members. Thus it requires the deletion of at least n edges from G in order to produce a disconnected graph or an empty graph. In other words, G is n-edge-connected, as required. □

Finally we give a result showing how the two connectivities are related. We let $\delta(G)$ denotes the smallest of all the vertex degrees in G.

Theorem 8.8 *Let G be a simple graph. Then*

$$\kappa(G) \leq \kappa_e(G) \leq \delta(G).$$

Proof Let v be a vertex with degree $\delta(G)$. Then deleting the $\delta(G)$ edges incident with v creates either the empty graph K_1 (if v is the only vertex in G) or a disconnected subgraph H, since v is an isolated vertex in H. Hence $\kappa_e(G) \leq \delta(G)$.

Now if $\kappa_e(G) = 0$ then either G is disconnected or empty and so $\kappa(G) = 0$. If $\kappa_e(G) = 1$ then G is connected and has a bridge, e say. In this case either $G = K_2$ or G has a cut vertex (namely one of e's ends). In either case $\kappa(G) = 1$. Thus we may now assume that $\kappa_e(G) \geq 2$. To simplify the notation we set $k = \kappa_e(G)$.

Then G has a set $\{e_1, \ldots, e_k\}$ of edges whose deletion disconnects G, but no set with less edges has this effect. Thus, deleting the first $k-1$ of these edges, $e_1, \ldots, e_{k-1}$, produces a connected subgraph H having the undeleted edge, e_k, as a bridge. Let $e_k = uv$. For $i = 1, \ldots, k-1$, it is now possible, since G is simple, to choose an end vertex u_i of the edge e_i, different from both u and v, such that $u_i \neq u_j$ for $i \neq j$. Let H' denote the subgraph obtained by deleting these $k-1$ vertices from G. If H' is disconnected then $\kappa(G) \leq k - 1 < k = \kappa_e(G)$ and we have our required result. If, however, H' is connected then, since it is a subgraph of H containing the edge $e_k = uv$ as a bridge, it is either isomorphic to K_2 or it has either u or v as a cut vertex. Thus a deletion of one further vertex will produce from H' either a disconnected graph or a K_1. This shows that $\kappa(G) \leq k = \kappa_e(G)$. Hence $\kappa(G) \leq \kappa_e(G) \leq \delta(G)$. □

Exercises for Section 8.3

8.3.1 Let u and v be two distinct vertices of the graph G. Prove that a set F of edges of G is $u-v$ separating if and only if every $u-v$ path has at least one edge belonging to F. (This is Theorem 8.3 (b).)

8.3.2 Verify that Theorems 8.4 and 8.6 are true for the vertices u and v of the graphs G of Figure 8.24.

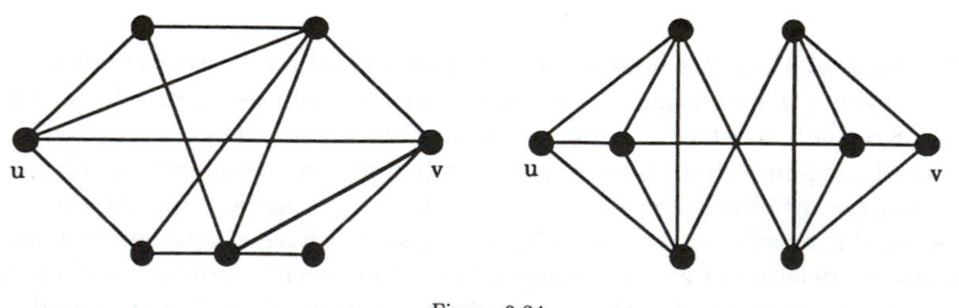

Figure 8.24

8.3.3 Verify that Theorems 8.5 and 8.7 are true for the complete bipartite graphs $K_{2,3}$ and $K_{3,4}$, the wheel W_5 and the 4-cube Q_4.

8.3.4 (In this exercise we show how matchings in a bipartite graph correspond to flows in an associated network.) Let G be a bipartite graph with bipartition $V(G) = X \cup Y$. Construct a network N from G by introducing two new vertices s and t, the source and sink of N, an arc from s to each of the vertices in X, an arc from each of these vertices in Y to t, and orienting the edges of G from vertices in X to those in Y. Assign capacity 1 to the arcs from s and also to the arcs to t and let all other arcs have infinite capacity (or a very large capacity).

(a) Let f be a flow from s to t in N and let M consist of all the edges xy in G such that the flow value $f(x, y)$ (on the corresponding arc) is positive. Prove that this defines a matching M in G.

(b) Conversely, let M be a matching in G. For each arc (u, v) in N, define $f(u, v) = 1$ if M saturates either u or v and $f(u, v) = 0$ otherwise. Prove that this defines a flow f on N with value equal to the number of edges in the matching M.

(c) Prove that the definitions of (a) and (b) establish a one-to-one correspondence between the set of matchings in G and the set of flows on N in such a way that maximum flows correspond to maximum matchings.

8.3.5 (In this exercise we sketch a proof of how Menger's Theorem can be used to prove Hall's Marriage Theorem (Theorem 4.3).) Let G be a bipartite graph with bipartition $V(G) = X \cup Y$ and suppose that $|N(S)| \geq |S|$ for every subset S of X. Adjoin to G two new vertices s and t, an edge joining s to each vertex of X and an edge joining t to each vertex of Y. (Compare this with the construction of the network in the previous exercise.) Denote this supergraph of G by G_1.

(a) Let $X = \{x_1, \ldots, x_n\}$. Prove that G has a matching M which saturates X if and only if G_1 has a set of n internally disjoint $s-t$ paths $P_{(1)}, \ldots, P_{(n)}$ (so that, after reordering the paths if necessary, $P_{(i)}$ must go through x_i for each i).

(b) Using Hall's condition that $|N(S)| \geq |S|$ for every subset S of X, prove that any $s-t$ separating set of vertices in G_1 has at least n vertices and hence that G has a matching which saturates X.

8.3.6 Let G be a plane graph and let G^* be its dual (as defined in Section 5.6). For each set of edges F in G let F^* denote the set of edges in G^* corresponding to those in F.

(a) Prove that a set of edges F in G forms a cycle in G if and only if the edge-deleted subgraph $G^* - F^*$ is disconnected but $G^* - F_1^*$ is connected for any proper subset F_1 of F.

(b) Prove that for any set F of edges in G, the edges of F^* form a cycle in G^* if and only if $G - F$ is disconnected but $G - F_1$ is connected for any proper subset F_1 of F.

Chapter 9

Ramsey Theory

9.1 A Party

Six people are at a party. Show that there are three people who all know each other or that there are three people who don't all know each other.

This problem was posed in a slightly different form in the 1953 Lowell Putnam Examination — a mathematics competition for university students in North America.

The easiest way to solve this problem is using graph theory but it is, in fact, one of the simplest illustrations of a theorem by F. P. Ramsey which was developed in the area of logic. Ramsey's theorem was rediscovered, in a combinatorial setting, by two Hungarian mathematicians, P. Erdös and G. Szekeres. Ramsey produced his version [52] in 1930, while the Erdös-Szekeres result [20] was five years later in 1935.

So let's go back to the party. What does the problem say in graph theory terms? Consider the complete graph $G = K_6$. Represent the six people by the six vertices of G. Now colour the edges joining two vertices blue, if the corresponding two people know each other. If two people don't know each other, colour the edge between the corresponding vertices red. If there are three people who know each other then this is represented by a blue triangle in K_6, on the vertices corresponding to these people. Similarly, if there are three people who don't know each other then this is represented by a red triangle. Figure 9.1 shows two possible parties.

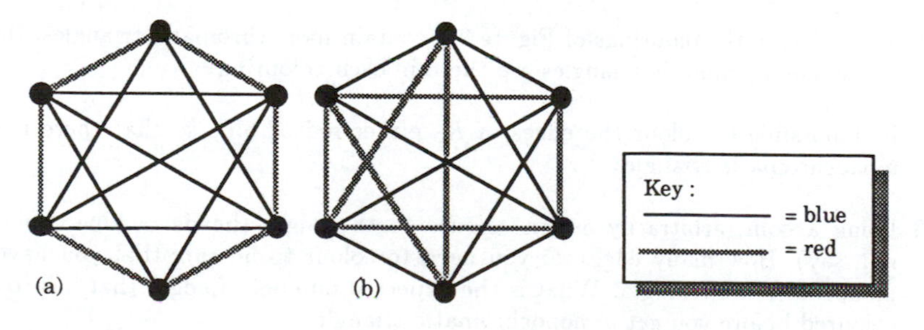

Figure 9.1: Two parties.

> Given an assignment of colours to all the edges of a graph G, a subgraph H of G is called **monochromatic** if all the edges of H have the same colour.

The party question now becomes: if we arbitrarily colour the edges of K_6 either blue or red, then must we have a monochromatic clique on three vertices, i.e., a monochromatic triangle?

We now give a simple argument to show that the answer to the question is yes.

Take an arbitrary vertex a in K_6. It has degree 5 in K_6. So when we colour the edges incident with a either red or blue, one colour must be used at least *three* times. Without loss of generality, assume that at least three edges are coloured blue as in Figure 9.2. (If three edges are coloured red we only need to interchange the words blue and red in the following discussion.)

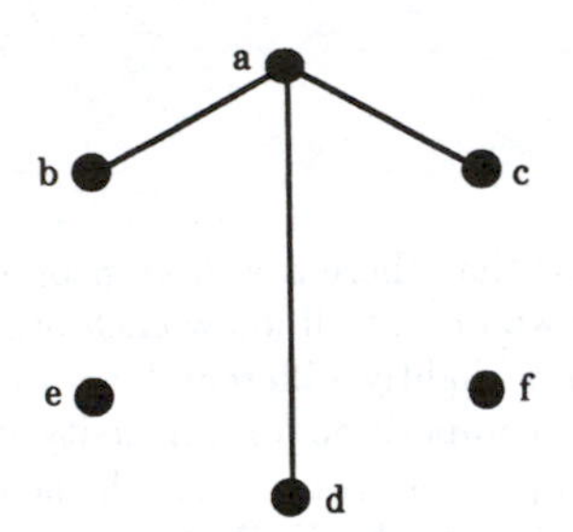

Figure 9.2: Three blue edges incident with the vertex a.

So assume that ab, ac, ad are coloured blue. (In other words, a and b know each other, a and c know each other and a and d know each other.) If any one of bc, bd or dc is coloured blue, then we have the required blue triangle. Hence we may suppose that bc, bd or cd are all coloured red. However, this then gives us a red triangle and the argument is completed.

Exercises for Section 9.1

9.1.1 Show that both colourings of Figure 9.1 contain monochromatic triangles. How many monochromatic triangles are there in each colouring?

9.1.2 Is it possible to colour the edges of K_5 either red or blue so that there is *no* monochromatic triangle?

9.1.3 Using a coin, arbitrarily assign colours to the edges (heads = blue, tails = red, say). How many edges do you need to colour to be *sure* that you have a monochromatic triangle? What is the expected number of edges that has to be coloured before you get a monochromatic triangle?

9.1.4 If every blue/red edge colouring of the edges of K_6 produces a monochromatic triangle, show that every blue/red edge colouring of the edges of K_7 also produces a monochromatic triangle.

9.1.5 Let n be a positive integer such that every blue/red edge colouring of the edges of K_n produces a monochromatic triangle. Show that every blue/red edge colouring of the edges of K_{n+1} also produces a monochromatic triangle.

9.1.6 Show that any blue/red edge colouring of K_6 actually produces at least *two* monochromatic triangles.

9.1.7 Arbitrarily colour the edges of K_7 blue or red. How many monochromatic triangles are guaranteed?

9.1.8 Suppose we arbitrarily colour the edges of K_8 either red or blue. Show that any vertex is on the end of at least four edges of the same colour.

9.1.9 (a) Colour the edges of K_{15} in blue, red or green. Show that any vertex is on the end of at least five edges of some colour.

(b) Colour the edges of K_n in one of m colours, where $n = mq + r$ with $0 \leq r < m$. How many edges of some colour can we guarantee are joined to any arbitrary vertex?

9.2 A Generalisation of the Party Problem

More generally, we might want to take any complete graph K_n and s colours. If we colour each edge of K_n arbitrarily in one of these s colours we might ask if we are forced to have a monochromatic triangle.

For instance, suppose we arbitrarily colour the edges of K_6 in blue, red or green. Must we have a monochromatic triangle or is it possible to arrange things so that there are *no* monochromatic triangles in such a colouring?

Try to answer this by yourself before continuing.

Before we go further, let us modify a definition which we gave in Section 6.6. There we defined an *edge colouring* of a nonempty graph G to be an assignment of colours to all the edges of G such that *adjacent edges are coloured differently.* In this Chapter we will drop the adjacent edges requirement, i.e., an **edge colouring** of G is simply an assignment of colours to all the edges of G, one colour per edge. If there are s different colours used in such a colouring then it is called a s-**edge colouring** of G.

Let us now try to construct a red, blue, green colouring of K_6, i.e., a 3-edge colouring of K_6, which avoids a monochromatic triangle. Clearly, to do this we have to use all three colours, otherwise the argument of Section 9.1 will give a monochromatic triangle. Now suppose that there is a vertex v of K_6 which has three blue edges incident with it, say the edges vx, vy and vz. Then, again as in Section 9.1, none of the three edges xy, xz and yz can be blue, they can not all be red and they can not all be green. Thus two of them are red and the other is green, say. Suppose that xy and xz are red while yz is green. In particular the vertex z is on the end of three edges which have different colours, namely vz, xz and yz.

Now suppose that the vertices of K_6 are given as a, b, c, d, e and f. As we have just seen, for our construction we can choose a vertex, say a, which is incident with three edges of different colours, say ab is blue, ad is red and af is green. Now colour ac blue and ae red, so that all the edges incident with a are now coloured. (See Figure 9.3.)

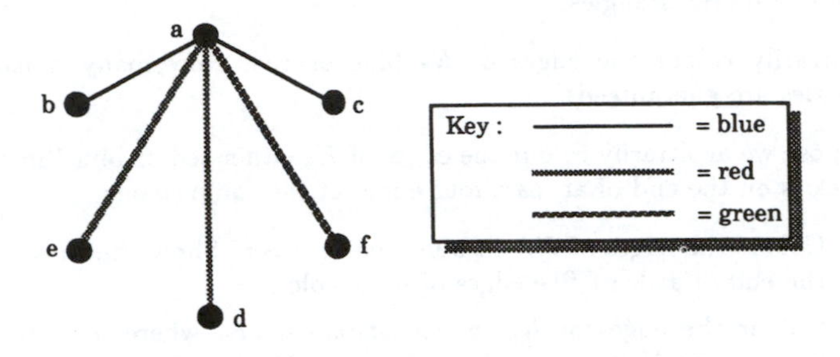

Figure 9.3: The colouring of the edges incident with a.

At the next stage of the construction, we clearly want to avoid bc being blue and de being red. So colour bc in red and de in green. Now, no matter how we colour the other edges incident with b, we can not get a monochromatic triangle with b as a vertex. So colour all three edges bd, be and bf blue.

Now look at c. Colouring cd, ce, cf blue gives no monochromatic triangles. With d we need a little caution. We can't colour df blue. But colouring it red is okay. Finally ef can't be blue but it can be red. This completes a red, blue, green colouring of K_6 which has no monochromatic triangle. We illustrate it in Figure 9.4.

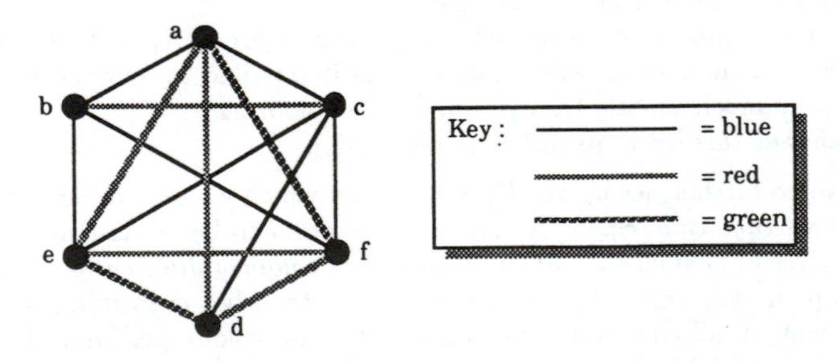

Figure 9.4: A 3-edge colouring of K_6 which has no monochromatic triangle.

Now, generalising Exercise 9.1.5, it is straightforward to see that if n is an integer such that every s-edge colouring of K_n has a monochromatic triangle then every s-edge colouring of K_{n+1} also has a monochromatic triangle, simply because any such colouring of K_{n+1} induces one for K_n. In view of the above, this raises the question:

What is the *smallest* value of n such that every 3-edge colouring of K_n must have a monochromatic triangle?

One way of tackling this question is to first show that for some n we *must* have a monochromatic triangle, and second to show that we can find a 3-edge colouring of K_{n-1} which has *no* monochromatic triangle. If we can find such an n then clearly it must give the smallest K_n in which every 3-edge colouring has a monochromatic triangle.

Now it turns out to be easy to show that K_{17}, when 3-edge coloured, must have a monochromatic triangle. The proof goes like this. (As before we work with the colours blue, red and green.)

Choose a vertex a of K_{17}. Since a has degree 16, if we 3-edge colour K_{17}, a must be an end of 6 edges which all have the same colour (by Exercise 9.1.9 (b)). Without loss of generality, assume that a is the end of the six blue edges ab, ac, ad, ae, af and ag. Then, if one of the edges joining any pair of the vertices b, c, d, e, f or g is blue, we have a monochromatic triangle and so we are finished. Otherwise the six vertices b, c, d, e, f and g form a K_6 in which the edges are 2-coloured, by red and green. By Section 9.1 we know that this clique K_6 must contain a monochromatic triangle. Hence K_{17} must contain a monochromatic triangle when the edges are 3-coloured.

If we can now show that the edges of K_{16} can be coloured in three colours without there being a monochromatic triangle, we will have found the answer to our "smallest n" question. Figure 9.5 gives such a colouring for K_{16}, with the blue edges given as solid lines, the red edges as the (thicker) shaded lines and (to make the figure clearer) the green edges missing.

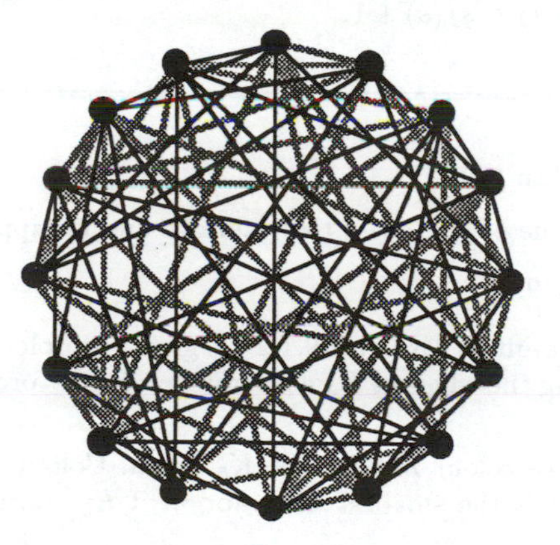

Figure 9.5: A 3-edge colouring (one colour invisible) of K_{16} which has no monochromatic triangles.

Now let's turn to four colours. If we want to find how big an n we need so that colouring the edges of K_n with *four* colours forces a monochromatic triangle, then we need to choose n sufficiently large so that we can use the fact that K_{17} works for *three* colours.

In this K_n, at a given vertex a, we try to find at least one colour α which has to be used on 17 edges. Then the 17 vertices adjacent to a by edges coloured α, either have an edge between them coloured α or they don't. If they do have an edge coloured α, then there is a triangle coloured α. If they don't have any edge coloured α, then there are only three edge colours used in the K_{17} that is left. We already know that this forces a monochromatic triangle.

So how big does n have to be to ensure that a vertex a has at least 17 edges coloured in one colour? Certainly if the degree of a is $4 \times 16 + 1$, that will be enough. So $n = 65$.

By the above argument, we know that for all $N \geq 65$, if we arbitrarily colour the edges of K_N in four colours, then a monochromatic triangle results. But is 65 the best possible? Can we colour the edges of K_{64} in four colours and avoid a monochromatic triangle? What is the smallest n here? At the moment no one knows the answer to this last question.

However, the above discussion gives a method of establishing an upper bound for the monochromatic triangle problem with $s+1$ colours, working from the known result with s colours.

Let us summarise what we have discovered so far. We want to know for what n, when we arbitrarily colour the edges of K_n in s colours, do we get a monochromatic triangle.

Let $f(s)$ be the smallest such n. Then we know that

$$f(2) = 6 \text{ and } f(3) = 17.$$

Using the argument that we have employed earlier in the $s = 3$ and $s = 4$ cases, we also get the following result.

Theorem 9.1 $f(s+1) \leq sf(s) + 1$.

Exercises for Section 9.2

9.2.1 What is the value of $f(1)$? Use this value to find an upper bound for $f(2)$.

9.2.2 Prove Theorem 9.1.

9.2.3 (a) Use the result of Exercise 9.1.2 to give a simpler proof of the fact that 3-colouring the edges of K_6 does not necessarily produce a monochromatic triangle.

(b) Suppose we colour m edges of K_6 green. Colour all other edges blue or red. What is the smallest value for m if K_6 contains no monochromatic triangles?

9.2.4 Show that the edges of K_7 can be coloured either blue, red or green so that it contains no monochromatic triangle.

9.2.5 Colour the edges of K_{17} arbitrarily with three colours. What is the minimum number of monochromatic triangles produced?

9.3 Another Generalisation of the Party Problem

But why are we hung up on triangles? Why not arbitrarily colour the edges of K_n in blue and red and ask when are we forced to have a monochromatic clique of *four* vertices, i.e., a monochromatic K_4?

Let us tackle this using the argument of the proof of Theorem 9.1. Note first that any vertex a of K_n has degree $n-1$. Thus, if we arbitrarily colour the edges of K_n in blue or red, then at a one colour must appear at least $\lceil\frac{1}{2}(n-1)\rceil = t$ times.[1] Suppose blue occurs most. Then we would need t to be big enough either to guarantee a blue triangle (which along with the blue edges from a gives a blue K_4) or a red K_4. On the other hand, if red occurs most, then we would need t to be big enough to guarantee a red triangle or a blue K_4.

This has got us into a slightly new situation. Before (in Section 9.2) we only had triangles to worry about. Now we seem to have to take account of triangles *and* K_4's. At least we do if we are to continue the argument we started above. So let's stop a minute and make a definition.

> Given two positive integers m and n we define the **Ramsey number** $r(m,n)$ to be the smallest p such that if we arbitrarily colour the edges of the complete graph K_p blue or red, then we must get a blue K_m or a red K_n as a subgraph.

From what we did in Section 9.1, we know that $r(3,3) = f(3) = 6$. The question we started this section with is: what is $r(4,4)$? More generally, do we know, or is there some way of finding, the value of $r(m,n)$ for small values of m and n?

First of all, does it make any sense for m (or n) to be 1? Since K_1 does not have any edges, colouring the edges of K_p blue or red (or pink or cerise for that matter) does not help us to get a K_1 with blue or red edges.

For this reason, we assume from now on that m and n are both at least two. Then the smallest value of $r(m,n)$ ought surely to be $r(2,2)$. What complete graph can we colour to be sure that we get either a blue edge or a red edge? Won't K_2 do? If the single edge is coloured blue, then we have a blue edge. If the single edge is coloured red, then we have a red edge. It follows that

$$r(2,2) = 2.$$

Can we now do $r(2,3)$? It is fairly easy to see that $r(2,3) \geq 3$. But is $r(2,3) = 3$? To see that it is, note that we can't arbitrarily colour edges of K_2 either blue or red and *force* either a blue K_2 or a red K_3 — we might just have used red on the one edge. Thus

$$r(2,3) = 3.$$

The same sort of argument shows that

$$r(2,4) = 4.$$

[1] For any number x the symbol $\lceil x\rceil$ denotes the largest *integer* greater than or equal to x.

We are clearly making progress!

The next number to try is probably $r(3,4)$. We try to tackle this using the Theorem 9.1 approach. Take some vertex a in our complete graph K_p, where $p = r(3,4)$, and let there be b blue edges and r red edges incident with a. (This means that $p = b+r+1$. Let B be the set of vertices adjacent to a by blue edges and R be the set of vertices adjacent to a by red edges.

Suppose that $b \geq r(2,4)$. Then in B we either have a blue edge or a red K_4. If we get the red K_4 we are done. If we get just a blue edge, then that blue edge, together with two of the edges incident with a form a blue K_3. So we must have a blue K_3 or a red K_4 and again we are done.

Now suppose that $r \geq r(3,3)$. Then in R we must have a blue K_3 or a red K_3. A blue K_3 is fine, so assume we get a red K_3. This red K_3, together with three red edges incident with a, give a red K_4. Again we have either a blue K_3 or a red K_4.

So how can we guarantee that either $b \geq r(2,4)$ or $r \geq r(3,3)$? Simply take $p = r(2,4) + r(3,3)$. If $b < r(2,4)$ and $r < r(3,3)$, then $b \leq r(2,4) - 1$ and $r \leq r(3,3) - 1$. Hence, since $p = b+r+1$, we have $p \leq (r(2,4)-1)+(r(3,3)-1)+1 = r(2,4)+r(3,3)-1$ which contradicts the fact that $p = r(2,4) + r(3,3)$.

Thus, if $p = r(2,4) + r(3,3)$ and we colour the edges of K_p blue or red, we must have either $b \geq r(2,4)$ or $r \geq r(3,3)$. Either way we then get our blue K_3 or our red K_4. Hence

$$r(3,4) \leq r(2,4) + r(3,3).$$

But is it possible that a p smaller than $r(2,4) + r(3,3)$ also does the job? Since, from above, $r(2,4) = 4$ and $r(3,3) = 6$ this is asking if $r(3,4) \leq 9$. In particular, can we colour K_9 so that there is no blue K_3 or red K_4? Well no, but the beginnings of such a colouring for K_8 are shown in Figure 9.6.

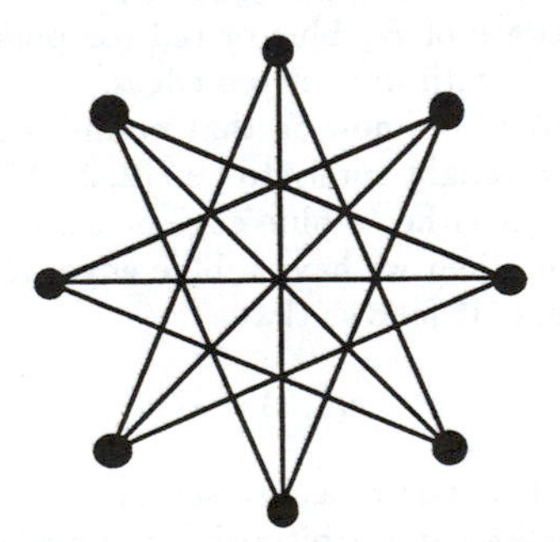

Figure 9.6: A partial 2-edge colouring of K_8, showing blue edges.

Suppose that the edges of K_8 shown in Figure 9.6 are blue. By the symmetry of the figure it is easy to see that there are no blue triangles. Now colour in the remaining edges red. There cannot be a red K_4. Any four vertices you choose have at least one blue edge between them. Hence $r(3,4) \geq 9$.

We now know that $10 \geq r(3,4) \geq 9$. We leave it as Exercise 9.3.6 to show that

$$r(3,4) = 9.$$

But we can make some progress with a more general situation. The argument used to show that $r(3,4) \leq r(2,4) + r(3,3)$ can be used to prove the next result.

Theorem 9.2 *For $m, n \geq 2$,*

$$r(m,n) \leq r(m-1,n) + r(m,n-1).$$

We can box in the number $r(m,n)$ by using Theorem 9.2 to obtain an upper bound for $r(m,n)$ and then using specific colourings to obtain a lower bound for $r(m,n)$. Hard work and special arguments (or a computer) can then give us $r(m,n)$ in the cases where the difference between the upper and lower bounds is small. This approach has produced the following numbers but beyond this there has been little progress.

$$r(3,4) = 9, \quad r(3,5) = 14, \quad r(3,6) = 18,$$

$$r(3,7) = 23, \quad r(3,8) = 28, \quad r(4,4) = 18.$$

Exercises for Section 9.3

9.3.1 Prove Theorem 9.2, i.e., that $r(m,n) \leq r(m-1,n)+r(m,n-1)$ for all $m, n \geq 2$.

9.3.2 Find estimates for $r(3,5)$, $r(4,5)$ and $r(5,5)$.

9.3.3 Prove that

$$r(m,n) \leq \binom{m+n-2}{m-1}.$$

9.3.4 (a) Why is $r(2,3) \neq 2$?

(b) Prove that $r(3,2) = 3$.

(c) Prove that $r(m,n) = r(n,m)$.

(d) Find $r(2,n)$ for all $n \geq 2$.

9.3.5 Show that $r(3,4) = 9$.

9.3.6 Try to show that $r(3,5) = 14$, $r(3,6) = 18$ and $r(4,4) = 18$.

9.3.7 Use Theorem 9.2 to show that $r(m,n)$ is finite for $m, n \geq 2$.

9.4 The Compleat Ramsey

Suppose that we decide to use s colours, called $1, \ldots, s$, to arbitrarily colour the edges of the complete graph K_p. We might combine the generalisation of the last two sections in the following question:

How big does p have to be to guarantee that this results in either a K_{n_1} in colour 1, or a K_{n_2} in colour 2, ..., or a K_{r_s} in colour s?

From what we have done so far it would seem that, provided we choose $n_i \geq 2$ for $i = 1, 2, \ldots, s$, then there probably is some *finite* value of p. With this optimism we have the following definition.

> Given s positive integers $n_1, \ldots, n_s$, all at least 2, we define the **(generalised) Ramsey number** $r(n_1, n_2, \ldots, n_s)$ to be the smallest integer p such that if we arbitrarily colour the edges of the complete graph K_p with s colours, called $1, 2, \ldots, s$, then we must get either a K_{n_1} coloured 1, a K_{n_2} coloured 2, ..., or a K_{n_s} coloured s (as a subgraph of K_p).

Our optimism is justified by the following result, known as Ramsey's Theorem. We omit its proof but those interested can consult Chapter 13 of Behzad, Chartrand and Lesniak-Foster [4]

Theorem 9.3 (Ramsey) *Given s positive integers $n_1, \ldots, n_s$, all at least 2, the Ramsey number*

$$r(n_1, n_2, \ldots, n_s)$$

is finite.

The difficulty with the generalised Ramsey numbers $r(n_1, n_2, \ldots, n_s)$ is that, although we know they exist, it is proving extremely difficult to find them.

In Section 9.2 we defined $f(s)$. Of course $f(s) = r(3, 3, \ldots, 3)$, where there are s threes in the last bracket. Theorem 9.1 gave us an inequality for $f(s)$. This inequality seems to be very poor and so does not provide us with a useful upper bound for $f(s)$.

The inequality of Theorem 9.2 also appears to become less and less useful as an upper bound for $r(m, n)$ as the values of m and n increase.

Because of the difficulty found in obtaining precise values for the Ramsey numbers, or indeed in obtaining better estimates for them, Ramsey Theory has generalised further. Much effort has been expended on finding the minimum value of p, so that when K_p is coloured, we are forced to have, not necessarily *complete* monochromatic subgraphs, but, say, monochromatic subtrees or monochromatic cycles or some other kind of monochromatic subgraph.

Some results of this type are listed below. But first we give another definition.

> Let G and H be two simple graphs. Then the **Ramsey number** $r(G, H)$ is defined to be the smallest integer p so that when the edges of K_p are coloured blue or red there is always either a blue G or a red H occurring as a subgraph of K_p.

Thus $r(m,n) = r(G,H)$ if we take $G = K_m$ and $H = K_n$. The following gives $r(G,H)$ for some common G and H.

(1) $r(K_m, T) = (m-1)(n-1)+1$, where T is a tree with n vertices.
(2) $r(sK_2, sK_2) = 2s+t-1$, for $s \geq t \geq 1$. (Here, and below, sG denotes the graph with s connected components each isomorphic to G.)
(3) $r(s_1K_2, s_2K_2, \ldots, s_cK_2) = s_1 + 1 + \sum_{i=1}^{c}(s_i - 1)$, when $s_1 \geq s_i$
(4) $r(sK_3, tK_3) = 3s+2t$, for $s \geq t \geq 1$ and $s \geq 2$.

For more information on results in Ramsey Theory, the reader should have a look at a recent article [29] in Scientific American by R.L. Graham and J.H. Spencer. Other references are cited there.

Exercises for Section 9.4

9.4.1 (a) Can the edges of K_7 be coloured blue, green or red so that K_7 is a disjoint union of monochromatic Hamiltonian cycles?

(b) Repeat (a) with K_9 and the colours blue, red, green and yellow.

(c) Show that the edges of K_{2n+1} can be coloured with n colours so that it is the disjoint union of n monochromatic hamiltonian cycles.

(d) What corresponding result holds for K_{2n}?

9.4.2 Show that if $s \geq t \geq 1$, then

$$ps + (q-1)t - 1 \leq r(sK_p, tK_q) \leq ps + (q-1)t + f(p,q).$$

Here $f(p,q) = r(t_0K_p, t_0K_q)$, where t_0 is chosen so that $t_0 \min\{p,q\} \geq 2r(K_p, K_q)$.

9.4.3 Prove that

(a) $r(2K_3, K_3) = 8$;

(b) $r(2K_3, 2K_3) = 10$;

(c) $r(C_4, C_4) = 6$;

(d) $r(C_4, C_5) = 7$;

(e) $r(K_{1,3}, K_{1,3}) = 6$;

(f) $r(P_5, P_5) = 6$.

Prove (a) and (b) directly without recourse to the general result in Section 9.4.

9.4.4 Show that $r(2,3,3) = 6$. Determine $r(2,3,3,3)$.

Chapter 10

Reconstruction

10.1 The Reconstruction Conjecture

The open question that we will describe in this final chapter was initially posed by Stanislaw Ulam, one of the scientists involved in the Manhattan Project which produced the first atomic bomb. The question was initially investigated by P. J. Kelly for his Ph.D. thesis in 1942. Although the question was raised about 40 years ago, it still remains unanswered. This chapter lists some of the progress made to date.

To describe the problem, we first introduce some terminology.

Mathematicians often describe a (finite) *set* X by listing its elements in the form

$$X = \{x_1, \ldots, x_n\}.$$

In doing so, there is an implicit agreement that the set X has exactly n different elements, namely $x_1, \ldots, x_n$. However, for the purposes of this chapter, we want to look at the less restrictive notion of a *collection* X. Here

$$X = \{x_1, \ldots, x_n\}$$

is a **collection** of the n elements $x_1, \ldots, x_n$, but *these elements need not be distinct*, in other words, some elements may be duplicated.[1] For example,

$$X = \{1, 2, 2, 3, 4, 4, 4\}$$

is a collection of seven elements in which there are only four distinct elements — the *set* of elements corresponding to X is $Y = \{1, 2, 3, 4\}$. To emphasise this distinction between *collection* and *set* we will use square brackets when listing the elements of a collection as opposed to the curly brackets usually used for sets. Thus we write the collection X above as

$$X = [x_1, \ldots, x_n].$$

Clearly we need not restrict ourselves to *finite* collections. For example, the collection

$$X = [1, 2, 2, 3, 3, 3, 4, 4, 4, 4, \ldots]$$

[1] Just as there are often duplicates in a stamp *collection*.

in which the element 1 occurs once, the element 2 occurs twice, the element 3 three times, etc., has the set of natural numbers as its "underlying" set.

The collections that we are particularly interested in here consist of graphs.

Let $\mathcal{G}$ and $\mathcal{H}$ be two finite collections of simple graphs, say

$$\mathcal{G} = [G_1, \ldots, G_m] \text{ and } \mathcal{H} = [H_1, \ldots, H_n].$$

We say that $\mathcal{G}$ and $\mathcal{H}$ are **equal**, writing this as usual by $\mathcal{G} = \mathcal{H}$, if they have the same number of elements, i.e., $m = n$, and if, after reordering one of the collections if necessary, the graphs G_i and H_i are isomorphic for each $i = 1, \ldots, m$.

We are now ready to describe what is considered by many to be the most important unsolved problem in graph theory.

Take any simple graph G with at least three vertices and let $\mathcal{G}$ denote the collection of all its vertex deleted subgraphs of the form $G - x$, i.e., $\mathcal{G} = [\,G - x : x \in V(G)\,]$. Now take another simple graph H, again with at least three vertices, and similarly let $\mathcal{H}$ denote the collection $[\,H - y : y \in V(H)\,]$.

The **Reconstruction Conjecture** says:

If the collections $\mathcal{G}$ and $\mathcal{H}$ are equal then G is isomorphic to H.

Why must the graphs G and H in the definition have at least three vertices? For the simple reason that there are non-isomorphic graphs G and H on two vertices for which $\mathcal{G} = \mathcal{H}$. Figure 10.1 shows two such graphs. However if we keep to our condition that there be at least three vertices then to date nobody knows of any non-isomorphic pair of graphs G and H for which $\mathcal{G} = \mathcal{H}$.

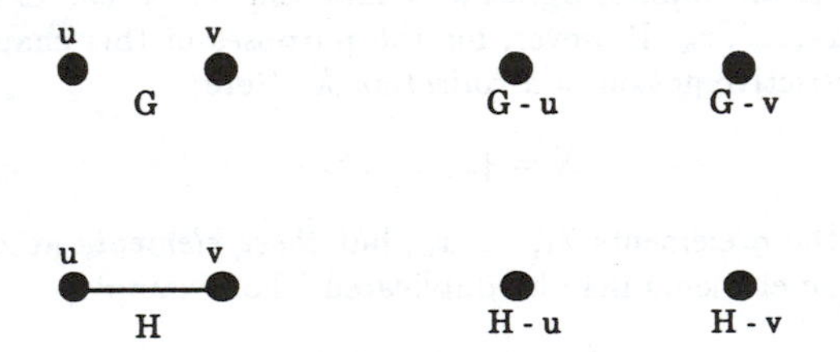

Figure 10.1: Two non-isomorphic graphs G and H with $\mathcal{G} = \mathcal{H}$.

Another way of looking at the Reconstruction Conjecture is that, given a collection of subgraphs of the form $\mathcal{G}$, then *exactly one* graph G can be uniquely recaptured from $\mathcal{G}$, where $\mathcal{G} = [G - v : v \in V(G)]$. In this case we say that G is **reconstructible (from $\mathcal{G}$)**.

What basic information can we derive from this collection of subgraphs $\mathcal{G}$? Can we determine the number of vertices of G? What about the number of edges?

Well, $|V(G)|$ is straightforward. This is simply the number of graphs in $\mathcal{G}$ because the deletion of each vertex produces a graph in the collection $\mathcal{G}$.

So what about $|E(G)|$? Let v be any vertex of G. We start counting the edges of G using the edges of $G-v$. From Figure 10.2, we see that

$$|E(G)| = |E(G-v)| + d(v).$$

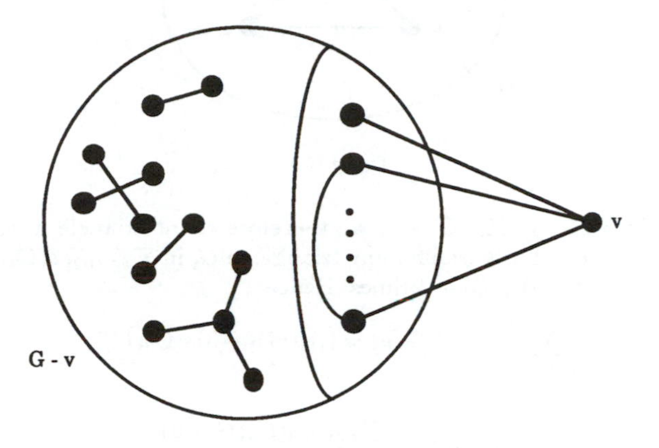

Figure 10.2: $|E(G)| = |E(G-v)| + d(v)$.

Hence

$$\sum_{v\in V(G)} |E(G)| = \sum_{v\in V(G)} |E(G-v)| + \sum_{v\in V(G)} d(v).$$

By Theorem 1.1, this gives $$\sum_{v\in V(G)} |E(G)| = \sum_{v\in V(G)} |E(G-v)| + 2|E(G)|.$$

Thus $$|V(G)||E(G)| = \sum_{v\in V(G)} |E(G-v)| + 2|E(G)|$$

and so $$|E(G)|(|V(G)|-2) = \sum_{v\in V(G)} |E(G-v)|.$$

From this we get

$$|E(G)| = \frac{\sum_{v\in V(G)} |E(G-v)|}{|V(G)|-2}.$$

Since $|V(G)| > 2$, we are not dividing the right hand side of the last equation by zero. Further, on that right hand side we know $\sum |E(G-v)|$, namely all the edges in the entire collection $\mathcal{G}$, and we know $|V(G)|$, namely the number of graphs in $\mathcal{G}$. Hence we can always find $|E(G)|$.

The argument we have just used to find $|E(G)|$ can be extended to find how often a particular graph H occurs as a non-spanning subgraph of G. To see this, we first find the number of triangles in G, given that $|V(G)| \geq 4$. For each $v \in G$, let

$$s(C_3, G) \text{ and } s(C_3, G-v)$$

denote the number of triangles in G and $G-v$, respectively.

From Figure 10.3, we can see that the triangle with vertices a, b and c lies in any $G-v$ except for $v = a, b$ or c. So triangle a, b, c is counted once in $s(C_3, G-v)$ unless

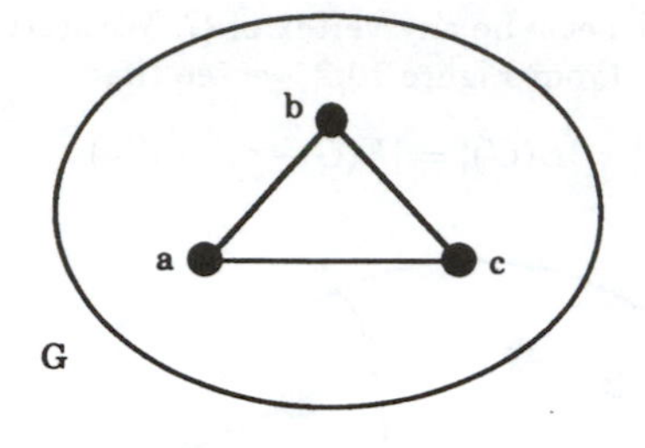

Figure 10.3

$v = a, b$ or c. In $\sum_{v \in V(G)} s(C_3, G - v)$ we therefore count triangle a, b, c a total of $|V(G)| - 3$ times. Since this is true for any triangle in G, in $\sum_{v \in V(G)} s(C_3, G - v)$ every triangle of G is counted $|V(G)| - 3$ times. Hence

$$\sum_{v \in V(G)} s(C_3, G - v) = (|V(G)| - 3)s(C_3, G),$$

so

$$s(C_3, G) = \frac{\sum_{v \in V(G)} s(C_3, G - v)}{|V(G)| - 3}.$$

Before reading further you should attempt Exercises 10.1.6, 10.1.7 and 10.1.8 — this should help to fully understand the proof of the following theorem.

Theorem 10.1 *Let $s(H, G)$ and $s(H, G - v)$ be the number of subgraphs of G and $G - v$, respectively, which are isomorphic to H, where $|V(H)| < |V(G)|$. Then*

$$s(H, G) = \frac{\sum_{v \in V(G)} s(H, G - v)}{|V(G)| - |V(H)|}.$$

Proof Take a particular copy H' of H in G. Since $|V(H')| = |V(H)| < |V(G)|$, H' occurs in $|V(G)| - |V(H')|$ of the subgraphs $G - v$. So in $\sum s(H, G - v)$, H' is counted $|V(G)| - |V(H')|$ times. But this is the case for every subgraph of G isomorphic to H. Hence

$$\sum_{v \in V(G)} s(H, G - v) = (|V(G)| - |V(H)|)s(H, G),$$

and the result follows. □

The result we proved earlier for edges is an easy consequence of Theorem 10.1.

Corollary 10.2

$$|E(G)| = \frac{\sum |E(G - v)|}{|V(G)| - 2}.$$

Proof Let $H = P_2$, the path consisting of a single edge. Then $s(H, G) = |E(G)|$ and $s(H, G - v) = |E(G - v)|$. Clearly $|V(H)| = 2$ and the Corollary follows by substituting these values into the equation of Theorem 10.1. □

Exercises for Section 10.1

10.1.1 Show that the Reconstruction Conjecture holds for each of the four collections $\mathcal{G}$ shown in Figure 10.4. To make the individual members of each collection clearer, we have displayed them as if drawn on a playing card. The problem is then to reconstruct G from its deck of cards $\mathcal{G}$.

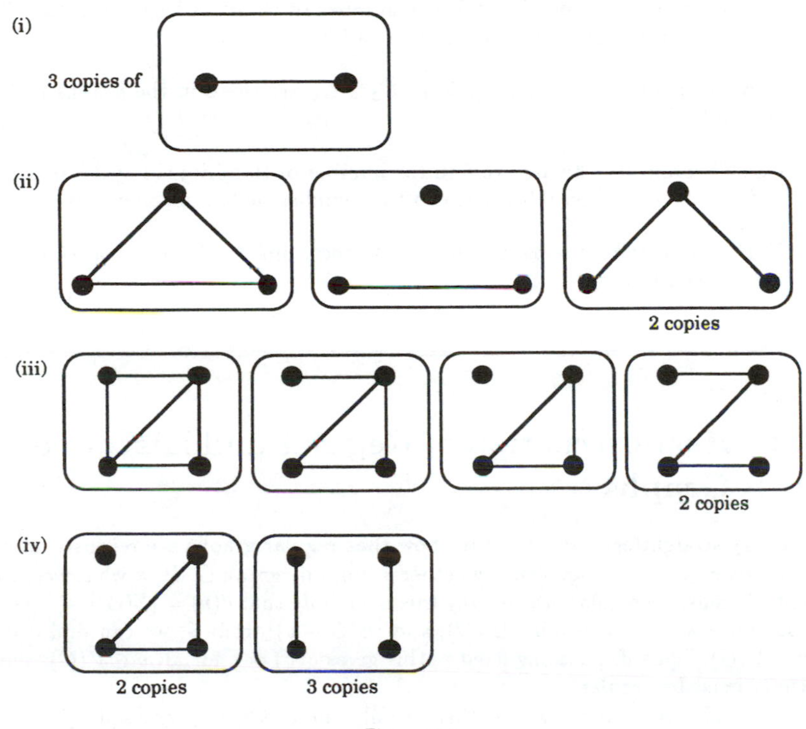

Figure 10.4

10.1.2 What is the smallest graph for which $\mathcal{G}$ is a non-empty set?

10.1.3 For the various collections $\mathcal{G}$ of Exercise 10.1.1, calculate $|V(G)|$, $|E(G)|$ and $d(v)$ for all $v \in V(G)$. Do these calculations make it easier to reconstruct G in each case?

10.1.4 Is G a regular graph if and only if the vertices in the graphs of $\mathcal{G}$ have only two different kinds of degree?

If the answer to the question is yes, prove it. If the answer to the question is no, how "close" to being true is it? Prove as much as you can.

10.1.5 How many triangles are there in the graphs which are reconstructible from the collections $\mathcal{G}$ in Exercise 10.1.1.

10.1.6 Use an argument similar to that used in the text to count triangles to determine $s(P_3, G)$, where $s(P_3, G)$ is the number of paths of length 2 in G. Check your answer on the graphs of Exercise 10.1.1.

10.1.7 Find an expression for $s(C_4, G)$, the number of cycles of length 4 in G. Check your answer on the graphs of Exercise 10.1.1.

10.1.8 How many subgraphs isomorphic to $C_3 \cup K_1$ are there in the graphs of Exercise 10.1.1?

10.1.9 Can Theorem 10.1 be used to find the number of Hamiltonian cycles in a graph G from the collection $\mathcal{G}$? If not, can this number be found some other way?

10.1.10 If H is a spanning subgraph of G can the number of copies of H in G be determined from $\mathcal{G}$?

10.2 Reconstruction of Regular and Disconnected Graphs

It is a very straightforward matter to show that regular graphs are reconstructible.

First, given $\mathcal{G}$, we can determine whether or not any graph G, H or whatever, which leads to $\mathcal{G}$, must be regular. To see why this is so, note that $d(v) = |E(G)| - |E(G-v)|$ and so, since we can determine $|E(G)|$ and $|E(G-v)|$ from $\mathcal{G}$, we can find $d(v)$ for all $v \in V(G)$. Thus if, for some fixed r, this gives $d(v) = r$ for all $v \in V(G)$, then of course G must be regular.

Our second stage is to show that there is only one G which gives rise to $\mathcal{G}$. For this, take any $v \in V(G)$. Then $G - v$ will have r vertices of degree $r-1$ produced when v was deleted from G. If we now add a new vertex v' and join it to these r vertices of degree $r-1$, we will have recovered the original graph G from $G - v$. Since there is only one way to add this vertex v' to $G - v$ then G is uniquely determined by $\mathcal{G}$.

We will now use the same two step procedure to reconstruct disconnected graphs. So first we show that we can **recognise** from $\mathcal{G}$ whether G, H or whatever is disconnected. Then we show that the disconnected graph G can be reconstructed uniquely from $\mathcal{G}$.

For the first step, let G be a disconnected graph. What are the graphs $G - v$ likely to look like? Will they all be disconnected? Have a go at Exercises 10.2.1 and 10.2.2 before you go further.

From Exercise 10.2.1 it is clearly *not* the case that if G is disconnected, then so are all the $G - v$. However, it might be the case that if G is disconnected, then *all but one* of the $G - v$ is disconnected.

From the results of Exercise 10.2.2 you should see that if G contains at least three components, then each $G - v$ contains at least two components. You can only lose a component in $G - v$ if v is an isolated vertex component in G. The same argument applies to a graph G with two components. Hence, we can prove the following result.

Theorem 10.3 *If G is disconnected with $|V(G)| > 2$, then at most one subgraph $G - v$ is connected.*

Now try Exercises 10.2.3–10.2.6 inclusive.

Suppose now that each graph $G - v$ in $\mathcal{G}$ is disconnected. Is there any chance that G itself is connected? If it were, then every vertex of G would be a cut vertex. This is impossible, since, by Theorem 2.20, G has at least two vertices which are not cut vertices. In fact, this also shows it to be impossible to have precisely one connected graph $G - v$ in $\mathcal{G}$. We can therefore improve upon Theorem 10.3 as follows:

Theorem 10.4 *Let $|V(G)| > 2$. Then G is disconnected if and only if at most one subgraph of the form $G - v$ is connected.*

This result enables us to *recognise* from $\mathcal{G}$ when G (or H) is disconnected, in other words, disconnected graphs are **recognisable**. The problem now is to find out how to reconstruct such a G uniquely. Now try Exercises 10.2.7–10.2.10 inclusive before you continue.

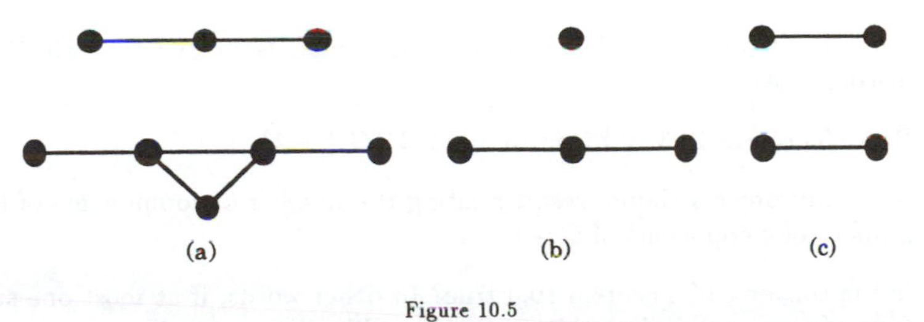

Figure 10.5

We use Exercise 10.2.8 to illustrate the approach used in this second stage of the reconstruction. In Exercise 10.2.8 (iii) the first two distinct graphs have the components of Figure 10.5 (a) in common. If we assume these are components of G, then the other two components of G must be obtainable from the two components in Figure 10.5 (b) or Figure 10.5 (c). But if the components of Figure 10.5 (a) are in G, then those of Figure 10.5 (b) and Figure 10.5 (c) must have come from deleting a vertex of G. So in each case one of these components is in G and the other is $J - v$ where J is a component of G.

The path of length two, P_3, is the largest of the components of Figure 10.5 (b) and Figure 10.5 (c) so it must be a component of G. It is clear that the other component is P_2. So we have reconstructed G.

This technique will work in general. Suppose G has at least three components. In $\mathcal{G}$, unless G is regular, there must be two distinct members $G - v$ and $G - w$ which have all but two components the same. These components are in G.

Of the four remaining components, the largest J in $G - w$ say, is in G. Now we can identify $J - v$ in $G - v$ and the remaining component of $G - v$ is the final required component of G.

If G has only two components (neither of which is an isolated vertex), then we use the argument of the last paragraph.

The above discussion is summarised in Theorem 10.5.

Theorem 10.5 *Disconnected graphs are reconstructible.*

Exercises for Section 10.2

10.2.1 Find $\mathcal{G}$ for the following graphs:

(a) $P_4 \cup P_4$,

(b) $C_4 \cup C_3$,

(c) $K_n \cup K_1$.

Do these results support the conjecture that G is disconnected if and only if each $G - v$ is disconnected?

10.2.2 Suppose G contains at least three components. Is it true that all $G - v$ are disconnected?

10.2.3 Prove Theorem 10.3. Why do we need $|V(G)| > 2$?

10.2.4 State and prove a similar result relating the number of components of G to the number of components of $G - v$.

10.2.5 Is the converse of Theorem 10.3 true? In other words, if at most one subgraph $G - v$ is connected, does this imply that G is disconnected?

10.2.6 Is the converse of Exercise 10.2.4 true?

10.2.7 Show that if G has one or more isolated vertices, then G is uniquely reconstructible.

Can you recognise from $\mathcal{G}$ when G has isolated vertices?

10.2.8 From each collection shown in Figure 10.6 reconstruct a unique graph G.

10.2.9 What strategies used in the last exercise will enable you to reconstruct a disconnected graph which has *no* isolated vertices? Will these strategies work if G has only two components?

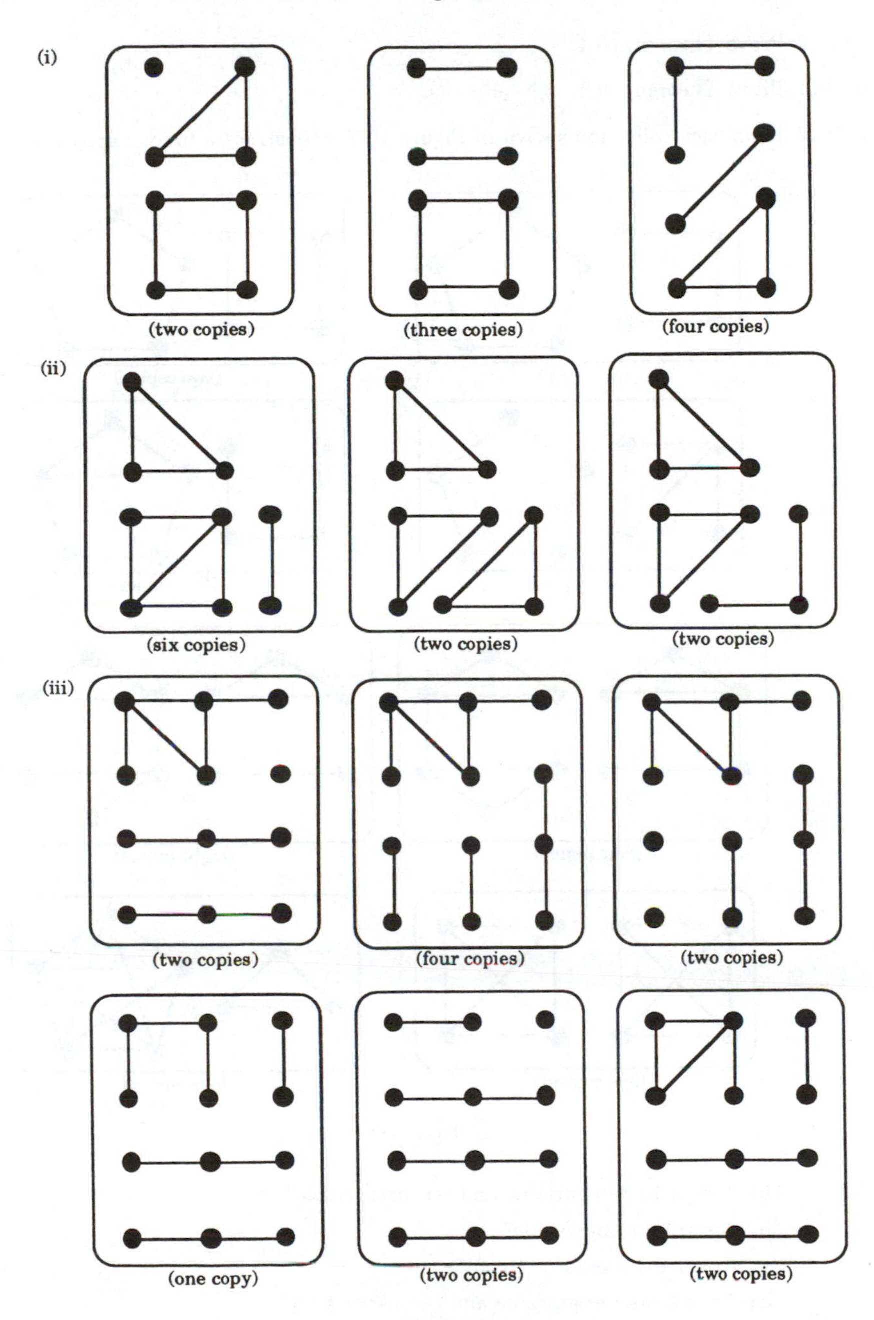
(i)
(two copies)
(three copies)
(four copies)
(ii)
(six copies)
(two copies)
(two copies)
(iii)
(two copies)
(four copies)
(two copies)
(one copy)
(two copies)
(two copies)

Figure 10.6

10.2.10 Prove Theorem 10.4.

10.2.11 Prove Theorem 10.5.

10.2.12 From each collection shown in Figure 10.7 reconstruct a unique graph G.

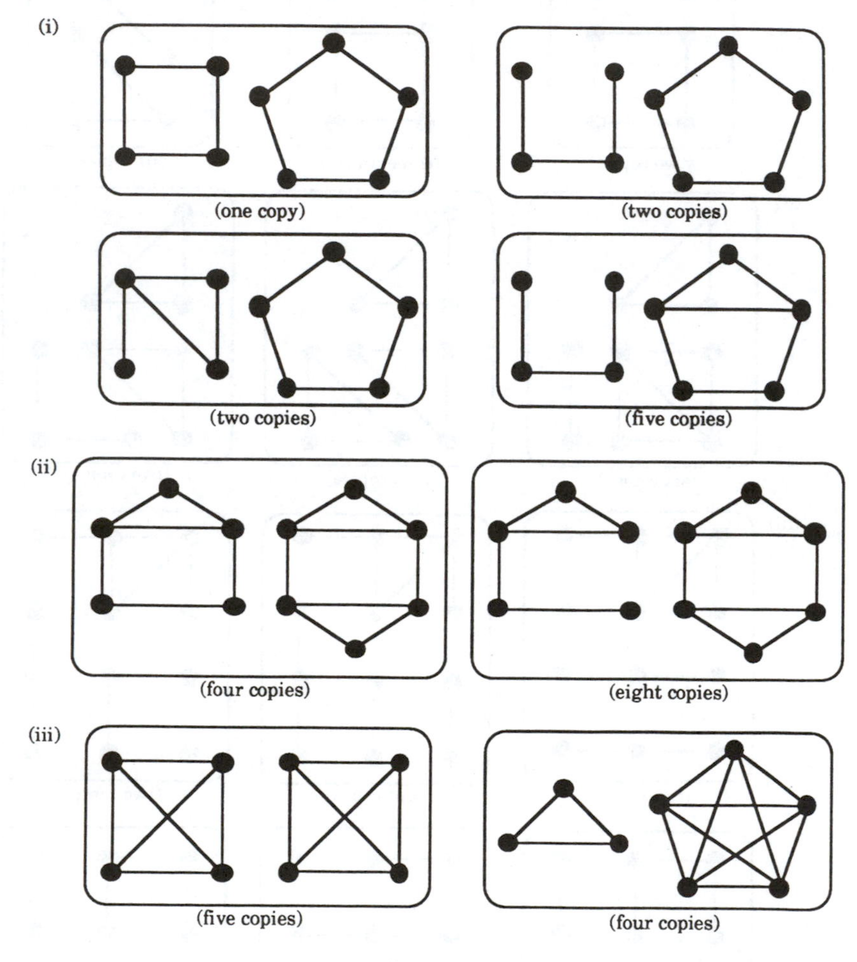

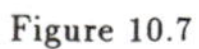
Figure 10.7

10.2.13 (a) Are paths recognisable and reconstructible?

(b) Are trees recognisable?

(c) Are trees reconstructible?

(d) Are forests recognisable and reconstructible?

10.3 Edge Reconstruction

Let G be the graph of Figure10.8 (a). Then its edge deleted subgraphs are shown in Figure 10.8 (b).

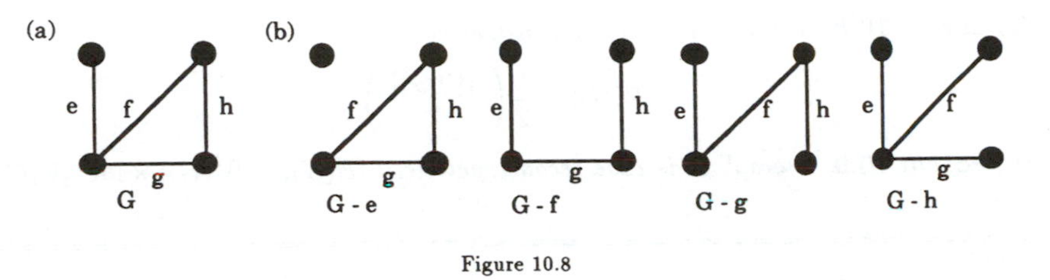

Figure 10.8

In view of the previous two sections a natural question to ask is can a graph be reconstructed uniquely from its edge deleted subgraphs?

> Take any simple graph G with at least four edges and let $\mathcal{G}_e$ denote the collection of all its edge deleted subgraphs of the form $G - e$, i.e., $\mathcal{G}_e = [G - x : e \in E(G)]$. Now take another simple graph H, again with at least four edges, and similarly let $\mathcal{H}_e$ denote the collection $[H - f : f \in E(H)]$.
>
> The **Edge Reconstruction Conjecture** says:
> If the collections $\mathcal{G}_e$ and $\mathcal{H}_e$ are equal then G is isomorphic to H.

Why must the graphs G and H in the definition have at least four edges? For the simple reason that there are non-isomorphic graphs G and H with less than four edges for which $\mathcal{G}_e = \mathcal{H}_e$. Figure 10.9 shows two such graphs. As the reader can easily check, the same is also true for the pair of graphs $K_1 \cup K_3$ and $K_{1,3}$, with three edges. However if we keep to our condition that there be at least four edges then to date nobody knows of any non-isomorphic pair of graphs G and H for which $\mathcal{G}_e = \mathcal{H}_e$.

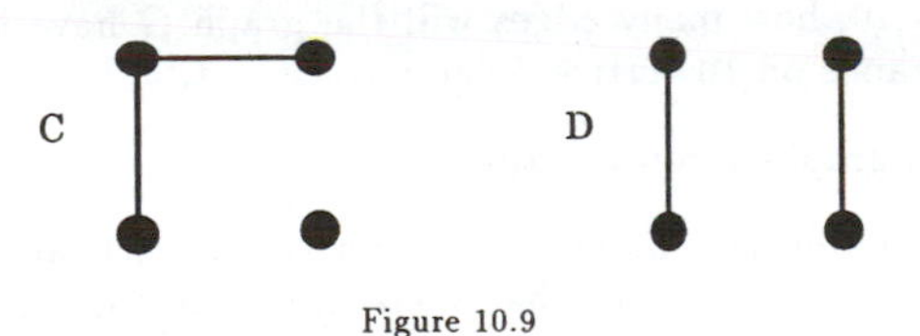

Figure 10.9

At first glance edge reconstruction appears to be easier than vertex reconstruction. Certainly deleting an edge seems to leave us with a lot more information than deleting a vertex does. It turns out that at present more progress has been made with the Edge Reconstruction Conjecture than with the Reconstruction Conjecture. Partly this is due to the following theorem (and its corollary) of Greenwell [30] published in 1971. We omit the proofs.

Theorem 10.6 (Greenwell, 1971) *If G is reconstructible and has no isolated vertices, then G is edge reconstructible.*

Corollary 10.7 *If G is reconstructible, then it is edge reconstructible.*

But in fact it has been shown that the Edge Reconstruction Conjecture *is* true except possibly for graphs with a relatively sparse collection of edges. More specifically, we have the following results, again omitting the proofs. The first is due to Lovász (1972) [42] and the second to Müller (1977) [46].

Theorem 10.8 *A graph G is edge reconstructible if*

$$|E(G)| > \frac{1}{2}\binom{|V(G)|}{2}.$$

Theorem 10.9 *A graph G is edge reconstructible if $|E(G)| > |V(G)| \times \log_2 |V(G)|$.*

Exercises for Section 10.3

10.3.1 Find all pairs of graphs with three or fewer edges which have the same collection of edge deleted subgraphs.

10.3.2 Reconstruct graphs from the collections of edge deleted subgraphs of Figure 10.10.

10.3.3 What graph parameters (such as $|E(G)|$, $d(v)$, etc) can be reconstructed from the edge deleted subgraphs of a graph?

10.3.4 Are regular graphs edge reconstructible? Are trees edge reconstructible?

10.3.5 Prove that an edge analogue of Theorem 10.1 holds.

10.3.6 Assuming Theorem 10.6, prove Corollary 10.7.

10.3.7 Give further classes of graphs which are edge reconstructible.

10.3.8 If $|V(G)| = 10$, how many edges will the graph G have if $2^{|E(G)|-1} > |V(G)|$? Are most graphs on 10 vertices edge reconstructible?

10.3.9 Are directed graphs reconstructible?

10.3.10 Recall that a tournament on n vertices is K_n with an arbitrary orientation assigned to each edge. Show that there are small tournaments which are non-reconstructible.

10.3.11 Show that all tournaments on five or more vertices are reconstructible or show that there are non-reconstructible tournaments on $2^m + 2^n$ vertices for m, n not both zero.

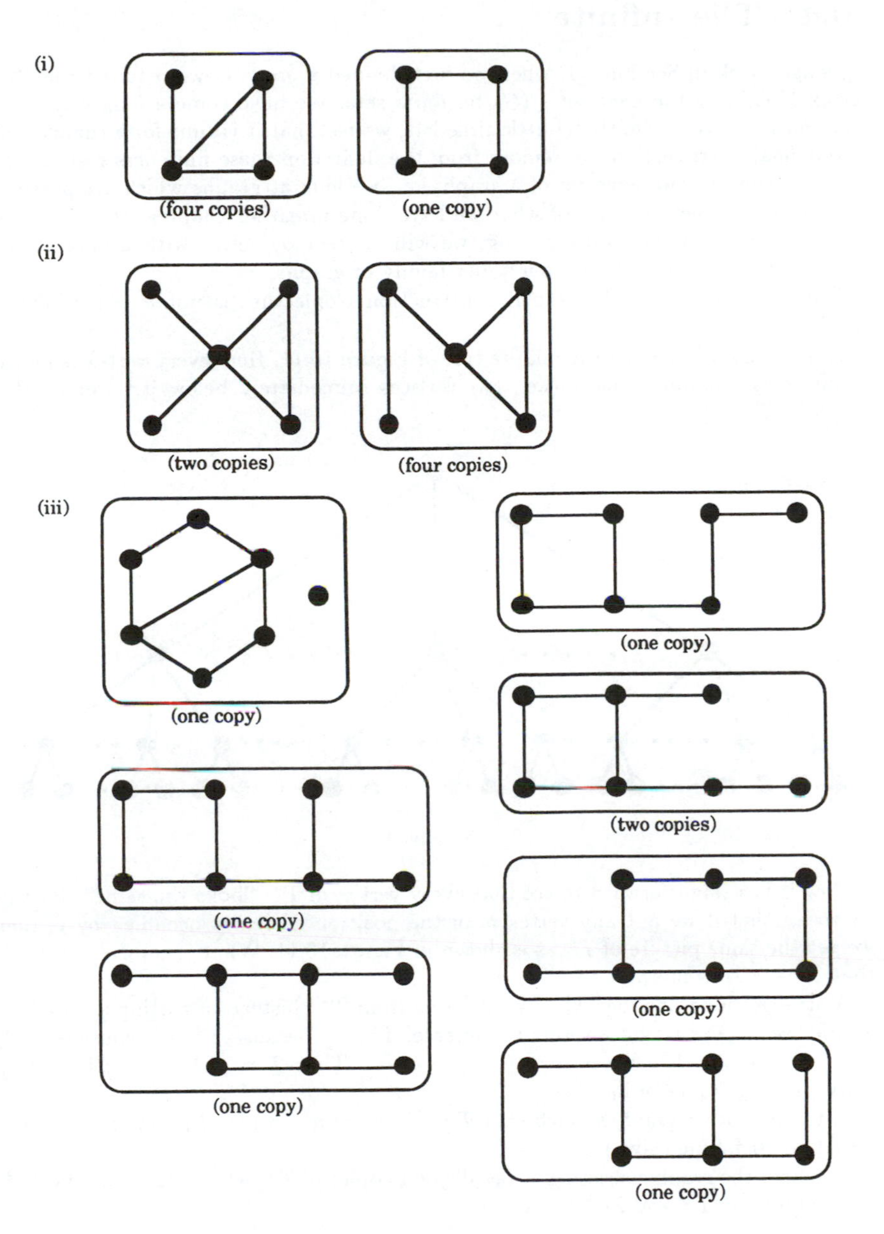
(i)
(four copies)
(one copy)
(ii)
(two copies)
(four copies)
(iii)
(one copy)
(one copy)
(two copies)
(one copy)
(one copy)
(one copy)
(one copy)

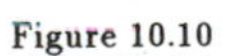
Figure 10.10

10.4 The Infinite

Ages ago, back in Section 1.1, when we first defined a *graph* G we required that the vertex $V(G)$ and the edge set $E(G)$ be *finite* sets. We have come a long way since then but now, with admittedly little time left, we feel that it is time for a change. So, in this final short section, we remove from the definition these finiteness restrictions on the vertex set and edge set of a graph, i.e., we look at graphs which are possibly *infinite*. While there is a lot of what we have done in earlier chapters that can now be re-examined in this wider setting, we will content ourselves with a look at the Reconstruction Conjecture for our wider family of graphs.

The only problem is that the Reconstruction Conjecture is not true for infinite graphs.

To see this, we consider the infinite tree of Figure 10.11. Here every vertex is joined to infinitely countably many (i.e., $\aleph_0$) vertices immediately below it.[2] Denote this graph by T^+.

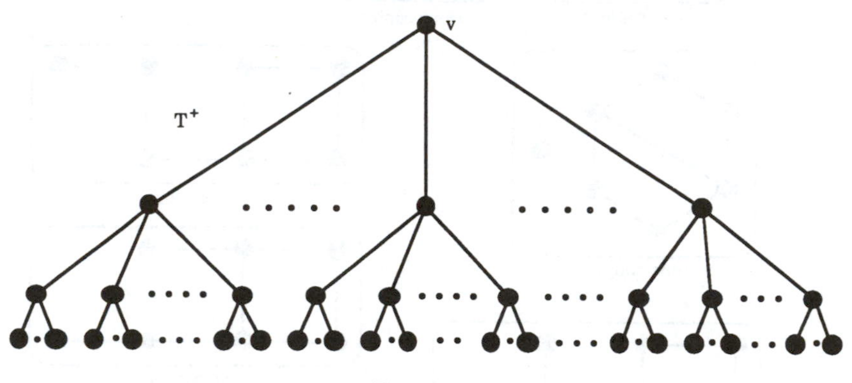

Figure 10.11

Now it is straightforward to see that every vertex of T^+ "looks the same". By this we mean that if we put any vertex v' in the position currently occupied by v, then we get the same picture of T^+ as is shown in Figure 10.11. What this means, then, is that $T^+ - v$ is isomorphic to $T^+ - v'$.

A strange thing happens when we delete v from T^+. Instead of getting a "smaller" graph, we appear to get an infinite increase! This is because $T^+ - v$ consists of $\aleph_0$ disjoint copies of T^+. We write $T^+ - v = \aleph_0 T^+$. Thus $\mathcal{T} = [T^+ - v : v \in V(T^+)]$ consists of $\aleph_0$ copies of $\aleph_0 T^+$.

Is there another graph T' such that $\mathcal{T} = [T' - w : w \in V(T')]$? (Now try Exercises 10.4.1 to 10.4.3 inclusive.)

Consider the graph consisting of *two* disjoint copies of T^+, which we denote by $2T^+$. First of all, are T^+ and $2T^+$ the same?

[2]The symbol $\aleph_0$, pronounced "aleph nought", is an infinite number, defined to be the number of elements in the set of natural numbers $\{1, 2, 3, \ldots\}$.

Clearly not. Given long enough, an ant could get from any vertex of T^+ to any other vertex using edges of T^+. The same can not be said for $2T^+$. In other words, T^+ is connected and $2T^+$ is not. So T^+ is *not* isomorphic to $2T^+$.

But is $\mathcal{T} = [(2T^+) - w : w \in V(2T^+)]$? First, we should note that $(2T^+) - w = T^+ \cup \aleph_0 T^+ = \aleph_0 T^+$. (This relies on infinity plus one being equal to infinity, or, more precisely, $\aleph_0 + 1 = \aleph_0$.) Moreover, this is regardless of which vertex w is removed from $2T^+$. So in fact $[(2T^+ - w : w \in V(2T^+)]$ *is* equal to $\mathcal{T}$.

Hence the Reconstruction Conjecture is false for infinite graphs.

Maybe the reason for the failure of the Reconstruction Conjecture with $T^+, 2T^+$ and so on, is that the vertices of these graphs have *infinite* degree. Perhaps order would be restored if we restricted our attention to graphs where every vertex has finite degree (but still permitting the vertex set to be infinite).

A graph G is said to be **locally finite** if every vertex of G has finite degree.

Of course, the class of locally finite graphs contains the class of finite graphs of our earlier sections. However, we will now describe a construction of two locally finite graphs which shows that the Reconstruction Conjecture is also false for locally finite graphs.

This construction involves a paricular locally finite graph which we now define.

The locally finite graph $P_{2\infty}$, called the **two-way infinite path**, is defined to have the countably infinite vertex set $\{\ldots, -3, -2, -1, 0, 1, 2, 3, \ldots\}$ and an edge between every pair of vertices of the form $i, i+1$, (so that each vertex has degree 2), as shown in Figure 10.12.

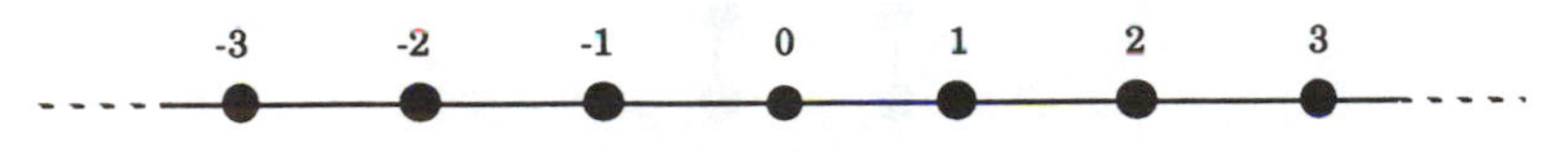

Figure 10.12: The two-way infinite path $P_{2\infty}$.

The construction also involves a series of locally finite graphs A_i, for $i = 1, 2, 3, \ldots$, the first three of which are shown in Figure 10.13. To form these, we essentially take i copies of $P_{2\infty}$ and join a new vertex u_i to an arbitrary vertex from each of the i copies of $P_{2\infty}$.

For each j, let A_{ij} be any subtree of A_i except $P_{2\infty}$. We then construct G to be the union of $\aleph_0$ copies of each A_{ij}.

Now if v is a vertex in a finite component of G then $G - v$ is isomorphic to G.

On the other hand, if v is in an infinite subtree of A_1, then $G - v = G$ unless $v = u_1$ (see Figure 10.13), in which case $G - v = G \cup P_{2\infty}$. This latter graph occurs $\aleph_0$ times since G contains $\aleph_0$ copies of A_1.

In the case that v is in an infinite subtree of A_2, then $G - v = G \cup 2P_{2\infty}$, $G \cup P_{2\infty}$ or G, depending on whether v is a vertex u_2 of Figure 10.13 in a copy of A_2 in G, or v corresponds to the vertex u_2 in a subtree T of A_2 where $T - v = T' \cup P_{2\infty}$ (T' being an arbitrary tree) or v does not correspond to u_2. (Here $2P_{2\infty}$ is the graph consisting of 2 (disjoint) copies of $P_{2\infty}$ and, more generally, $rP_{2\infty}$ consists of r copies of $P_{2\infty}$ for any $r \geq 2$.

Similarly we see that if v is in an infinite subtree of A_3, $G - v = G \cup 3P_{2\infty}$, $G \cup 2P_{2\infty}$, $G \cup P_{2\infty}$ or G. Further each of these graphs occurs $\aleph_0$ times.

Hence we find that $\mathcal{G}$ consists of $\aleph_0$ copies of $G \cup rP_{2\infty}$ for $r \geq 0$.

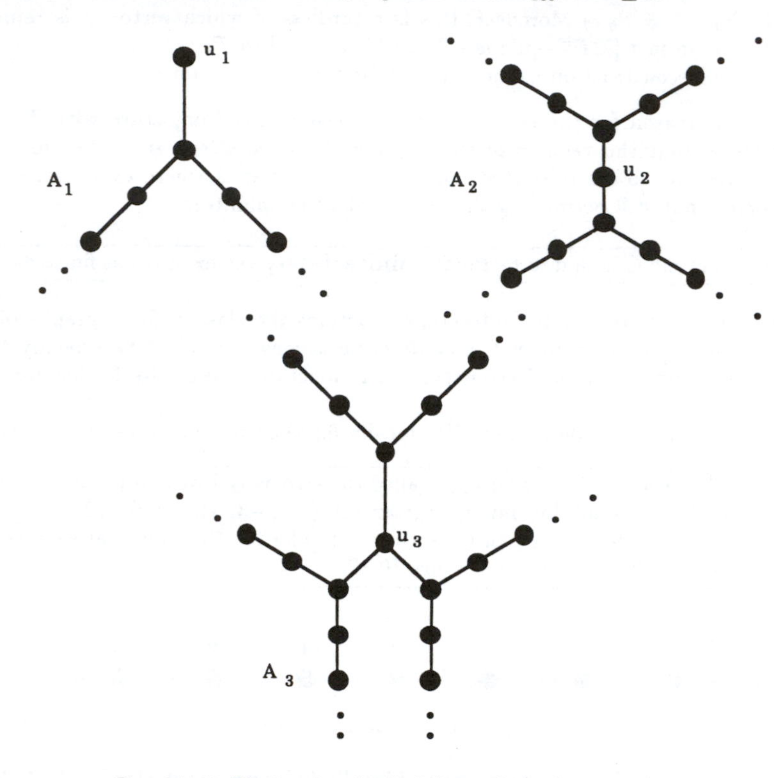

Figure 10.13: The first three locally finite graphs in the sequence A_i.

Now take the graph $H = G \cup P_{2\infty}$. We know that $H - w = G \cup (r+1)P_{2\infty}$ if $w \in V(G)$ for $r \geq 0$. If $w \in V(P_{2\infty})$, then, since $P_{2\infty} - w = 2P_\infty$, $H_w = G$. Now this means that $H - w = G \cup rP_{2\infty}$ for $r \in \mathcal{N} \cup \{0\}$ and that each such vertex deleted subgraph occurs $\aleph_0$ times. Hence $\mathcal{H} = \mathcal{G}$.

The two examples that we have given of non-reconstructible infinite graphs are

$$\text{(i) } T^+ \text{ and } 2T^+, \text{ and (ii) } G \text{ and } G \cup P_{2\infty}.$$

In both cases the first graph of the pair is a subgraph of the second. Further, the second graph is an induced subgraph of the first. Is this the reason that infinite graphs are not reconstructible?

Finally, let us broaden our definition of reconstructibility.

Two graphs G_1 and G_2 are called **H-isomorphic** if G_1 is an induced subgraph of G_2 and G_2 is an induced subgraph of G_1.

A graph G is called **H-reconstructible** if every reconstruction of G is H-isomorphic to G.

The German mathematician Rudolph Halin conjectures that *all* non-reconstructible graphs are H-isomorphic. This is still an open question.

Exercises for Section 10.4

10.4.1 (a) Draw the infinite regular tree R of degree 3.

(b) Is $R - v$ isomorphic to $R - v'$ for $v, v' \in V(R)$?

(c) Is $R - v$ isomorphic to $3R$?

(d) Is the collection of vertex deleted subgraphs for R the same as that for $2R$?

10.4.2 Is $2T^+ - w$ isomorphic to $(2T^+) - w'$ for all $w, w' \in V(2T^+)$?

10.4.3 (a) Is $\mathcal{T} = [(3T^+) - x : x \in V(3T^+)]$?

(b) Find infinitely many graphs not isomorphic to T^+ whose collection of vertex deleted subgraphs is $\mathcal{T}^+$.

10.4.4 $\aleph_0$ is just one "size" of infinity. Show that the Reconstruction Conjecture is false for *every* infinity.

10.4.5 Can you find any other counterexamples to the Reconstruction Conjecture for infinite graphs?

10.4.6 Find two other examples to show that locally finite graphs are not reconstructible.

10.4.7 Show that if Halin's conjecture is true, then so is the Reconstruction Conjecture.

10.4.8 Show that T^+ and $3T^+$ are H-isomorphic.

10.4.9 Let P and Q be finite graphs. If P and Q are H-isomorphic, is P isomorphic to Q?

Bibliography

[1] M.O. Albertson and J.P. Hutchinson. *Discrete Mathematics with Algorithms.* Wiley, New York, 1988.

[2] K. Appel and W. Haken. The solution of the four-colour map problem. *Sci. Amer.*, 237:108–121, 1977.

[3] K. Appel, W. Haken, and J. Koch. Every planar map is four-colourable. *Illinois J. Math.*, 21:429–567, 1977.

[4] M. Behzad, G. Chartrand, and L. Lesniak-Foster. *Graphs and Digraphs.* Prindle, Weber & Schmidt, Massachusetts, 1962.

[5] C. Berge. Two theorems on graph theory. *Proc. Nat. Acad. Sci. U.S.A.*, 43:842–844, 1957.

[6] N.L. Biggs, E.K. Lloyd, and R.J. Wilson. *Graph Theory 1736-1936.* Oxford University Press, Oxford, 1976.

[7] J. A. Bondy and U. S. R. Murty. *Graph Theory with Applications.* Elsevier/MacMillan, New York/London, 1976.

[8] J.A. Bondy and V. Chvátal. A method in graph theory. *Discrete Math.*, 15:111–136, 1976.

[9] R.L. Brooks. On colouring the nodes of a network. *Proc. Cambridge Philos. Soc.*, 37:194–197, 1941.

[10] J.C. Butcher. The Fibonacci sequence, chromatic numbers and slam bidding. *Bull. Inst. Math. Appl.*, 26:129–132, 1990.

[11] P. Camion. Chemins et circuits hamiltoniens des graphes complets. *C.R. Acad. Sci. Paris*, 249:2151–2152, 1959.

[12] A. Cayley. A theorem on trees. *Quart. J. Math.*, 23:376–378, 1889.

[13] G. Chartrand. *Graphs as Mathematical Models.* Prindle, Weber & Schmidt, Massachusetts, 1978.

[14] G. B. Dantzig and D. R. Fulkerson. On the max-flow, min-cut theorem of networks. In *Linear Inequalities and Related Systems*, pages 215–221. Annals of Math. Study 38, Princeton University Press, 1956.

[15] N. G. de Bruijn. A combinatorial problem. *Indag. Math.*, 8:461–467, 1946.

[16] E.W. Dijkstra. A note on two problems in connexion with graphs. *Numer. Math.*, 1:269–271, 1959.

[17] G.A. Dirac. A property of 4-chromatic graphs and some remarks on critical graphs. *J. London Math. Soc.*, 27:85–92, 1952.

[18] G.A. Dirac. Some theorems on abstract graphs. *Proc. London Math. Soc.*, 2:69–81, 1952.

[19] E. Egerváry. On combinatorial properties of matrices (Hungarian). *Mat. Fiz. Lapok*, 38:16–28, 1931.

[20] P. Erdös and Szekeres G. A combinatorial problem in geometry. *Compositio Math.*, 2:463–470, 1935.

[21] L. Euler. Solutio problematis ad geometriam situs pertinentis. *Comment. Academiae Sci. I. Petropolitanae*, 8:128–140, 1736.

[22] L. Euler. Demonstratio nonnullarum insignium proprietatum quibus solida hedris planis inclusa sunt praedita. *Novi Comm. Acad. Sci. Imp. Petropol.*, 4:140–160, 1758.

[23] L. R. Ford, Jr. and D. R. Fulkerson. Maximal flow through a network. *Canad. J. Math.*, 8:399–404, 1956.

[24] L. R. Ford, Jr. and D. R. Fulkerson. A simple algorithm for finding maximal network flows and an application to the Hitchcock problem. *Canad. J. Math.*, 9:210–218, 1957.

[25] P.A. Fowler. The Königsberg bridges — 250 years later. *Amer. Math. Monthly*, 95:42–43, 1988.

[26] A. Gibbons. *Algorithmic Graph Theory.* Cambridge, Cambridge, 1989.

[27] I. J. Good. Normal recurring decimals. *J. Lond. Math. Soc.*, 21:167–169, 1946.

[28] R. Gould. *Graph Theory.* Benjamin/Cummings, Menlo Park, 1989.

[29] R.L. Graham and Spencer J.H. Ramsey theory. *Sci. Amer.*, 263:112–117, 1990.

[30] D.D. Greenwell. Reconstructing graphs. *Proc. Amer. Math. Soc.*, 30:431–433, 1971.

[31] E.J. Grinberg. Plane homogeneous graphs of degree three without hamiltonian circuits. *Latvian Math. Yearbook*, 4:51–58, 1968.

[32] P. Hall. On representation of subsets. *J. London Math. Soc.*, 10:26–30, 1935.

[33] P.J. Heawood. Map colour theorems. *Quart. J. Math.*, 24:332–338, 1890.

[34] C. Hierholzer. Uber die Möglichkeit, einen Liniensug ohne Wiederholung und ohne unterbrechnung zu umfahren. *Math, Ann.*, 6:30–32, 1873.

[35] J. E. Hopcroft and R. E. Tarjan. Algorithm 447: Efficient algorithms for graph manipulation. *Comm. ACM*, 16:372–378, 1973.

[36] A.B. Kempe. On the geographical problem of the four colors. *Amer. J. Math.*, 2:193–200, 1879.

[37] D. König. Graphs and matrices (Hungarian). *Mat. Fiz. Lapok*, 38:116–119, 1931.

[38] J.B. Kruskal. On the shortest spanning subtree of a graph and the traveling salesman problem. *Proc. Amer. Math. Soc.*, 7:48–50, 1956.

[39] M.-K. Kuan. Graphic programming using odd or even points. *Chinese Math.*, 1:273–277, 1962.

[40] H.W. Kuhn. The Hungarian method for the assignment problem. *Naval Res. Logist. Quart.*, 2:83–97, 1955.

[41] K. Kuratowski. Sur le problème des courbes gauches en topologie. *Fund. Math.*, 16:271–283, 1930.

[42] L. Lovász. A note on the line reconstruction problem. *J. Combinatorial Theory B*, 13:309–310, 1972.

[43] D.W. Matula, G. Marble, and J.D. Isaacson. Graph colouring algorithms. In R.C. Read, editor, *Graph Theory and Computing*, pages 109–122. Academic Press, 1972.

[44] K. Menger. Zur allgemeinen Kurventheorie. *Fund. Math.*, 10:95–115, 1927.

[45] J.L. Mott, A. Kandel, and T.P. Baker. *Discrete Mathematics for Computer Scientists.* Reston Pub. Co., Virginia, 1983.

[46] V. Müller. The edge reconstruction hypothesis is true for graphs with more than $n \cdot \log_2 n$ edges. *J. Combinatorial Theory B*, 22:281–283, 1977.

[47] J. Munkres. Algorithms for the assignment and transportation problems. *J. Soc. Indust. Appl. Math.*, 5:32–38, 1957.

[48] J. Mycielski. Sur le coloriage des graphes. *Colloq. Math.*, 3:161–162, 1955.

[49] O. Ore. *Graphs and Their Uses (revised and updated by Robin J. Wilson).* The Math. Assoc. of America, Providence, 1990.

[50] A.D. Polimeni and H.J. Straight. *Foundations of Discrete Mathematics.* Brooks/Cole Pub. Co., Monterey, California, 1985.

[51] R.C. Prim. Shortest connection networks and some generalizations. *Bell System Tech. J.*, 36:1389–1401, 1957.

[52] F. Ramsey. On a problem of formal logic. *Proc. London Math. Soc.*, 30:264–286, 1930.

[53] L. Rédei. Ein Kombinatorischer Satz. *Acta. Litt. Szeged*, 7:39–43, 1934.

[54] H.E. Robbins. A theorem on graphs, with an application to a problem of traffic control. *Amer. Math. Monthly*, 46:281–283, 1939.

[55] F.S. Roberts. *Applied Combinatorics.* Prentice-Hall, Englewood Cliffs, New Jersey, 1984.

[56] D. Seinsche. On a property of the class of n-colourable graphs. *J. Combinatorial Theory B*, 16:191–193, 1974.

[57] D.K. Smith. *Network Optimisation Practice, A Computational Guide.* Ellis Horwood, Chichester, 1982.

[58] M.M. Sysło, N. Deo, and J.S. Kowalik. *Discrete Optimization Algorithms with Pascal Programs.* Prentice-Hall, Englewood Cliffs, New Jersey, 1983.

[59] C. Thomassen. Kuratowski's Theorem. *J. Graph Theory*, 5:225–241, 1981.

[60] C. Thomassen. A link between the Jordan Curve Theorem and the Kuratowski Planarity Criterion. *Amer. Math. Monthly*, 97:216–218, 1990.

[61] A. Tucker. *Applied Combinatorics.* Wiley, New York, 1980.

[62] V.G. Vizing. On an estimate of the chromatic class of a p-graph. *Diskret. Analiz.*, 3:25–30, 1964.

[63] D.J.A. Welsh and M.B. Powell. An upper bound for the chromatic number of a graph and its application to timetabling problems. *Comput. J.*, 10:85–86, 1967.

[64] H. Whitney. Congruent graphs and the connectivity of graphs. *Amer. J. Math.*, 54:150–168, 1932.

[65] R.J. Wilson. *Introduction to Graph Theory, Third Edition.* Longman, Harlow, 1985.

[66] R.J. Wilson and J.J. Watkins. *Graphs: An Introductory Approach.* Wiley, New York, 1989.

Index

Absent colour, 213
Acyclic graph, 47
Adjacency matrix, 35
Adjacent vertices, 7
Adjacent edges, 13
Algorithm,
  Breadth First Search, 69, 71
  Closest Insertion, 117
  Dijkstra's, 72
  Fleury's, 89
  Ford and Fulkerson, 274, 276
  Fusion, 42
  Hierholzer's, 95
  Hopcroft and Tarjan, 256
  Hungarian, 136
  Kruskal's, 63
  Kuhn-Munkres, 147
  Largest-First Sequential, 201
  Prim's, 65
  Simple Sequential, 199
  Smallest-Last Sequential, 202
  Two-Optimal, 110
  Welsh and Powell, 201
Alternating path, 122
Alternating tree, 136
Ancestor, 259
Arc, 230
  external, 283
  forward, 268
  internal, 283
  reverse, 268
Arc set, 230
Articulation point, 78
Assignment problem, 143
Augmenting path, 122

Berge's Theorem, 124
BFS Technique, 69
Bipartite graph, 10
  complete, 10
Bipartition, 10
Birkhoff diamond, 197
Bondy and Chvatal's Theorem, 105
Breadth First Search Technique, 69
Breakthrough, 275
Bridge, 53
Brooks' Theorem, 194

C_n, 29
$c(G)$, 104
Capacity
  of an arc, 261
  constraint, 262
Cayley's Theorem, 59
$\chi(G)$, 192
$\chi_1(G)$, 213
Chinese postman problem, 96
Chromatic graph
  k-, 192
  k-edge, 213
Chromatic index, 192
  edge, 213
Chromatic number, 192, 213
$cl(G)$, 208
Clique, 208
Clique number, 208
Closed walk, 26
Closest Insertion Algorithm, 117
Closure, 104
Collection, 303
Colour,
  absent, 213
  present, 213

Colourable graph
  k-, 192
  k-edge, 213
Colouring, 192
  k-, 192
  k-edge, 213
  edge, 213
  vertex, 192
Complement
  of a graph, 23
  of a digraph, 236
Complete graph, 10
Complete bipartite graph, 10
Complete k-partite graph, 199
Complete tripartite graph, 34
Component, connected, 28
Concatenation, 28
Condensation, 237
Connected component, 28
Connected digraph
  strongly, 232
  unilaterally, 235
  weakly, 232
Connected graph, 28
  n-, 80
Connected vertices, 27
Connectivity, 79
  edge, 287
  vertex, 79
Contracted edge, 60
Contractible edge, 175
Contraction of an edge, 174
Converse of a digraph, 238
Convex face, 179
Convex solid, 169
Covering, 134
$cr(G)$, 179
Critical graph, k-, 205
Critical planar graph, 161
Crossing number, 179
Cube, k-, 16
Curve, Jordan, 158
Cut, 264
Cut edge, 53
Cut vertex, 78
Cycle, 29
  even, 29
  k-, 29
  odd, 29

d_λ, 147
$d(\varphi)$, 165
$d(v)$, 14
de Bruijn diagram, 242
de Bruijn sequence, 241
Defect, 147
Degree of a vertex, 14
Degree of a face, 165
$\delta(G)$, 287
$\Delta(G)$, 194
Diagonal, 182
Diameter, 33
Diconnected digraph, 232
Digraph, 230
Digraph isomorphism, 232
Dijkstra's Algorithm, 72
Dirac's Theorem on Hamiltonian graphs, 102
Dirac's Theorem on critical graphs, 206
Directed cycle, 231
Directed edges, 230
Directed Euler tour, 239
Directed Euler trail, 239
Directed graph, 230
Directed Hamiltonian cycle, 249
Directed Hamiltonian path, 248
Directed path, 231
Directed spanning walk, 235
Directed tour, 231
Directed trail, 231
Directed walk, 231
Disconnected graph, 28
Disjoint subgraphs, 20
Distance between vertices
  in a graph, 32
  in a digraph, 252
Dot, 2
Double dual, 186
Dual, 185
Duplicated edge, 96

E_λ, 144

Eccentricity, 33
Edge, 2, 230
    cut, 53
    directed, 230
Edge chromatic graph, k-, 213
Edge chromatic index, 213
Edge chromatic number, 213
Edge colourable graph, k-, 213
Edge colouring, k-, 213
Edge colouring, 213, 293
Edge connected, n-, 287
Edge connectivity, 287
Edge deleted subgraph, 18
Edge disjoint subgraphs, 20
Edge Reconstruction Conjecture, 313
Edge set, 2
Edges,
    adjacent, 13
    parallel, 7
Empty graph, 10
End vertex, 2
Equal collections, 304
Equality subgraph, 145
Euler digraph, 239
Euler graph, 83
Euler Polyhedron Formula, 169
Euler's Formula, 163
Euler's Theorem, 85
Euler tour, 83
    directed, 239
Euler trail, 83
    directed, 239
Eulerian graph, 83
Euler's Formula, 163
Even vertex, 14
Even cycle, 29
Exterior of a Jordan curve, 158
Exterior face, 162
External arc, 283

f-incrementing walk, 269
f-saturated vertex, 268
f-unsaturated vertex, 268
f-unsaturated tree, 274
$f(G)$, 162
$f(s)$, 296

Face of a plane graph, 162
    exterior, 162
    interior, 162
Face colourable, 220
Feasible vertex labelling, 144
First Theorem of Graph Theory, 14
Five Colour Problem, 222
Five Colour Theorem, 223
Fleury's Algorithm, 89
Flow, 262
    conservation, 262
    maximal, 265
    revised, 270
    zero, 274
Flow value, 263
Ford and Fulkerson Algorithm, 274, 276
Forest, 51
Forward arc, 268
Four Colour Conjecture, 220
Four Colour Theorem, 226
Fusing vertices, 41
Fusion, 41
Fusion Algorithm, 42

G_λ, 145
Good diagram, 242
Graph, 2
Graph isomorphism, 8
Grinberg graph, 183
Grinberg's Theorem, 182
Grötzsch graph, 210

H-isomorphic, 319
H-reconstructible, 319
Hall's Marriage Theorem, 130
Hamiltonian graph, 100
Hamiltonian circuit, 100
Hamiltonian cycle, 100
    directed, 249
Hamiltonian digraph, 249
Hamiltonian path, 99, 248
Head, 230
Hierholzer's Algorithm, 95
Hopcroft and Tarjan Algorithm, 256
Hungarian method, 136

$I(u)$, 262

$i(W)$, 268
$id(v)$, 238
Identifying vertices, 41
Incidence matrix of a graph, 39
Incident edge and vertex, 13
Increment, 268
Incrementing walk, 269
Indegree, 238
Induced subgraph, 20
Initial vertex, 230
Interior of a Jordan curve, 158
Interior face, 162
Intermediate vertex in a network, 261
Internal arc, 283
Internal vertices, 25
Internally disjoint paths, 80
Intersection of graphs, 21
Isolated vertex, 7
Isomorphic graphs, 8
    H-, 319
Isomorphic digraphs, 232
Isomorphism
    of graphs, 8
    of digraphs, 232
Isthmus, 53

Join of two vertices, 2, 230
Join of graphs, 24
Jordan curve, 158
Jordan Curve Theorem, 158

k-partite graph, 198
    complete, 199
$K_{m,n}$, 10
K_n, 10
$K_{r,s,t}$, 34
$\kappa(G)$, 80
$\kappa_e(G)$, 287
Kempe chain, 194, 213
    argument, 194
Königsberg bridge problem, 213
Kruskal's Algorithm, 63
Kuhn-Munkres Algorithm, 147
Kuratowski's Theorem, 178

Labelled based on, 275
Labelling, feasible vertex, 144
Largest-First Sequential Algorithm, 201
Latin square, 214
Length of a walk, 25, 231
Locally finite graph, 317
Loop, 2

M-alternating path, 122
M-alternating tree, 136
M-augmenting path, 122
M-saturated vertex, 121
M-saturates v, 121
M-unsaturated vertex, 121
Map, 220
Marriage problem, 129
Marriage Theorem, 130
Matching, 121
    maximum, 122
    optimal, 144
    perfect, 122
Matrix, adjacency, 35
Matrix, incidence, 39
Max-Flow, Min-Cut Theorem, 269
Maximal flow, 265
Maximal non-Hamiltonian, 101
Maximum vertex degree, 194
Maximum matching, 122
Menger's Theorem, 283
    edge version of, 287
Minimal spanning tree, 62
Minimum covering, 134
Monochromatic subgraph, 292
Multigraph, 7
Mycielski's Theorem, 209

$N(S)$, 129
$N(v)$, 7
Neighbour, 7
Neighbour set, 129
Neighbourhood set, 7
Network, 261
Node, 2

$O(u)$, 262
$od(v)$, 238
Odd vertex, 14
Odd cycle, 29
$\omega(G)$, 28

Open walk, 26
Optimal assignment problem, 143
Optimal circuit, 110
Optimal matching, 144
Optimal tree, 62
Ordered pair, 229
Orientable graph, 254
Orientation, 232
Orienting edges of a graph, 254
Origin
  of an arc, 230
  of a directed walk, 232
  of a walk, 25
Outdegree, 238

P_n, 26
$P_{2\infty}$, 317
Parallel edges, 7
Path, 26
  alternating, 122
  augmenting, 122
Perfect matching, 122
Personnel assignment problem, 135
Petersen graph, 32
Planar graph, 157
Plane graph, 157
Platonic bodies, 170
Platonic solids, 170
Point, 2
Polyhedral graph, 169
Polyhedron, 169
  regular, 170
Present colour, 213
Prim's Algorithm, 65
Proper subgraph, 17

Q_n, 16

$r(m,n)$, 297
$r(n_1,n_2,\ldots,n_s)$, 300
Radius, 33
Ramsey number, 297, 300
  generalised, 300
Randomly traceable graph, 95
Reachable vertex, 232
Recognisable property of a graph, 309
Reconstructible graph, 304
  H-, 319
Reconstruction Conjecture, 304
Regular
  digraph, 245
  graph, 15
  polyhedron, 170
Reverse arc, 268
Revised flow, 270

Saturated vertex, 268
Scan, 275
Score, 252, 253
  second level, 252
Score sequence, 253
Self-complementary graph, 23
Self-dual graph, 187
Separates, 282
Separating set
  of edges, 282
  of vertices, 282
Shift register sequence, 242
Simple
  digraph, 233
  graph, 7
Simple Sequential Algorithm, 199
Sink, 261
Smallest-Last Sequential Algorithm, 202
Source, 261
Spanning subgraph, 17
Spanning supergraph, 17
Spanning tree, 57
Square of a graph, 33
Star graph, 51
Strong component, 233
Strongly connected digraph, 232
Subdigraph, 233
Subdivision of a graph, 173
Subgraph, 17
  proper, 17
Supergraph, 17
  spanning, 17
Symmetric matrix, 35

Tail of an arc, 230
Teleprinter's problem, 245
Terminal vertex of an arc, 230

Terminus
 of an arc, 230
 of a directed walk, 232
 of a walk, 25
Tour, 83
 directed, 231
 Euler, 83, 239
Tournament, 246
Trail, 26
 directed, 231
 Euler, 83, 239
Transfer along augmenting path, 125
Transitive digraph, 253
Travelling salesman problem, 110
Tree, 47
 alternating, 136
 f-unsaturated, 274
 spanning, 57
Triangulation, 223
Trivial graph, 10
Trivial walk, 25
Two-Optimal Algorithm, 110
Two-way infinite path, 317

Underlying graph of a digraph, 230
Underlying simple graph, 18
Unicyclic graph, 57
Unilaterally connected digraph, 235
Union of two graphs, 21
Unsaturated tree, 274
Unsaturated walk, 268

Value of flow, 263
Vertex, 2
 end, 2
 even, 14
 initial, 230
 isolated, 7
 odd, 14
 terminal, 230
Vertex colouring, 192
Vertex connectivity, 79
Vertex deleted subgraph, 18
Vertex set, 2
Vizing's Theorem, 216

W_n, 32

Walk, 25
 closed, 26
 incrementing, 269
 open, 26
 trivial, 25
 saturated, 268
 unsaturated, 268
Weakly connected digraph, 232
Weight, 62
Weighted graph, 62
Welsh and Powell Algorithm, 201
Wheel, 32
Whitney's Theorem, 80
Word, 241

Zero flow, 274